Advanced Research on Biological Control of Plant Disease or Microbial Interactions

Advanced Research on Biological Control of Plant Disease or Microbial Interactions

Editors

Beibei Ge
Tomislav Cernava
Kyungseok Park

MDPI

Basel • Beijing • Wuhan • Barcelona • Belgrade • Novi Sad • Cluj • Manchester

Editors
Beibei Ge
Institute of Plant Protection
Chinese Academy of
Agricultural Sciences
Beijing
China

Tomislav Cernava
Institute of Environmental
Biotechnology
Graz University of
Technology
Graz
Austria

Kyungseok Park
Research Institute of
International Agriculture
Technology and Information
Ansung
Korea, South

Editorial Office
MDPI AG
Grosspeteranlage 5
4052 Basel, Switzerland

This is a reprint of articles from the Special Issue published online in the open access journal *Microorganisms* (ISSN 2076-2607) (available at: www.mdpi.com/journal/microorganisms/special_issues/biological_control_disease_interactions).

For citation purposes, cite each article independently as indicated on the article page online and as indicated below:

Lastname, A.A.; Lastname, B.B. Article Title. *Journal Name* **Year**, *Volume Number*, Page Range.

ISBN 978-3-7258-1556-2 (Hbk)
ISBN 978-3-7258-1555-5 (PDF)
doi.org/10.3390/books978-3-7258-1555-5

Contents

Preface

Plant diseases are caused by a large number of plant pathogens, with fungi, bacteria, and viruses being the main causes of the loss of crop yield and quality worldwide. Although several control strategies have been developed to reduce the negative effects of plant diseases, biological control remains an environmentally friendly and cost-effective green technique in environmental protection and agricultural production; it generally uses selected bioresources, including beneficial microorganisms. This Special Issue will comprehensively explore the general principles and advances in the "Biological Control of Plant Disease or Microbial interactions", in relation to the modes of action and applications of biocontrol agents in the control of plant diseases, the interactions between plant pathogens and biocontrol agents, and biological control agents and their mechanisms.

Beibei Ge, Tomislav Cernava, and Kyungseok Park
Editors

microorganisms MDPI

Review

Can Entomopathogenic Nematodes and Their Symbiotic Bacteria Suppress Fruit Fly Pests? A Review

Jorge Toledo [1,*], Brenda M. Morán-Aceves [2], Jorge E. Ibarra [3] and Pablo Liedo [1]

1 El Colegio de la Frontera Sur, Tapachula 30700, Chiapas, Mexico; pliedo@ecosur.mx
2 Laboratorio de Reproducción de Hongos Entomopatógenos, Asociación de Cañeros del Soconusco, A.C. CNPR, Huixtla 30640, Chiapas, Mexico; breyma16ibt@gmail.com
3 Departamento de Biotecnología y Bioquímica, Centro de Investigaciones y de Estudios Avanzados (CINVESTAV-IPN), Irapuato 36500, Guanajuato, Mexico; jorge.ibarra@cinvestav.mx
* Correspondence: jtoledo@ecosur.mx

Abstract: Fruit flies (Diptera: Tephritidae) are serious pests that affect fruit production and marketing. Both third instar larvae and pupae are biological stages that persist in the soil until adult emergence. Entomopathogenic nematodes (ENs) are biological control agents that are used to control agricultural pests in greenhouse or field conditions. Several studies have been carried out under laboratory and field conditions showing how ENs can be applied within an area-wide integrated pest management approach to control fruit fly species in orchards and backyard fruit trees. In this review, we analyze how soil physical characteristics and biotic factors affect the performance of these biological control agents. Of the reviewed papers, more than half evaluated the influence of soil texture, humidity, temperature, and other factors on the performance of infective juveniles (IJs). Abiotic factors that significantly influence the performance of IJs are temperature, humidity, and texture. Among the biotic factors that affect IJs are fungi, bacteria, mites, insects, and earthworms. We conclude that ENs have the potential to be applied in the drip area of fruit trees that are infested by fruit flies and contribute to their suppression. This approach, in conjunction with an area-wide pest management approach, may contribute to pest suppression and increase the sustainability of agroecosystems.

Keywords: fruit flies; entomopathogenic nematodes; biological control agent; sustainable approach

Citation: Toledo, J.; Morán-Aceves, B.M.; Ibarra, J.E.; Liedo, P. Can Entomopathogenic Nematodes and Their Symbiotic Bacteria Suppress Fruit Fly Pests? A Review. *Microorganisms* **2023**, *11*, 1682. https://doi.org/10.3390/microorganisms11071682

Academic Editors: Beibei Ge, Tomislav Cernava and Kyungseok Park

Received: 23 April 2023
Revised: 6 June 2023
Accepted: 9 June 2023
Published: 28 June 2023

1. Introduction

The family Tephritidae (Diptera) includes species that constitute the most important fruit and vegetable pests worldwide. The family comprises ~4900 described species within 481 genera. Six of these genera (*Anastrepha*, *Bactrocera*, *Ceratitis*, *Dacus*, *Rhagoletis*, and *Zeugodacus*) contain ~70 species of agricultural importance [1–5]. These pests cause considerable economic losses because of direct damage to a wide range of crops [4,5]. In numerous countries, the development of fruticulture has been restricted by strict quarantines imposed by countries that import the produce and by the high costs associated with efforts directed towards the prevention, containment, suppression, and eradication of these pests.

To prevent or minimize the harmful effects of tephritid pests, growers must comply with health and safety standards required by the market, applying several cultural, chemical, legal, and biological control methods, including the sterile insect technique (SIT), which are applied using an area-wide approach [6,7]. Parasitoids can be released as part of the biological control of fruit flies, and this method can be enhanced by the application of entomopathogenic nematodes (ENs) and fungi [8–16]. It has been demonstrated that ENs have the potential to attack and infect the larvae of a wide range of fruit fly species and other pests and may thus represent an effective additional strategy when applied to the control of fruit flies in both conventional and organic agricultural production systems of tropical and subtropical fruits [14,17,18]. These ENs are produced at large scale and thus represent a cost-effective option for application as a suppression strategy mainly in domestic backyard fruit trees and marginal areas where the pest takes refuge outside of the crop season [10,14].

Entomopathogenic nematodes are organisms that naturally inhabit the soil and can persist in this habitat due to the wide range of hosts they can attack [19]. The principal route of infection by infective juveniles (IJs) is via natural apertures in the host, such as the spiracles [20], mouth, and anus [21], although some species can penetrate softer tissues such as the intersegmental joints. The juvenile stage, known as the "dauer" larva, is the infective phase that enters the host's body and releases its symbiotic bacteria, causing a type of septicemia (putrefaction of fatty tissue) and consequent death of the host within two to three days following infection [22]. The juveniles feed on the putrefied tissue and continue their development until reaching the adult stage and reproduce within the host. The IJs of the family Heterorhabditidae are associated with bacteria of the genus *Photorhabdus*, and those of the family Steinernematidae with the genus *Xenorhabdus*, both of which belong to the family Gammaproteobacteria [22].

The use of *Steinernema* nematode species to pest suppression is already commonplace and has produced satisfactory results [17] against a number of pests. Their application has been reported in the suppression of populations of the root weevil (*Diaprepes abbreviatus* L.), a pest of citrus trees [23], the Japanese beetle, *Popillia japonica* Newman, and the sweet potato weevil, *Cylas formicarius* (F.) (Coleoptera: Apionidae) [24,25]. They can also effectively infect and kill larvae of the order Lepidoptera when applied in humid environments with moderate temperatures [26]. Similarly, it has been reported that various species of *Heterorhabditis* are effective and can be considered for the biological control of pest insects [27]. The infective capacity of nematodes on dipteran pests has been shown against larvae of *Musca domestica* (L.) [28], the cabbage root fly (Anthomyiidae, *Delia radicum* L.) [29], the onion fly (Anthomyiidae, *Delia antiqua* (Meigen), and the mushroom sciarid fly (Sciaridae, *Lycoriella auripila* Winnertz) [30].

In the case of fruit flies, nematodes mainly infect second- and third-instar larvae (early and late), as well as the adult phase during emergence, while infection in the pupal stage is low or null [9,31,32]. To obtain satisfactory results, it is necessary to optimize the application method of these biological control agents, considering the prevailing abiotic and biotic factors present in the regions where they will be applied, the susceptible stage of the pest, and the phenology of the crops that require protection. Moreover, it has been shown that their greatest potential may be mainly against pest populations of fruit trees in marginal areas, abandoned orchards (unmanaged), greenhouses, commercial orchards, and orchards under organic production. Recent reports have already documented the potential of ENs as biological control agents against fruit flies and other fruit pests [33,34]. Therefore, it is necessary to analyze the advantages they offer and their limitations for successful application under field conditions. The use of these biological control agents will contribute to improving pest management strategies and reducing the use of pesticides in the control of fruit flies in order to produce fruits that meet market quality and food safety requirements.

The main objective of this review was to document the potential offered by ENs of the families Heterorhabditidae and Steinernematidae by analyzing their interaction with biotic and abiotic factors, in terms of developing an effective application strategy that can function as an additional element in the area-wide management of fruit flies.

2. EN–Fruit Fly Interaction

2.1. Fruit Flies

Studies on the effect of ENs on fruit flies have been conducted on the third-instar larvae of different fruit fly species, such as *Anastrepha fraterculus*, *A. ludens*, *A. obliqua*, *A. serpentina*, *A. suspensa*, *Ceratitis capitata*, *C. rosa*, *Bactrocera dorsalis*, *B. litrofons*, *B. oleae*, *B. tryoni*, *B. zonata*, *Dacus ciliatus*, *D. curcubitae*, *D. dorsalis*, *Rhagoletis indifferens*, *R. cerasi*, *R. pomonella*, and *Zeugodacus cucurbitae* usually provided by laboratories that produce them using artificial larval diets according to the specifications established for each species and *serpentina*, *A. suspensa*, *Ceratitis capitata*, *C. rosa*, *Bactrocera dorsalis*, *B. litrofons*, *B. oleae*, *B. tryoni*, *B. zonata*, *Dacus ciliatus*, *D. curcubitae*, *D. dorsalis*, *Rhagoletis indifferens*, *R. cerasi*, *R. pomonella*, and *Zeugodacus cucurbitae* usually provided by laboratories that produce them using artificial larval diets according to the specifications established for each species and laboratory [9,35–43]. For some species, third instar larvae have been obtained from fruits infested by gravid wild females in controlled conditions [44,45] and others from infested fruits collected in field conditions for obtaining the larvae [10,46].

2.2. Nematode Rearing

Some nematodes evaluated in bioassays and applied directly in the field are reproduced through a rearing process, while others are obtained from commercial products distributed to control other pests. The rearing and multiplication of nematodes are conducted in vivo and in vitro following the criteria established in different methodologies. However, they have been conditioned to the requirements of each of the various studies conducted.

In vivo rearing has been conducted by inoculating greater wax moth, *Galleria mellonella* (L.) larvae, where infective juveniles (IJs) were collected using White traps [21]. The IJs produced by this process were used in laboratory bioassays, pilot tests, and for application in small agricultural cooperatives [47]. When a suitable host does not exist or when experimental conditions require it, in vitro rearing methods can be used. In vitro rearing methods allow for the development of nematodes without requiring a host, and they are also used for rearing aposymbiotic nematodes (without symbionts) [28].

In vitro production is conducted in both solid and liquid media. Monoxenic cultures are produced in solid media, where fermentation of the symbiont bacteria is first carried out, followed by the inoculation of IJs for their subsequent reproduction. When production using this process first began, Petri dishes were used with production media based on dog food (kibble), pig kidney, beef blood, and nutrient agars [48]. The Bedding method was subsequently developed [49], adopting the use of Erlenmeyer flasks with polyurethane foam, which were later replaced by sterilizable bags and a sterile air pumping system [47] (Figure 1).

Figure 1. Production of *S. carpocapsae* using Bedding's method. Foam cubes were imbedded in a paste made of cooked chicken livers, sterilized, and inoculated with *Xenorhabditis* sp. Five days later, the bacterial culture was inoculated with IJs. After several generations, IJs were migrated to the flask's wall and collected.

Production in liquid media is conducted using monoxenic cultures. The concentration of symbiont bacteria prior to inoculation with IJs in the production medium is a crucial factor in the productivity of IJs, and in vitro production in liquid media is the most cost-effective process [50]. Although the industrial production of IJs of some species of nematodes is taking place, greater efficiency is still required through the standardization of processes and methods, since the processes are still trial and error.

In laboratory conditions, in vivo and in vitro techniques that are frequently used for rearing nematodes have been replicated as described and reported by several authors and are adapted to the requirements of each laboratory [28]. In in vivo breeding systems, all laboratories use White traps to harvest IJs that develop in live insects (generally *G. mellonella*) used as hosts. The harvested IJs are used in experimental evaluations and to maintain the nematode colonies [51,52]. Two types of media have been described in the in vitro rearing system, liver–kidney and lipids, which are used for the growth and successful reproduction of entomopathogenic nematodes with or without symbiotic bacteria. The liver–kidney agar system involves IJs to mature and reproduce without being associated with a host and is used to rear aposymbiotic nematodes [53]. However, a disadvantage of this rearing system is that it is easily contaminated due to the nutrients it includes, which promote the development of unwanted microorganisms. The lipid agar system is prepared by mixing the medium with nutrient broth, agar, and yeast extract in a 2 L Erlenmeyer flask, then adding distilled water, $MgCl_2{\bullet}6H_2O$, corn oil, and corn syrup. Finally, ~20 mL volumes of sterile agar are placed in Petri dishes. The disadvantage of this method is that the inoculation of IJs and the harvest of nematodes are carried out under a laminar flow hood to reduce the risk of contamination of the medium, and the use of the nematodes produced depends exclusively on the requirements of the laboratory and/or project.

2.3. Commercial Production of ENs

The technology for the mass rearing of IJs has already been developed in several countries, including Mexico [54,55]. Based on standard requirements, namely, those conditioned by supply and demand, it is possible to reproduce IJs of all species at a mass scale [56]. Mass rearing is carried out in large bioreactors and managed to reduce costs and produce large quantities of ENs, ranging from 50,000 to 100,000 L in culture media. This approach is already used in Europe, North America, and Asia [57]. For the specific use of ENs in pest management programs, it should be noted that, although nematodes are easily produced both in vivo and in vitro in various complex semi-solid organic media, the cost of mass production using these methods represents a major constraint to their commercialization [57]. Therefore, development strategies and methodologies undergo slight adaptations according to the process adopted in each site. There are also factors that need to be improved, such as: (1) temperature, which is the easiest factor to control during the process; (2) oxygen concentration, temperature, and pH of the substrate used during the development of IJs; (3) humidity and aeration are more difficult factors to keep constant as the magnitude of mass rearing increases; and (4) the host density and inoculation rate should be estimated for each nematode species and host to optimize production [58]. Once these factors are controlled and standardized, labor costs can be reduced because the processes will be mechanized. In addition, with the mechanization of the process, a more stable and higher quality product should be obtained [59].

Efforts have also been intensified to develop new methods of mechanization of the process to improve in vivo rearing. One example is the "LOTEK" breeding system, which consists of perforated trays that allow for the automated collection of IJs for inoculation, incubating and harvesting IJs in 48 h, and a 97% carcass extraction efficiency [60]. The bases for the design of new separation processes have also been established, and the physical properties of the components have been described [61]. Despite the efforts that have been made, the technology still needs to be improved to optimize the separation process and scale up to a mass rearing system, which will contribute to substantially speed up the process, as in most cases it depends on sedimentation of the IJs by gravity. However, the technology that is being currently developed has been allowed to overcome the setback caused by the mass reproduction of IJs, and thus the priority now is to define a strategy with the greatest potential for field application, and thereby ensure the highest probability of success in terms of pest suppression. Additionally, an in vitro mass rearing method of IJs using solid monoxenic culture with four species of nematodes within the genera *Steinernema* and *Heterorhabditis* (*S. pakistanense*, *S. asiaticum*, *S. feltiae*, and *H. indica*), was described [62]. The method started by homogenizing and placing fresh chicken viscera on a porous foam (polyester polyurethane) substrate. Subsequently, they inoculated 10 mL of *Xenorhabdus* symbiotic bacteria culture in yeast extract (YS) broth. The production of IJs obtained for each species was: between 5 and 7 million per 500 mL flask for *S. pakistanense*, 4 and 5 million for *S. asiaticum*, 1 and 2 million for *S. feltiae*, and between 5 and 7 million for *H. indica*. The authors concluded that this large-scale production process may generate enough IJs with the quality required for applications under field conditions.

2.4. Evaluation of ENs on Fruit Flies

(a) Bioassays

The process begins by counting the number of IJs in a concentrated homogenized suspension, and then 1 mL is added to make a total volume of 100 mL with sterile water. The suspension is homogenized, and 1 mL is evenly distributed in one quadrant of a Petri dish. A stereoscopic microscope is used for quantification, and at least 10 counts (replicates) are performed per batch. The estimated number of IJs in a known volume is then used to make and adjust dilutions to the required densities for evaluation in the experiment.

The efficacy of these microbial control agents has been demonstrated by various laboratory bioassays and field trials. Several studies have been recently conducted against different fruit fly life stages (larvae, pupae, and adults), either exposed directly or indirectly via substrates [32,37,41,63–66]. These tests have been conducted in small experimental arenas, using sand, clay loam, or vermiculite as substrate. Other factors are also considered in these evaluations, including nematode origin, moisture content, temperature, and soil depth of hosts. Each experiment has been carried out independently from each other. The different developmental stages of the fruit fly species evaluated in bioassays are either from mass rearing or from infested fruits under forced laboratory conditions and from larvae that remain in infested mango, orange, and apple fruits [14,44,67]. The moisture content used in the different soil textures or substrates has been in the range of 10–30%.

(b) Experiments simulating natural conditions

Experiments that evaluate the efficacy of IJs under field conditions are relatively scarce; however, results from few studies that have been carried out indicate the potential of these organisms for the control of fruit fly larvae [14,68]. Therefore, the promotion of this strategy has merit, although it should be noted that evaluations have been conducted in experimental arenas of relatively small dimensions (100, 150, 400, 640, 710, and 2500 cm^2) and in soils with controlled moisture content [10,13,14]. However, in a field study where IJs were applied directly on the soil in the drip area of trees using equipment adapted to a tractor, the results indicate the potential of IJs as biological control agents given the significant reduction in the number of adult flies emerging from these treated soils [46]. Similarly, Ref. [44] applications of IJs at a rate of 180 IJs/cm^2 on infested fruits in experimental arenas with a surface area of 132 cm^2 (12 × 11 cm), where each experimental unit contained 4 kg of sandy clay soil with a moisture content of 25% (w/v ratio) and three infested fruits were placed in the treated soil, have been tested [44]. Ten days after treatment, the soil was sieved to separate the larvae and to determine the infection level. Larvae that remained inside the fruits were collected with entomological forceps and observed to determine the infection. All the larvae and pupae were placed in Petri dishes containing moistened filter paper and dissected under a stereoscopic microscope to verify the presence of IJs and thus estimate the infection.

(c) Effect of IJs combined with pesticides and other biological control agents

In almost every case, the use of IJs combined with pesticides and other biological control agents has produced synergistic results when evaluated against fruit pests. When fruit flies were evaluated with IJs plus Spinosad, this resulted in a higher mortality of larvae of *C. capita* than with IJs or insecticide tested alone [69]. Moreover, when IJs have been applied with other insecticides, mortality in *C. capitata* pupae was higher or similar compared to IJs applied alone [66]. This synergistic effect provides greater efficacy to these control agents and prolongs the useful life of insecticides by delaying the development of resistance in the pests. Nematodes can also be applied in combination with entomopathogenic fungi such as *Beauveria bassiana* and *Metarhizium anisopliae*, since their efficacy is not reduced, and this method has shown to have a greater impact on the larvae, pupae, and adults of *Bactrocera zonata* and *B. darsolis* [70]. Additionally, the use of entomopathogenic fungi and nematodes separately was assessed against the larvae of *C. capitata* and showed that both biological control agents have a similar efficacy at the evaluated concentrations [71]. The combined applications of *B. bassiana* with *H. bacteriophora*, *H. megidis*, *S. feltiae* or *S. carpocapsae* against *C. capitata* resulted in synergistic interactions under laboratory conditions [72,73]. Additionally, when IJs of *S. feltiae* were applied alone and in combination with the conidia of three species of entomopathogenic fungi (*B. bassiana* (Bals.-Criv.) Vuill., *M. robertsii* JF Bisch, SA Rehner and Humber, and *Isaria fumosorosea* Wize) against larvae of *B. dorsalis*, higher mortality was recorded when they were applied together. Similarly, when two species of nematodes (*S. carpocapsae* and *S. riobrave*) were evaluated in combination with *B. bassiana*, *M. bruneum*, and *I. javanica*, a synergistic effect against *R. pomonella* larvae was observed [41]. However, there are other cases where there has been no indication of synergy, but there was also no effect of the fungi on the IJs [74].

(d) Environmental conditions

Entomopathogenic nematodes are natural inhabitants of the soil that live and persist in that environment, attacking many species of insect hosts that spend at least one stage of their life cycle in the soil. However, EN populations are affected by various biotic and abiotic factors, among which the main abiotic factors are extreme temperatures, low moisture content, and soil texture and pH. The main biotic factors are nematophagous fungi and acari, as well as some bacteria that can cause illnesses in nematodes. In the laboratory bioassays discussed in this review, Petri dishes and plastic containers of different sizes have been utilized as experimental arenas with filter paper, vermiculite, or soils of different textures as substrates. When filter paper is used, larvae or pupae used for tests do not serve as the diet or substrate. In contrast, when other substrate types, such as soils, are used, they must be sterilized beforehand in autoclaves to ensure the absence of any natural enemies of IJs. The textures of these soil types used as substrates are sandy, sandy loam, sandy clay, and clay. In general, the mean temperature used in these evaluations is 25 ± 1 °C. However, the range of temperatures reported is 15 to 35 °C. In studies conducted under field conditions, temperature is not controlled, but a range of 25 to 28 °C has generally been recorded [14]. In the case of environmental humidity, evaluations have been conducted in a range of 50 to 80% relative humidity (RH), with a mean value of $70 \pm 5\%$ R.H in most studies. However, interactions with abiotic factors, such as environmental temperature, soil texture and moisture content, and UV light, have also been analyzed. It is important to define the optimal soil moisture content, since it has been shown that high values of this parameter limit the displacement of IJs of *H. bacteriophora*. A soil moisture content of $\geq$30% has been observed to result in lower infection levels on *A. obliqua* larvae [75–78]. These results reflect the potential of these organisms against cryptic pests such as fruit flies, and a strategy that reflects the most effective application method must thus be defined, as well as the dose (density of IJs) required for application in the field.

2.5. Infective Capacity of Nematodes in Soil

The mobility of IJs is closely related to the texture and humidity of the soil. It has been observed that the infective capacity of *H. bacteriophora* on larvae of *A. obliqua* is higher in soils with a moisture content of 20, 25, and 30%. Despite slight variations in the levels of larval mortality, no significant difference was found among these moisture content values. In contrast, another study found that soils with a moisture content of 15% resulted in 40% mortality, while a moisture content of 10% resulted in mortality of 8%, which was the lowest value observed across all the different moisture content levels evaluated [74]. The same pattern was observed in *A. ludens* and *A. serpentina*, where the highest larval mortality by *H. bacteriophora* and *S. feltiae* was found in sandy clay soil with a moisture content of 15% and a temperature of 26 ± 1 °C [31,77,79,79]. Similarly, a mortality of 40% was observed in *A. obliqua* larvae in soil with a moisture content of 15%, while a moisture content of 25% resulted in 50% mortality [44]. Additionally, when IJs were applied directly on the soil with infested fruits placed on it and where IJs were applied directly on the infested fruits, this resulted in 12.3% and 16.6% larval mortality, respectively, and no significant differences were found between the two treatments [44].

There is evidence that when IJs are applied directly on infested fruits, there is an acceptable level of infection in both the larvae that are still inside the fruits and those that have left the fruits to pupate on the ground. However, the highest mortality by EN infection has been observed when IJs were applied on the soil at a density of 180 IJs/cm^2, where a mortality of 27.5% was recorded in larvae recovered from the soil and extracted from the fruits. Infected juveniles applied directly on the fruits resulted in a mortality of 24.6%, both in larvae that had emerged to pupate in the soil and those extracted from the fruits. Similarly, IJ performance differs between different soil textures, as has been observed using *A. ludens* and *A. obliqua* fruit flies, where the best performance was recorded in sandy clay soil [31,32,79,80]. Every day, new species with potential as biological control agents against larvae and pupae of *C. capitata* are reported, such as *H. noenieputensis* and *S. yirgalemense* [81].

The effect of these two application methods was evaluated under field conditions on 36 fruits infested by *A. obliqua*, where a total of 423 pupae were obtained from soil treated with IJs that were applied on infested fruit, and 52 infected larvae were obtained from the fruits, which determined a 12.3% infection in larvae. However, when IJs were sprayed directly on the infested fruits, a total of 570 pupae were obtained from the soil, and 95 larvae were extracted from the fruits, resulting in a 16.7% infection in larvae obtained from fruits [44]. A similar strategy was proposed to control *B. dorsalis* and other species of fruit flies with ENs in sub-Saharan Africa [21].

3. Factors Affecting Infectivity

(a) Abiotic factors

During the development of the fruit flies' life cycle, larvae and pupae are biological phases that occur in soils with different textures and moisture content and at different depths [82–84], which are optimal environments for the survival of EN IJs and for effectively attacking fruit fly larvae. For IJs to provide greater control of these pests, it is necessary to determine the influence of the factors that may limit their infective capacity. It is also crucial to consider the physical aspects that may limit the performance of IJs in the sites where they will be applied, among which the main ones are those associated with soil, such as humidity, texture, temperature, pH, and solar radiation. Therefore, it is necessary to determine the interaction between ENs and abiotic factors [14,32,40,85,86]. The first report of naturally occurring attacks by the nematode *Neoaplectana* (*Steinernema*) sp. was against *R. pomonella* larvae in Massachusetts, USA [87]. Subsequently, the nematodes were experimentally tested [41,42].

Soil texture is closely associated with soil moisture content, and this factor has been found to directly impact the movement of IJs of *H. indica* [31,32,88]. Thus, the performance of IJs is directly dependent on soil texture, as observed in IJs infecting the larvae of *A. obliqua*, which performed better in soil with sandy clay texture compared to sandy and sandy loam soils, all with a moisture content of 10% [75]. However, the foraging strategy of the nematodes must also be considered. For example, the "dauer" larvae of some nematode species, such as those of the genus *Heterorhabditis*, detect their host while moving randomly through the soil, while other species, such as *S. carpocapsae* and *S. feltiae*, and other *Steinernema* spp., ambush their host in the upper soil layers, and other species use both strategies (*Heterorhabditis* spp.). In addition, higher larvae infestation levels have been recorded in soils with a moisture content of 20, 25, and 30%, but the infective capacity of IJs has been observed to decrease with lower soil humidity. It has been reported that the IJs of *H. bacteriophora* have a greater capacity to move towards their host in sandy soils when the substrate has a higher moisture content [31,80]. Soil texture, moisture content, depth, and compression all influence the mobility of IJ hosts, as is the case of the larvae of several fruit fly species, where the soil depth where pupation occurs has been found to be related to these factors [82,89]. Soil texture may thus favor or limit the infective capacity of IJs [90].

Soil texture also influences the infection level caused by IJs to a great extent. Evaluations of different soil textures under laboratory conditions showed that soils with a sandy clay texture exhibit the greatest displacement of IJs of *H. bacteriophora* and *S. feltiae*, and the IJs in these soils also cause higher larval mortality compared to other soil textures [31,79]. Some EN species show a behavior similar to that observed in *H. indica* [91], and an evaluation of the vertical displacement and infective capacity of *Neoplectana* (*Steinernema*) *carpocapsae* in substrates with four different textures (pure silica sand, sandy clay, sandy loam, and clay) showed mortalities of 62.9, 58.6, 41.9, and 13.2%, respectively. These results indicate that the infective capacity of IJs when attacking and infecting the pupae of *G. mellonella* appears to decrease as the clay content of the substrate increases.

Entomopathogenic nematodes have a wide host range but, due to their excellent foraging ability, they have a low impact on non-target organisms. They also have a high reproductive rate, and can thus be mass-produced at a low cost. Their application in the field is compatible with various control methods already used for other agricultural pests [92].

Under laboratory conditions, satisfactory results have been reported in evaluations with larvae and recently emerged adults of different fruit fly species, such as *C. capitata*, *B. oleae*, *B. dorsalis*, *A. ludens*, *A. fraterculus*, *A. serpentina*, and *A. obliqua* [14,88,93–98]. Moreover, the species *S. riobrave* has been shown to cause high mortality in the third-stage larvae of *A. ludens* [99]. Additionally, bioassay-based characterizations of other ENs species are being carried out to determine their potential as biological control agents; as in the case of *Oscheius tipulae* tested against *C. capitata*, ~65% and 75% mortality was recorded with the application of 250 and 500 IJs per larva, respectively [100].

However, the persistence and efficacy of ENs are conditioned by abiotic factors such as ultraviolet light, extreme temperatures, and soil moisture content, pH, and texture, among others [101–103], and by biotic factors associated with the habitat. These include the availability and abundance of hosts and the interaction with their natural enemies in the form of pathogenic microorganisms (fungi) and microarthropods, such as hematophagous mites [103]. These cause a significant reduction in the abundance of IJs, raising questions regarding their efficacy. Adverse temperatures affect the oxygen exchange, persistence, survival, and dispersal of nematodes, and thus limit their infective capacity [104]. In addition, nematodes are also sensitive to low soil moisture content and extreme temperatures, and therefore, all these abiotic factors must be considered fundamental to ensure higher survival in the field for successful pest suppression [104,105].

A study under field conditions using concentrations of 5000, 1500, 500, and 150 IJs/cm^2 showed that *S. feltiae* has a high infective capacity as it caused high mortality in the larvae of *C. capitata* [95]. However, other strains of this nematode species have caused high mortality at a lower IJ density. It has also been observed that a density of approximately 1000 IJs of *S. feltiae* per 3 dL of soil causes an average mortality of 58% under controlled conditions, while a density of 10,000 IJs per 3 dL of soil causes a mortality of 96%. This is a significantly higher mortality than that observed in the presence of entomopathogenic fungi and bacteria in soils without nematodes [74]. However, the foraging strategies of IJs, which are classified into ambushers and cruisers, may also influence the selection of nematodes for pest control programs.

Nematode concentration is closely associated with the foraging activity of species of the genus *Heterorhabditis* [105] since a mortality of more than 50% was reported in the larvae of five white grub species (Coleoptera: Scarabeidae) following the application of 30,000 IJs of *Heterorhabditis* sp. in soil [106]. It should be noted that this amount of soil is lower than that used in evaluations with fruit flies, but this example helps to emphasize the need to apply higher densities of IJs to achieve higher mortality in soil pests.

It has also been observed that many larvae removed from fruits exposed to IJs have not been infected, which suggests that the larvae were late third instars, and the nematodes were not able to infect them at this developmental stage [44]. This occurs because, once the puparium is formed, the hard exoskeleton acts as a protective cover, as spiracles, which are natural openings for infection, practically close and are surrounded by small spinnerets that act as a barrier that prevents penetration by IJs. Therefore, if the IJs encounter their host in the pupal stage, they cannot penetrate inside its body [79].

IJs have high ability to infect fruit fly larvae in sandy clay soil, with higher mortality when the moisture content is between 15 and 25%. It has been reported that inoculations of 1.2 to 2.5 billion nematodes/ha produce satisfactory results in terms of controlling larvae of coleopteran and dipteran pests [101].

To apply IJs for pest suppression purposes, the first step is to consider the availability of the infective agents and the cost associated with their production. However, if we consider that large-scale production has already been developed to produce large quantities of IJs at low cost, it only remains to define and optimize a viable route for their application in the environments where the pest is located, mainly where an area-wide integrated management program is carried out. Infective juveniles have been applied in this way to regulate populations of larvae or pupae of the citrus root weevil, *Diaprepes abbreviates*, and the plum curculio (*Conotrachelus nenuphar* (Herbst) [23,107].

Field studies on the use of IJs against fruit flies are limited and an adequate application strategy has yet to be defined. However, the results obtained regarding the use of IJs against the main fruit fly species [108], and the prevailing interest in generating more sustainable strategies suggest that there is an increasing use of IJs for pest control, and their application for the control of fruit flies should therefore be promoted in orchards, such as mango, citrus, guava, sapodilla, where these flies cause damage. Infective juveniles of ENs should be applied on the soil mainly in the drip area of "trap trees" within marginal areas, which exhibit high levels of damage caused by fruit flies, in line with the criteria used to control the plum weevil, *C. nenuphar* [107]. The knowledge generated provides the basic elements to integrate a strategy for the application of ENs in the field and to reduce the use of pesticides for pest control. However, the use of IJs will be enhanced with additional knowledge about the potential they offer to suppress fruit fly populations under field conditions in different environments and that will help to define the best form of application in the field. This approach could be used against fruit flies and contribute to the production of fruits and vegetables with less pesticide residues.

(b) Biotic factors

Among the main biotic factors that affect the efficacy of nematodes, fungi and bacteria are the most important. An antagonistic effect of these organisms has been observed on nematodes, as they cause lethal diseases and thus significantly reduce the number of effective IJs per soil area [109]. Nematophagous organisms such as mites and insects may limit the survival of entomopathogenic nematodes applied in the field [103]. The reduction in IJs by natural enemies and their limited dispersal capacity questions the effectiveness of these organisms as biological control agents, and thus it may sometimes be necessary to carry out additional applications for better pest control [110].

The constant presence of hosts is an important factor for the persistence of IJs in the soil. In this sense, tropical and subtropical environments can constantly provide insect hosts that guarantee a permanent population of IJs [111]. Therefore, considering that fruit fly larvae use the soil to pupate and that mango, citrus, guava, and sapodilla, among other orchards, are located in tropical and subtropical regions where temperatures are warm and humidity is high, combined with a predominance of soils with clayey loam and sandy clayey textures, a control program for these pests can be implemented by applying nematodes as an additional control agent, which would result in less environmental pollution caused by the conventional pesticides that are currently applied.

fly larvae use the soil to pupate and that mango, citrus, guava, and sapodilla, among other orchards, are located in tropical and subtropical regions where temperatures are warm and humidity is high, combined with a predominance of soils with clayey loam and sandy clayey textures, a control program for these pests can be implemented by applying nematodes as an additional control agent, which would result in less environmental pollution caused by the conventional pesticides that are currently applied.

Considering the advantages and disadvantages that entomopathogenic nematodes face against fruit flies, areas of opportunity are identified to make their use more effective in a management program (Figure 2). An analysis was performed to determine the strengths, opportunities, weaknesses, and threats (SWOT) on the use of EPNs. Then, we can see that nematodes face several weaknesses, but there are also several opportunities that we should investigate to improve their performance, mainly under field conditions.

Figure 2. Factors that promote and limit the use of entomopathogenic nematodes for their application against fruit flies.

4. Future Perspective

To date, a total of 70 scientific articles have been published in 15 countries, which report different results that generally indicate a high level of ENs infection against fruit fly larvae and pupae demonstrating their potential as control agents against these pests. Additionally, a total of 32 species of ENs have been evaluated against 18 fruit fly species (Table 1). The study of this topic will continue in a dynamic way, focused mainly against the most economically important species such as *C. capitata* [112–117], *B. oleae* [118,119], *B. dorsalis* [120], *R. cerasi* and *A. fraterculus* [121]. The exploration of new species of nematodes continues and bioassays are carried out against the different biological stages of fruit flies [115]. It is also necessary to explore the most effective routes for its application in the field [107].

Currently, more than a dozen commercial EN strains have been developed and commercialized, globally, using 13 EN species. As mentioned above, ENs can be produced in vivo (by infecting susceptible insects), in vitro on solid media, and in vitro liquid media. The last two techniques require the culture of the symbiotic bacteria, before inoculating the IJs. The latter technique is the most efficient and cost/benefit accepted, in spite of the requirement of initial large investments. Still, [illegible] can be made to improve the production efficiency and biocontrol potential. For example, in vitro production can be improved by optimizing the culture media and by expanding the fermentation processing of the symbiotic bacteria. Additionally, the development of programs focused on finding improved nematode strains and the stabilization of beneficial traits, related to the obtain a stable production of ENs, is highly recommended. These are some of the main constraints in the productions of commercial EN products [122].

On the other hand, the short shelf-life of their commercial products may be the most important limitation to expand the commercial use of ENs. ENs must be prepared in an adequate formulation that should keep IJs under a humid and aeriated environment to keep them alive for an acceptable time. Many efforts have been carried out on this constrain and ideas such as the use of clay pellets with humid cores or, more recently, the use of gels balls (i.e., alginate capsules), have improved the shelf-life of these products. However, more research is required to make a real advancement, such as the development of methods focused to obtain dormant individuals which can be recovered before use. The specific use of ENs in fruit fly control has the interesting alternative of using fruits infested with fruit fly larvae, previously treated with IJs, and then, scattered under the dripping area of fruit trees [59].

Table 1. Species of entomopathogenic nematodes evaluated against several species of fruit flies in different experimental arenas.

Fruit Flies Species	Nematodes Entomopathogenic Species Tested	N^1	References
Anastrepha fraterculus	*Heterorhabditis bacteriophora* *H. chongmingensis* *H. amazonensis* *Heterorhabditis* sp. *Steinernema feltiae* *S. carpocapsae* *S. glaseri* *S. riobrave* *S. rarum* *Steinernema* sp. *Oscheius* sp.	1 1 1 1 2 2 1 1 1 2 2	Barbosa-Negrisoli et al., 2009 [97] Foekel et al., 2016 [37] Foekel et al., 2017 [38] Chaneiko et al., 2021 [65]
A. ludens	*H. bacteriophora* *S. carpocapsae* *S. feltiae* *S. glaseri* *S. riobrave*	1 1 1 1 1	Lezama-Gutiérrez et al., 1996 [99]. Lezama-Gutiérrez et al., 2006 [80]. Toledo et al., 2005 [32]. Toledo et al., 2006 [14]. Toledo et al., 2014 [76]
A. serpentina	*H. bacteriophora*	1	Toledo et al., 2006 [98]
A. suspensa	*H. bacteriophora* *H. indica* *H. floridensis* *H. heliothidis* *H. zealandica* *S. carpocapsae* *S. feltiae* *S. glaseri* *S. riobrave* *S. rarum* *S. diaprepesi*	1 1 1 1 1 1 1 2 1 1 1	Beavers & Calkin., 1984 [93]. Have et al., 2017 [9] Have et al., 2018 [10]
A. obliqua	*H. bacteriophora* *S. carpocapsae*	1 1	Toledo et al., 2005 [31] Toledo et al., 2009 [75]
Bactrocera dorsalis	*H. indica* *H. taysearae* *Heterorhabditis* sp. *S. feltiae* *S. taysearae* *S. kandii*	1 1 1 1 1 2	Godjo et al., 2018 [13] Godjo et al., 2021 [120] Menzler-Hokkanen et al., 2022 [74]. Usman et al., 2021 [43] Wakil et al., 2022 [70]
B. latifrons	*S. siamkayai*	1	Ta-Oun et al., 2022 [78].
B. oleae	*H. bacteriophora* *H. marelatus* *S. carpocapsae* *S. feltiae* *S. glaseri* *S. riobrave*	2 1 2 1 1 1	Sirjani et al., 2009 [123]. Torrini et al., 2017 [118]. Torrini et al., 2020 [119].

Table 1. *Cont.*

Fruit Flies Species	Nematodes Entomopathogenic Species Tested	N^1	References
B. tryoni	*H. bacteriophora* *S. carpocapsae* *S. feltiae*	1 1 2	Langford et al., 2014 [86]. Aryal et al., 2022 [124] Attalla & Eweis, 2002 [63] Usman et al., 2021 [43].
B. zonata	*H. bacteriophora* *S. carpocapsae* *S. feltiae*	2 2 1	Mohmoud & Mohamed-Osman 2007 [125] Fetoh et al., 2011 [64] Nouh & Hussein 2014 [126]
Ceratitis capitata	*H. bacteriophora* *H. indica* *H. zealandica* *H. baujardi* *H. noenieputensis* *H. megidis* *H. downesi* *Heterorhabditis* sp. *S. (Neoplactana) carpocapsae* *S. carpocapsae* *S. feltiae* *S. riobrave* *S. brazilense* *S. weiseri* *S. yirgalemense* *S. khoisanae* *Oscheius tipulae*	6 1 1 1 1 1 1 2 1 7 6 1 1 1 1 1 1	Almeida et al., 2007 [71]. Shaurub et al., 2015 [114]. Minas et al., 2016 [68]. Nouh & Hussein 2014 [126] Abdel-Razek & Abd-Elgawad 2021 [69] Kaprananas et al., 2021 [45] Yağcı1 et al., 2021 [116] Lindegren & Vail 1986 [94] Lindegren et al., 1990 [95] Kapranas et al., 2023 [117] Gazit et al., 2000 [67] Kepenekci & Susurluk 2006 [127] Karagoz et al., 2009 [8]. Malan & Manrakhan 2009 [40]. James et al., 2018 [81].
C. rosa	*H. bacteriophora* *H. zealandica* *S. khoisanae*	1 1 1	Malan & Manrakhan 2009 [40]
Dacus cialiatus	*H. bacteriophora* *S. carpocapsae* *S. riobrave*	3 3 1	Fetoh & El-Gendi 2006 [36] Fetoh et al., 2011 [64] Kamali et al., 2013 [85]
D. curcubitae	*S. feltiae*	1	Lindegren & Vail 1986 [94]
D. dorsalis	*S. feltiae*	1	Lindegren & Vail 1986 [94]
Rhagoletis cerasi	*H. bacteriophora* *H. marelatus* *S. carpocapsae* *S. feltiae*	1 1 1 1	Kepenekci & Susurluk 2006 [127] Kepenekci et al., 2015 [121]
R. indifferens	*H. bacteriophora* *H. marelatus* *S. carpocapsae* *S. feltiae* *S. riobrave* *S. intermedium*	1 1 1 1 1 1	Yee & Lacey 2003 [128] Patterson & Lacey 1999 [129]
R. pomonella	*H. bacteriophora* *H. indica* *S. (Neoaplectana)* sp. *S. carpocapsae* *S. feltiae* *S. riobrave*	1 1 1 2 1 2	Poinar Jr et al., 1977 [53] Usman et al., 2020a [41] Usman et al., 2020b [42]

N^1 = Frecuency of tested.

5. Conclusions

Based on the results reported to date, the larval phase appears to be the most susceptible biological stage to attacks by IJs, followed by the adult stage, while the pupal stage is the least susceptible one. The most frequent studies conducted to date have focused on determining whether IJs of different species have the potential to infect and kill fruit flies.

Considering the results obtained in different studies carried out under both laboratory and field conditions, it has been demonstrated that ENs have the potential to infect and kill larvae and adults of fruit flies to be considered effective biological control agents of these pests. Infective juveniles could be applied using a backpack sprayer, boom sprayer, microjet irrigation systems, trunk-sprayers, subsurface injection, or baits. However, it is necessary to know the physical and chemical properties of the soil where the fruit orchards are established for an optimal IJ performance. In this way, IJs can be more efficient at searching for and finding the different life stages of fruit flies that may exist in the tree drip area where they are less likely to be affected by conventional chemical insecticides. It has also been observed that the efficacy of these organisms is directly related to the texture, moisture content, and temperature of the substrate and, to a lesser extent, to host species and age. Interestingly, IJs applied directly on the soil can cause a high mortality of larvae when these leave the fruit to pupate in the soil and while they are still inside the fruit. Although the mortality caused by nematodes in larvae inside the fruit is lower, the application may be effective in domestic backyard trees, abandoned orchards, or fruit fly host trees located in marginal areas where no other control methods are used. Further studies are required to develop an effective method of application in the field that will provide the highest level of pest suppression.

Author Contributions: Conceptualization, writing draft original and structure the idea, J.T.; writing—review and editing, B.M.M.-A.; writing—original draft preparation, P.L.; review and editing the final version, J.E.I. All authors have read and agreed to the published version of the manuscript.

Funding: Funding came from core budgets of El Colegio de la Frontera Sur and Centro de Investigaciones Avanzadas—IPN.

Data Availability Statement: No new data were created, or all data cited are available. Therefore, authorization is not required.

Acknowledgments: We acknowledge Trevor Williams (Instituto de Ecología, A.C. -INECOL-) for their support and advise in the early stages of our research on ENs and for the suggestions made to the draft. To Azucena Oropeza (ECOSUR) for technical support.

Conflicts of Interest: The authors declare no conflict of interest.

References

1. Norrbom, A.L.; Kim, K.C. *A List of the Reported Host Plants of the Species of Anatrepha (Díptera: Tephritidae)*; United States Department of Agriculture, APHIS: Washington, DC, USA, 1988; pp. 81–52.
2. Drew, R.A.I.; Hancock, D.L. The *Bactrocera dorsalis* complex of fruit flies (Diptera: Tephritidae: Dacinae) in Asia. *Bull. Entomol. Res. Suppl. Ser.* **1994**, *2*, 1–68. [CrossRef]
3. Hernández-Ortiz, V. *El Género Anastrepha Schiner en México (Diptera: Tephritidae): Taxonomía, Distribución y sus Plantas Huéspedes*; Instituto de Ecología, A.C. & la Sociedad Mexicana de Entomología: Xalapa, Mexico, 1992; p. 162.
4. White, I.M.; Elson-Harris, M.M. *Fruit Flies of Economic Significance: Their Identification and Bionomics*; CAB International: Wallingford, UK, 1992; p. 601. [CrossRef]
5. Ekesi, S.; De Meyer, M.; Mohamed, S.A.; Virgilio, M.; Borgemeister, C. Taxonomy, ecology, and management of native and exotic fruit fly species in Africa. *Annu. Rev. Entomol.* **2016**, *61*, 219–238. [CrossRef]
6. Reyes, J.; Santiago, G.; Hernández, P. The Mexican fruit fly programme. In *Area-Wide Control of Fruit Flies and Other Insect Pests*; Tan, K.H., Ed.; Penerbit Universiti Sains Malaysia: Penang, Malaysia, 2000; pp. 377–380.
7. Enkerlin, W.; Ruelas, J.G.; Pantaleon, R.; Litera, C.S.; Cortés, A.V.; López, J.Z.; Dávila, D.O.; Gerardo, P.M.; Villarreal, L.S.; Roldán, E.C.; et al. The Moscamed Regional Programme: Review of a success story of area-wide sterile insect technique application. *Entomol. Exp. Appl.* **2017**, *164*, 188–203. [CrossRef]
8. Karagoz, M.; Gulcu, B.; Hazir, C.; Kaya, H.K.; Hazir, S. Biological control potential of Turkish entomopathogenic nematodes against the Mediterranean fruit fly *Ceratitis capitata*. *Phytoparasitica* **2009**, *37*, 153–159. [CrossRef]

9. Heve, W.K.; El-Borai, F.E.; Carrillo, D.; Duncan, L.W. Biological control potential of entomopathogenic nematodes for management of Caribbean fruit fly, *Anastrepha suspensa* Loew (Tephritidae). *Pest Manag. Sci.* **2017**, *73*, 1220–1228. [CrossRef] [PubMed]
10. Heve, W.K.; El-Borai, F.E.; Carrillo, D.; Duncan, L.W. Increasing entomopathogenic nematode biodiversity reduces efficacy against the Caribbean fruit fly *Anastrepha suspensa*: Interaction with the parasitoid *Diachasmimorpha longicaudata*. *J. Pest Sci.* **2018**, *19*, 799–813. [CrossRef]
11. Heve, W.K.; El-Borai, F.E.; Johnson, E.G.; Carrillo, D.; Crow, W.T.; Duncan, L.W. Responses of *Anastrepha suspensa*, *Diachasmimorpha longicaudata*, and sensitivity of guava production to *Heterorhabditis bacteriophora* in fruit fly integrated pest management. *J. Nematol.* **2018**, *50*, 261–272. [CrossRef]
12. Heve, W.K.; Adjadeh, T.A.; Billahc, M.K. Overview and future research needs for development of effective biocontrol strategies for management of *Bactrocera dorsalis* Hendel (Diptera: Tephritidae) in sub-Saharan Africa. *Pest Manag. Sci.* **2021**, *77*, 4224–4237. [CrossRef] [PubMed]
13. Godjo, A.; Zadji, L.; Decraemer, W.; Willems, A.; Afouda, L. Pathogenicity of indigenous entomopathogenic nematodes from Benin against mango fruit fly (*Bactrocera dorsalis*) under laboratory conditions. *Biol. Control* **2018**, *117*, 68–77. [CrossRef]
14. Toledo, J.; Rasgado, M.A.; Ibarra, J.E.; Gómez, A.; Liedo, P.; Williams, T. Infection of *Anastrepha ludens* following soil aplications of *Heterorhabditis bacteriophora* in a mango orchard. *Entomol. Exp. Appl.* **2006**, *119*, 155–162. [CrossRef]
15. Toledo, J.; Flores, S.; Campos, S.; Villaseñor, A.; Enkerlin, W.; Liedo, P.; Valle, Á.; Montoya, P. Pathogenicity of three formulations of *Beauveria bassiana* and efficacy of autoinoculation devices and sterile fruit fly males for dissemination of conidia for the control of *Ceratitis capitata*. *Entomol. Exp. Appl.* **2017**, *164*, 340–349. [CrossRef]
16. Montoya, P.; Toledo, J. Estrategias de control biológico. In *Moscas de la Fruta: Fundamentos y Procedimientos Para su Manejo*; Montoya, P., Toledo, J., Hernández, E., Eds.; S y G Editores: México City, Mexico, 2020; pp. 169–182.
17. Dolinski, C.; Lacey, L.A. Microbial control of arthropod pests of tropical tree fruits. *Neotrop. Entomol.* **2007**, *36*, 161–179. [CrossRef]
18. Dolinski, C. Entomopathogenic nematodes against the main guava insect pests. *BioControl* **2016**, *61*, 325–335. [CrossRef]
19. Kaya, H.K. Entomogenous nematodes for insect control in IPM systems. In *Biological Control in Agricultural IPM Systems*; Hoy, M.A., Herzog, D.C., Eds.; Academic Press: New York, NY, USA, 1985; pp. 283–302.
20. Triggiani, O.; Poinar, G.O., Jr. Infection of adult Lepidoptera by *Neoaplectana carpocapsae* (Nematoda). *J. Invertebr. Pathol.* **1976**, *27*, 413–414. [CrossRef]
21. Woodring, J.L.; Kaya, H.K. *Steinernematid and Heterorhabditid Nematodes: A Handbook of Biology and Techniques*; Southern Cooperative Series Bulletin 331; Arkansas Agricultural Experiment Station: Stuttgart, AR, USA, 1988.
22. Lewis, E.E. Behavioral ecology. In *Entomopathogenic Nematology*; Gaugler, R., Ed.; CAB International: Wallingford, UK, 2002; pp. 205–224.
23. Stuart, R.J.; El-Borai, F.E.; Duncan, L.W. From augmentation to conservation of entomopathogenic nematodes: Trophic cascades, habitat manipulation and enhanced biological control of *Diaprepes abbreviatus* root weevils in Florida citrus groves. *J. Nematol.* **2008**, *40*, 73–84.
24. Jansson, R.K.; Lecrone, S.H.; Gaugler, R. Field efficacy and persistence of entomopathogenic nematodes (Rhabditida: Steinernematidae, Heterorhabditidae) for control of sweetpotato weevil (Coleoptera: Apionidae) in southern Florida. *J. Econ. Entomol.* **1993**, *86*, 1055–1063. [CrossRef]
25. Klein, M.G. Biological control of scarabs with entomopathogenic nematodes. In *Nematodes and Biological Control of Insect Pest*; Bedding, R., Akhurst, R., Kaya, H., Eds.; CSRIO: East Melbourne, Australia, 1993; pp. 49–57.
26. Glazer, Y.; Navon, A. Activity and persistence of entomoparasitic nematodes tested againt *Heliothis armigera* (Lepidoptera: Noctuidae). *J. Econ. Entomol.* **1990**, *83*, 1795–1800. [CrossRef]
27. Adams, B.J.; Fodor, A.; Koppenhofer, H.S.; Stackebrandt, E.; Stock, S.P.; Klein, M.G. Biodiversity and systematics of nematode-bacterium entomopathogens. *Biol. Control* **2006**, *37*, 32–49. [CrossRef]
28. Mullens, B.A.; Meyer, J.A.; Cyr, T.L. Infectivity of insect-parasitic nematodes (Rhabditida: Steinernematidae, Heterorhabditidae) for larvae of some Manure-breeding flies (Diptera: Muscidae). *Environ. Entomol.* **1987**, *16*, 769–773. [CrossRef]
29. Chen, S.; Han, X.; Moens, M. Effect of temperature on the pathogenicity of entomopathogenic nematodes (Steinernema and Heterorhabditis spp.) to *Delia radicum*. *BioControl* **2003**, *48*, 713–724. [CrossRef]
30. Navarro, M.J.; Gea, F.J. Entomopathogenic nematodes for the control of phorid and sciarid flies in mushroom crops. *Pesqui. Agropecuária Bras.* **2014**, *49*, 11–17. [CrossRef]
31. Toledo, J.; Martinez, C.; Liedo, P.; Ibarra, J.E. Susceptibilidad de larvas de *Anastrepha obliqua* Macquart (Diptera: Tephritidae) a Heterorhabditis bacteriophora (Poinar) (Rhabditidae: Heterorhabditidae) en condiciones de laboratorio. *Vedalia* **2005**, *12*, 11–22.
32. Toledo, J.; Ibarra, J.E.; Liedo, P.; Gomez, A.; Rasgado, M.A.; Williams, T. Infection of *Anastrepha ludens* (Diptera: Teprhitidae) larvae by *Heterorhabditis bacteriophora* (Rhabditida: Heterorhabditidae) under laboratory and field conditions. *Biocontrol Sci. Technol.* **2005**, *15*, 627–634. [CrossRef]
33. Diksha; Mahajan, E.; Singh, S.; Kaur-Sohal, S. Potential biological control agents of *Zeugodacus cucurbitae* (Coquillett): A review. *J. Appl. Entomol.* **2022**, *146*, 917–929. [CrossRef]
34. Shaurub, E.H. Review of entomopathogenic fungi and nematodes as biological control agents of tephritid fruit flies: Current status and a future vision. *Entomol. Exp. Appl.* **2023**, *171*, 17–134. [CrossRef]

35. Orozco-Dávila, D.; Quintero, L.; Hernández, E.; Solís, E.; Artiaga, T.; Hernández, R.; Ortega, C.; Montoya, P. Mass rearing and sterile insect releases for the control of *Anastrepha* spp. pests in Mexico—A review. *Èntomol. Exp. Appl.* **2017**, *164*, 176–187. [CrossRef]
36. Fetoh, B.E.A.; El-Gendi, S.S. Impact of entomopathogenic nematodes on different stages of the pumpkin fly, *Dacus ciliatus* as a new approach in its biological control. *Fayoum J. Agric. Res. Dev.* **2006**, *20*, 19–25. [CrossRef]
37. Foelkel, E.; Monteiro, L.B.; Voss, M. Virulence of nematodes against larvae of the south American fruit fly in laboratory using soil from Porto Amazonas, Paraná, Brazil, as substrate. *Ciência Rural* **2016**, *46*, 405–410. [CrossRef]
38. Foelkel, E.; Voss, M.; Monteiro, L.B.; Nishimura, G. Isolation of entomopathogenic nematode in an apple orchard in southern Brazil and its virulence to *Anastrepha fraterculus* (Diptera: Tephritidae) larvae, under laboratory conditions. *Braz. J. Biol.* **2017**, *77*, 22–28. [CrossRef]
39. Abbas, M.S.T.; Nouh-Gehan, M.; Abdel-Samad, S.S.M.; Negm, A.A. Infectivity of the entomopathogenic nematodes as bio-control agents to *Spodoptera littorals, Ceratitis capitata* and *Bactrocera zonata*. *Egypt. J. Biol. Pest Control* **2016**, *26*, 609–613.
40. Malan, A.P.; Manrakhan, A. Susceptibility of the Mediterranean fruit fly (*Ceratitis capitata*) and the Natal fruit fly (*Ceratitis rosa*) to entomopathogenic nematodes. *J. Invertebr. Pathol.* **2009**, *100*, 47–49. [CrossRef]
41. Usman, M.; Gulzar, S.; Wakil, W.; Wu, S.; Piñero, J.C.; Leskey, T.C.; Nixon, L.J.; Oliveira-Hofman, C.; Toews, M.D.; Shapiro-Ilan, D. Virulence of entomopathogenic fungi to *Rhagoletis pomonella* (Diptera: Tephritidae) and interactions with entomopathogenic nematodes. *J. Econ. Entomol.* **2020**, *113*, 2627–2633. [CrossRef] [PubMed]
42. Usman, M.; Gulzar, S.; Wakil, W.; Piñeiro, J.C.; Leskey, T.C.; Nixon, L.J.; Oliveira-Hofman, C.; Wu, S.; Shapiro-Ilan, D. Potential of entomopathogenic nematodes against the pupal stage of the apple maggot *Rhagoletis pomonella* (Walsh) (Diptera: Tephritidae). *J. Nematol.* **2020**, *52*, e2020-79. [CrossRef] [PubMed]
43. Usman, M.; Wakil, W.; Shapiro-Ilan, D.I. Entomopathogenic nematodes as biological control agent against *Bactrocera zonata* and *Bactrocera dorsalis* (Diptera: Tephritidae). *Biol. Control* **2021**, *163*, 104706. [CrossRef]
44. Herrera–Aguilar, J. Mortalidad de Larvas de *Anastrepha obliqua* (Diptera. Tephritidae) con el Nematodo *Heterorhabditis basc-teriophora* en Suelo Areno-Arcilloso. Bachelor's Thesis, Facultad de Ciencias Agrícolas, Universidad Autónoma de Chiapas, Huehuetan, Chiapas, México, 2013; 62p.
45. Kapranas, A.; Chronopoulou, A.; Lytra, I.C.; Peters, A.; Milonasa, P.G.; Papachristos, D.P. Efficacy and residual activity of commercially available entomopathogenic nematode strains for Mediterranean fruit fly control and their ability to infect infested fruits. *Pest Manag. Sci.* **2021**, *77*, 3964–3969. [CrossRef]
46. Herz, A.; Köppler, K.; Vogt, H.; Elias, E.; Katz, P.; Peters, A. Biological control of the cherry fruit fly, *Rhagoletis cerasi* L. (Diptera: Tephritidae) by use of entomopathogenic nematodes: First experiences towards practical implementation. In Proceedings of the Conference Ecofruit—12th International Conference on Cultivation Technique and Phytopathological Problems in Organic Fruit-Growing, Weinsberg, Germany, 31 January–2 February 2006; Fördergemeinschaft Ökologischer Obstbau eV (FÖKO): Weinsberg, Germany, 2006; pp. 67–72.
47. Shapiro-Ilan, D.I.; Gaugler, R.; Tedders, W.L.; Brown, I.; Lewis, E.E. Optimization of inoculation for in vivo production of entomopathogenic nematodes. *J. Nematol.* **2002**, *34*, 343–350. [PubMed]
48. Wouts, W.M. Mass production of the entomogenous nematode, *Heterorhabditis heliothidis* (Nematoda: Heterorhabditidae) on artificial media. *J. Nematol.* **1981**, *13*, 467–469.
49. Bedding, R.A. Low cost in vitro mass production of *Neoaplectana* and *Heterorhabditis* species (Nematoda) for field control of insect pests. *Nematologica* **1981**, *27*, 109–114. [CrossRef]
50. Ehlers, R.U. Mass production of entomopathogenic nematodes of plant protection. *Appl. Microbiol. Biotechnol.* **2001**, *56*, 623–633. [CrossRef]
51. Lindegren, J.E.; Valero, K.A.; Mackey, B.E. Simple in vivo production and storage methods for *Steinernema carpocapsae* infective juveniles. *J. Nematol.* **1993**, *25*, 193–197.
52. Peters, A.; Han, R.; Yan, X.; Leite, L.G. Production of entomopathogenic nematodes. In *Microbial Control of Insect and Mite Pests*; Academic Press: Cambridge, MA, USA, 2017; pp. 157–170. [CrossRef]
53. Poinar, G.O., Jr.; Thomas, G.M. Significance of *Achromobacter nematophilus* sp. nov. (Achromobacteriaceae: Eubacteriales) asso-ciated with a nematode. *Int. Bull. Bacteriol. Nomencl. Taxon.* **1966**, *15*, 249–252.
54. Maciel-Vergara, G.; Rodríguez-Hernández, A.I.; Chavarría-Hernández, N. Cultivo monoxénico sumergido del nematodo ento-mopatógeno, *Steinernema carpocapsae* CABA01, en biorreactor airlift con recirculación interna. *Bio Tecnol.* **2010**, *14*, 11–24.
55. Cho, C.H.; Whang, K.S.; Gaugler, R.; Yoo, S.K. Submerged monoxenic culture medium development for *Heterorhabditis bacteriophora* and its symbiotic bacterium *Photorhabdus luminescens*: Protein sources. *J. Microbiol. Biotechnol.* **2011**, *21*, 869–873. [CrossRef]
56. Dunn, M.D.; Belur, P.D.; Malan, A.P. A review of the in vitro liquid mass culture of entomopathogenic nematodes. *Biocontrol Sci. Technol.* **2021**, *31*, 1–21. [CrossRef]
57. Devi, G. Mass production of entomopathogenic nematodes—A review. *Int. J. Environ. Agric. Biotechnol.* **2018**, *3*, 1032–1043. [CrossRef]
58. Shapiro-Ilan, D.I.; Gaugler, R. Production technology for entomopathogenic nematodes and their bacterial symbionts. *J. Ind. Microbiol. Biotechnol.* **2002**, *28*, 137–146. [CrossRef] [PubMed]
59. Shapiro-Ilan, D.I.; Han, R.; Qiu, X. Production of entomopathogenic nematodes. In *Mass Production of Beneficial Organisms*; Morales-Ramos, J.A., Rojas, M.G., Shapiro-Ilan, D.I., Eds.; Academic Press: San Diego, CA, USA, 2014; pp. 321–355. [CrossRef]

60. Gaugler, R.; Brown, I. Apparatus and Method for Mass Production of Insecticidal Nematodes. U.S. Patent 6,474,259, 30 April 2001.
61. Young, J.M.; Dunnill, P.; Pearce, J.D. Separation characteristics of liquid nematode cultures and the design of recovery operations. *Biotechnol. Prog.* **2002**, *18*, 29–35. [CrossRef] [PubMed]
62. Tabassum, K.A.; Shahina, F. In vitro mass rearing of different species of entomopathogenic nematodes in monoxenic solid culture. *Pak. J. Nematol.* **2004**, *22*, 167–175.
63. Attalla, F.A.; Eweis, M.A. Preliminary investigation on the utilization of entomopathogenic nematodes as biological control agents against the peach fruit fly, *Bactrocera zonata* (Saunders) (Diptera: Tephritidae). *Egypt. J. Agric. Res.* **2002**, *80*, 1045–1053.
64. Fetoh, B.E.S.A.; Abdel-Gawad, A.A.; Shalaby, F.F.; Elyme, M.F. Pathogenic and lethal effects of the entomopathogenic nematodes on the peach fruit fly, *Bactrocera zonata* (Saunders) and the cucurbit fruit fly, *Dacus ciliatus* (Loew) (Diptera: Tephritidae). *Egypt. J. Agric. Res.* **2011**, *89*, 463–476. [CrossRef]
65. Chaneiko, S.M.; de Brida, A.L.; Bernardi, D.; Leite, L.G.; Garcia, F.R.M. Biological activity of entomopathogenic nematodes on *Anastrepha fraterculus* (Diptera: Tephritidae). *Biosci. J.* **2021**, *37*, e37047. [CrossRef]
66. Jean-Baptiste, M.C.; Lima de Brida, A.; Bernardi, D.; da Costa Días, S.; Pazini, J.B.; Leite, L.G.; Siciliano-Wilcken, S.R.; Mello-Garcia, F.R. Effectiveness of entomopathogenic nematodes against *Ceratitis capitata* (Diptera: Tephritidae) pupae and nematode compatibility with chemical insecticides. *J. Econ. Entomol.* **2021**, *114*, 248–256. [CrossRef]
67. Gazit, Y.; Rosler, Y.; Glazer, I. Evaluation of entomopathogenic nematodes for the control of Mediterranean fruit fly (Diptera: Tephritidae). *Biocontrol Sci. Technol.* **2000**, *10*, 157–164. [CrossRef]
68. Minas, R.S.; Moreira-Souza, R.; Dolinski, C.; da Silva-Carvalho, R.; da Silva-Burla, R. Potential of entomopathogenic nematodes (Rhabditida: Heterorhabditidae) to control Mediterranean fruit fly (Diptera: Tephritidae) soil stages. *Nematoda* **2017**, *3*, e02016. [CrossRef]
69. Abdel-Razek, A.S.; Abd-Elgawad, M.M.M. Spinosad combined with entomopathogenic nematode for biocontrol of the Mediterranean fruit fly (*Ceratitis capitata* [Wiedemann]) on citrus. *Egypt J. Biol. Pest Control* **2021**, *31*, 112. [CrossRef]
70. Wakil, W.; Usman, M.; Piñero, J.C.; Wu, S.; Toews, M.D.; Shapiro-Ilane, D.I. Combined application of entomopathogenic nematodes and fungi against fruit flies, *Bactrocera zonata* and *B. dorsalis* (Diptera: Tephritidae): Laboratory cups to field study. *Pest Manag. Sci.* **2022**, *78*, 2779–2791. [CrossRef]
71. Almeida, J.A.M.; Batista-Filho, A.; Oliveira, F.C.; Raga, A. Pathogenicity of the entomopathogenic fungi and nematode on medfly *Ceratitis capitata* (Wied.) (Diptera: Tephritidae). *BioAssay* **2009**, *2*, 7. [CrossRef]
72. Gava, C.A.T.; Paranhos, B.A.J. Combining the virulent *Beauveria bassiana* (Balsam) Vuillemin LCB289 and nematode strains to control pupae of *Ceratitis capitata* Wiedemann. *Biocontrol Sci. Technol.* **2023**, *33*, 383–396. [CrossRef]
73. Medina, P.; Corrales, E.; González-Nuñez, M.; Smagghe, G.; Viñuela, E. Effects of *Beauveria bassiana*, *Heterorhabditis bacteriophora*, *H. megidis* and *Steinernema feltiae* on the Mediterranean fruit fly *Ceratitis capitata* and the very sensitive braconid *Psyttalia concolor* in the lab. *IOBC/Wprs Bull.* **2008**, *35*, 113–121.
74. Menzler-Hokkanen, I.; Ruhanen, H.; Hokkanen, H.M.T. Mortality of the oriental fruit fly, *Bactrocera dorsalis*, during pupation in insect pest suppressive soils. *Entomol. Exp. Appl.* **2022**, *170*, 565–572. [CrossRef]
75. Toledo, J.; Williams, T.; Pérez, C.; Liedo, P.; Valle, J.F.; Ibarra, J.E. Abiotic factors affecting the infectivity *of Steinernema carpocapsae* (Rhabditida: Steinernematidae) on larvae of *Anastrepha obliqua* (Macquart) (Diptera: Tephritidae). *Biocontrol Sci. Technol.* **2009**, *19*, 887–898. [CrossRef]
76. Toledo, J.; Sánchez, J.E.; Williams, T.; Gómez, A.; Montoya, P.; Ibarra, J.E. Effect of soil moisture in relation to *Heterorhabditis bacteriophora* persistence and efficacy against *Anastrepha ludens* larvae. *Fla. Entomol.* **2014**, *97*, 528–533. [CrossRef]
77. Rohde, C.; Moino, A., Jr.; Da Silva, M.A.T.; Carvalho, F.D.; Ferreira, C.S. Influence of soil temperature and moisture on the infectivity of entomopathogenic nematodes (Rhabditida: Heterorhabditidae, Steinernematidae) against larvae of *Ceratitis capitata* (Wiedemann) (Diptera: Tephritidae). *Neotrop. Entomol.* **2010**, *39*, 608–611. [CrossRef]
78. Ta-un, P.; Ehlers, R.-U.; Nimkingrat, P. Effects of soil texture and moisture on the host searching abilities of *Steinernema siamkayai* against *Bactrocera latifrons*. *Nematology* **2022**, *24*, 559–570. [CrossRef]
79. Toledo, J.; Gurgúa, J.L.; Liedo, P.; Ibarra, J.E.; Oropeza, A. Parasitismo de larvas y pupas de la mosca mexicana de la fruta, *Anastrepha ludens* (Loew) (Diptera: Tephritidae) por el nematodo *Steinernema feltiae* (Filipjev) (Rhabditidae: Steinernematidae). *Vedalia* **2001**, *8*, 27–36.
80. Lezama-Gutiérrez, R.; Molina-Ochoa, J.; Pescador-Rubio, A.; Galindo-Velasco, E.; Ángel-Sahagún, C.A.; Michel-Aceves, A.C.; González-Reyes, E. Efficacy of Steinernematid nematodes (Rhabditida: Steinernematidae) on the suppression of *Anastrepha ludens* (Diptera: Tephritidae) larvae in soil of differing textures: Laboratory and field trials. *J. Agric. Urban Entomol.* **2006**, *23*, 41–49.
81. James, M.; Malan, A.P.; Pia Addison, P. Surveying and screening South African entomopathogenic nematodes for the control of the Mediterranean fruit fly, *Ceratitis capitata* (Wiedemann). *Crop Prot.* **2018**, *105*, 41–48. [CrossRef]
82. Jackson, C.G.; Long, J.P.; Klungness, L.M. Depth of pupation in four species of fruit flies (Diptera: Tephritidae) in sand with and without moisture. *J. Econ. Entomol.* **1998**, *91*, 138–142. [CrossRef]
83. Hodgson, P.J.; Sivinski, J.; Quintero, G.; Aluja, M. Depth of pupation and survival of fruit fly (*Anastrepha* sppTephritidae) Pupae in a Range of Agricultural Habitats. *Environ. Entomol.* **1998**, *27*, 1310–1314. [CrossRef]
84. Montoya, P.; Flores, S.; Toledo, J. Effect of rainfall and soil moisture on survival of adults and immature stages of *Anastrepha ludens* and *A. obliqua* (Diptera: Tephritidae) under Semi-field Conditions. *Fla. Entomol.* **2008**, *91*, 643–650. [CrossRef]

85. Kamali, S.; Karimi, J.; Hosseini, M.; Campos-Herrera, R.; Duncan, L.W. Biocontrol potential of the entomopathogenic nematodes *Heterorhabditis bacteriophora* and *Steinernema carpocapsae* on cucurbit fly, *Dacus ciliatus* (Diptera: Tephritidae). *Biocontrol Sci. Technol.* **2013**, *23*, 1307–1323. [CrossRef]
86. Langford, E.A.; Nielsen, U.N.; Johnson, S.N.; Riegler, M. Susceptibility of Queensland fruit fly, *Bactrocera tryoni* (Froggatt) (Diptera: Tephritidae), to entomopathogenic nematodes. *Biol. Control* **2014**, *69*, 34–39. [CrossRef]
87. Poinar, G.O., Jr.; Thomas, G.; Prokopy, R.J. Microorganisms associated with *Rhagoletis pomonella* (Diptera: Tephritidae) in Massachusetts. *Proc. Entomol. Soc. Ont.* **1977**, *108*, 19–22.
88. Toledo, A.J. Evaluación de Algunos Agentes Entomopatogenos Para el Control Microbiano de tres Especies de Moscas de la Fruta (Diptera: Teprhitidae) de Importancia Económica. Ph.D. Thesis, Facultad de Ciencias, Universidad Nacional Autónoma de México, México City, Mexico, 2002; 124p.
89. Alyokhin, A.V.; Mille, C.; Messing, R.H.; Duan, J.J. Selection of pupation habitats by oriental fruit fly larvae in the laboratory. *J. Insect Behav.* **2001**, *14*, 57–67. [CrossRef]
90. Portillo-Aguilar, C.; Villani, M.G.; Tauber, M.J.; Tauber, C.A.; Nyrop, J.P. Entomopathogenic nematode (Rhabditida: Heterorhabditidae and Steinernematidae) response to soil texture and bulk density. *Environ. Entomol.* **1999**, *28*, 1021–1035. [CrossRef]
91. Georgis, R.; Poinar, G.O., Jr. Effect of soil texture on the distribution and infectivity of *Neoaplectana carpocapsae* (Nematoda: Steinernematidae). *J. Nematol.* **1983**, *15*, 308–311.
92. Kaya, H.K. Constraints associated with commercialization of entomogenous nematodes. In *Fundamental and Applied Aspects of Invertebrate Pathology*; Samson, R.A., Vlak, J.M., Peters, D., Eds.; Foundation of the Fourth International Colloquium of Invertebrate Pathology: Wageningen, The Netherlands, 1986; pp. 661–664.
93. Beavers, J.B.; Calkin, C.O. Susceptibility of *Anastrepha suspensa* (Diptera: Tephritidae) to Steinernematid and Heterorhabditid namatodes in laboratory studies. *Environ. Entomol.* **1984**, *13*, 137–139. [CrossRef]
94. Lindegren, J.E.; Vail, P.V. Susceptibility of Mediterranean fruit fly, melon fly and oriental fruit fly (Diptera: Tephritidae) to the entomogenous nematode *Steinernema feltiae* in laboratory tests. *Environ. Entomol.* **1986**, *15*, 465–468. [CrossRef]
95. Lindegren, J.E.; Wong, T.T.; McInnis, D.O. Response of Mediterranean fruit fly (Diptera: Tephritidae) to the entomogenous nematode *Steinernema feltiae* in field tests in Hawaii. *Environ. Entomol.* **1990**, *19*, 383–386. [CrossRef]
96. Gurgua, J.L.; Liedo, P.; Ibarra, J.E.; Oropeza, A.; Toledo, J. Efecto del suelo y la temperatura en el parasitismo del nematodo entomopatógenos, *Steinernema feltiae* (Filipjev) (Rhabditidae: Steinernematidae) en la mosca mexicana de la fruta. In Proceedings of the XXII National Congress of Biological Control, Montecillo, Estado de México, Mexico, 28–29 October 1999; pp. 185–188.
97. Barbosa-Negrisoli, C.R.C.; Garcia, M.S.; Dolinski, C.; Negrisoli, A.S., Jr.; Bernardi, D.; Nava, D.E. Efficacy of indigenous entomopathogenic nematodes (Rhabditida: Heterorhabditidae, Steinernematidae), from Rio Grande do Sul Brazil, against *Anastrepha fraterculus* (Wied.) (Diptera: Tephritidae) in peach orchards. *J. Invertebr. Pathol.* **2009**, *102*, 6–13. [CrossRef]
98. Toledo, J.; Rojas, R.; Ibarra, J.E. Efficiency of *Heterorhabditis bacteriophora* (Nematoda: Heterorhabditidae) on *Anastrepha serpentina* (Diptera: Tephritidae) larvae under laboratory conditions. *Fla. Entomol.* **2006**, *89*, 524–526. [CrossRef]
99. Lezama, G.R.; Molina, O.J.; González, R.M.; Trujillo de la Luz, A.; Rebolledo, O.D. Susceptibilidad de larvas de *Anastrepha ludens* (Diptera: Tephritidae) a diversos nematodos entomopatógenos (Steinernematidae y Heterorhabditidae). *Vedalia* **1996**, *3*, 31–33.
100. Loulou, A.; M'saad-Guerfali, M.; Muller, A.; Bhat, A.H.; Abolafia, J.; Machado, R.A.R.; Kallel, S. Potential of *Oscheius tipulae* nematodes as biological control agents against *Ceratitis capitata*. *PLoS ONE* **2022**, *17*, e0269106. [CrossRef] [PubMed]
101. Gaugler, R. Ecological consideration in the biological control of soil-habitating insect with entomopathogenic nematode. *Agric. Ecosyst. Environ.* **1988**, *24*, 351–360. [CrossRef]
102. Geden, C.J.; Axtell, R.C. Effect of temperature on nematode *Steinernema feltiae* (Nematoda: Steinernematidae) treatment of soil for control of laser mealworm (Coleoptera: Tenebrionidae) in turkey house. *J. Econ. Entomol.* **1988**, *81*, 800–803. [CrossRef]
103. Epsky, N.D.; Walter, D.E.; Capinera, J.L. Potential role of nematophagous microartropods as biotic mortality factors of entomogenous nematodes (Rhabditida: Steinernematidae, Heterorhabditidae). *J. Econ. Entomol.* **1988**, *81*, 821–825. [CrossRef]
104. Kung, S.P.; Gaugler, R. Effects of soil temperature, moisture, and relative humidity on entomopathogenic nematode persistence. *J. Invertebr. Pathol.* **1991**, *57*, 242–249. [CrossRef]
105. Georgis, R.; Gaugler, R. Predictability in biological control using enthomophathogenic nematodes. *J. Econ. Entomol.* **1991**, *84*, 713–720. [CrossRef]
106. Koppenhöfer, A.M.; Grewal, P.S.; Fuzy, E.M. Virulence of the entomopathogenic nematodes *Heterorhabditis bacteriophora*, *Heterorhabditis zealandica*, and *Steinernema scarabaei* against five white grub species (Coleoptera: Scarabaeidae) of economic importance in turfgrass in North America. *Biol. Control* **2006**, *38*, 397–404. [CrossRef]
107. Shapiro-Ilan, D.I.; Wright, S.E.; Leskey, T.C.; Tuttle, A.F.; Cooley, D.R.; Leskey, T.C. Using entomopathogenic nematodes for biological control of plum curculio, *Conotrachelus nenuphar*: Effects of irrigation and species in apple orchards. *Biol. Control* **2013**, *67*, 123–129. [CrossRef]
108. Belien, T. Entomopathogenic nematodes as biocontrol agents of insect pests in orchards. *CABI Rev.* **2018**, *13*, 1–11. [CrossRef]
109. Poinar, G.O., Jr.; Jansson, H.B. Infection of *Neoaplectana* and *Heterhabditis* (Rhabditiae; Nematoda) with the predatory fungi, *Monacrospporium ellipsoporum* and *Artrobotrys oliogospora* (Monilliales: Deuteromyces). *Rev. De Nematol.* **1986**, *9*, 241–244.
110. Wright, R.J.; Agudelo-Silva, F.; Georgis, R. Soil applications of steinernematid and heterorhabditid nematodes for control of Colorado potato beetles, *Leptinotarsa decemlineata* (Say). *J. Nematol.* **1987**, *19*, 201–206.

111. Capinera, J.L.; Epsky, N.D. Potential for biological control of soils insects in the Caribbean basin using entomopathogenic nematode. *Fla. Entomol.* **1992**, *75*, 525–532. [CrossRef]
112. Da Silva, A.C.; Batista–Filho, A.; Leite, L.G.; Tavares, F.M.; Raga, A.; Schmidt, F.S. Efeito de nematoides entomopatogênicos na mortalidade da mosca-do-Mediterrâneo, *Ceratitis capitata*, e do gorgulho-da-goiaba, *Conotrachelus psidii*. *Nematol. Bras.* **2010**, *34*, 31–40.
113. Rohde, C.; Mertz, N.R.; Junior, A.M. Entomopathogenic nematodes on control of Mediterranean fruit fly (Diptera: Tephritidae). *Rev. Caatinga* **2020**, *33*, 974–984. [CrossRef]
114. Shaurub, E.H.; Soliman, N.A.; Hashem, A.G.; Abdel-Rahman, A.M. Infectivity of four entomopathogenic nematodes in relation to environmental factors and their effects on the biochemistry of the Medfly *Ceratitis capitata* (Wied.) (Diptera: Tephritidae). *Neotrop. Entomol.* **2015**, *44*, 610–618. [CrossRef]
115. Mokrini, F.; Laasli, S.E.; Benseddik, Y.; Joutei, A.B.; Blenzar, A.; Lakhal, H.; Sbaghi, M.; Imren, M.; Özer, G.; Paulitz, T.; et al. Potential of Moroccan entomopathogenic nematodes for the control of the Mediterranean fruit fly, *Ceratitis capitata* Wiedemann (Diptera: Tephritidae). *Sci. Rep.* **2020**, *10*, 19204. [CrossRef]
116. Yağcı1, M.; Tuğba Akdeniz Fırat, T.A.; Dolunay-Erdoğuş, F.; Şahin, M. Virulence of four entomopathogenic nematode against different stages of the Mediterranean fruit fly, *Ceratitis capitata* Wiedemann (Diptera: Tephritidae). *Egypt. J. Biol. Pest Control* **2021**, *31*, 126. [CrossRef]
117. Kapranas, A.; Anna Chronopoulou, A.; Peters, A.; Antonatos, S.; Lytra, I.; Milonas, P.; Papachristos, D. Early and off-season biological control of medfly with entomopathogenic nematodes: From laboratory experiments to successful field trials. *Biol. Control* **2023**, *179*, 105173. [CrossRef]
118. Torrini, G.; Mazza, G.; Benvenuti, C.; Roversi, P.F. Susceptibility of olive fruit fly, *Bactrocera oleae* (Diptera: Tephritidae) pupae to entomopathogenic nematodes. *J. Plant Prot. Res.* **2017**, *57*, 318–320. [CrossRef]
119. Torrini, G.; Mazza, G.; Benvenuti, C.; Simoncini, S.; Landi, S.; Frosinini, R.; Rocchini, A.; Roversi, P. FEntomopathogenic nematodes as potential biocontrol agents against *Bactrocera oleae* (Diptera: Tephritidae). *Biocontrol Sci. Technol.* **2020**, *30*, 909–919. [CrossRef]
120. Godjo, A.; Chabi, N.; Zadji, L.; Dossou, P.; Batcho, O.; Baimey, H.; Bonou, W.; Sinzogan, A.A.C.; Bokonon-Ganta, A.; Decraemer, W.; et al. Evaluation of the ability of indigenous nematode isolates of *Heterorhabditis taysearae* and *Steinernema kandii* to control mango fruit fly *Bactrocera dorsalis* under laboratory, semi-field and field conditions in Northern Benin. *Crop Prot.* **2021**, *149*, 105754. [CrossRef]
121. Kepenekci, I.; Hazir, S.; Özdem, A. Evaluation of native entomopathogenic nematodes for the control of the European cherry fruit fly *Rhagoletis cerasi* L. (Diptera: Tephritidae) Larvae Soil. *Turk. J. Agric. For.* **2015**, *39*, 74–79. [CrossRef]
122. Koppenhöfer, A.M.; Shapiro-Ilan, D.I.; Hiltpold, I. Entomopathogenic nematodes in sustainable food production. *Front. Sustain. Food Syst.* **2020**, *4*, 125. [CrossRef]
123. Sirjani, F.O.; Lewis, E.E.; Kaya, H.K. Evaluation of entomopathogenic nematodes against the olive fruit fly, *Bactrocera oleae* (Diptera: Tephritidae). *Biol. Control* **2009**, *48*, 274–280. [CrossRef]
124. Aryal, S.; Nielsen, U.N.; Sumaya, N.H.; Wilson, C.; Riegler, M. Virulence, penetration rate and reproductive potential of entomopathogenic nematodes from eastern Australia in Queensland fruit fly, *Bactrocera tryoni*. *Biol. Control* **2022**, *169*, 104871. [CrossRef]
125. Mahmoud, M.F.; El-Naeim-Mohamed-Osman, M.A. Use of the nematode *Steinernema feltiae* Cross N 33 as a biological control agent against the peach fruit fly, *Bactrocera zonata*. *Tunis. J. Plant Prot.* **2007**, *110*, 109–115.
126. Nouh, G.M.; Hussein, M.A. The role of entomopathogenic nematodes as biocontrol agents against some Tephritid flies. *Adv. Biol. Resear.* **2014**, *8*, 301–306.
127. Kepenekci, I.; Susurluk, A. Infectivity of two Turkish isolates of *Steinernema feltiae* (Rhabditida: Steinernematidae) against *Rhagoletis cerasi* and *Ceratitis capitata*. *Nematol. Mediterranea* **2006**, *34*, 95–97.
128. Yee, W.L.; Lacey, L.A. Stage-specific mortality of *Rhagoletis indifferens* (Diptera: Tephritidae) exposed to three species of *Steinernema* nematodes. *Biol. Control* **2003**, *27*, 349–356. [CrossRef]
129. Patterson-Stark, J.E.; Lacey, L.A. Susceptibility of western cherry fruit fly (Diptera: Tephritidae) to five species of entomopathogenic nematodes in laboratory studies. *J. Invertebr. Pathol.* **1999**, *74*, 206–208. [CrossRef]

microorganisms

MDPI

Review

Nematophagous Fungi: A Review of Their Phosphorus Solubilization Potential

Marcos Vera-Morales [1,2,*], Segundo E. López Medina [1], Jaime Naranjo-Morán [3], Adela Quevedo [2] and María F. Ratti [2,4]

1 Escuela de Postgrado, Universidad Nacional de Trujillo, Jr. San Martin 392, Trujillo 13007, Perú
2 Escuela Superior Politécnica del Litoral, ESPOL, Centro de Investigaciones Biotecnológicas del Ecuador, CIBE, Campus Gustavo Galindo Km. 30.5 vía Perimetral, Guayaquil EC090112, Ecuador
3 Laboratorio de Biotecnología Vegetal, Ingeniería en Biotecnología, Facultad Ciencias de la Vida, Campus María Auxiliadora, Universidad Politécnica Salesiana (UPS), Km 19.5 Vía a la Costa, Guayaquil P.O. Box 09-01-2074, Ecuador
4 Escuela Superior Politécnica del Litoral, ESPOL, Facultad de Ciencias de la Vida, FCV, Campus Gustavo Galindo Km. 30.5 vía Perimetral, Guayaquil EC090608, Ecuador
* Correspondence: mxvera@espol.edu.ec

Abstract: Nematophagous fungi (NF) are a group of diverse fungal genera that benefit plants. The aim of this review is to increase comprehension about the importance of nematophagous fungi and their role in phosphorus solubilization to favor its uptake in agricultural ecosystems. They use different mechanisms, such as acidification in the medium, organic acids production, and the secretion of enzymes and metabolites that promote the bioavailability of phosphorus for plants. This study summarizes the processes of solubilization, in addition to the mechanisms of action and use of NF on crops, evidencing the need to include innovative alternatives for the implementation of microbial resources in management plans. In addition, it provides information to help understand the effect of NF to make phosphorus available for plants, showing how these biological means promote phosphorus uptake, thus improving productivity and yield.

Keywords: biocontrol; nematophagous fungi; phosphorus; solubilization

Citation: Vera-Morales, M.; López Medina, S.E.; Naranjo-Morán, J.; Quevedo, A.; Ratti, M.F. Nematophagous Fungi: A Review of Their Phosphorus Solubilization Potential. *Microorganisms* **2023**, *11*, 137. https://doi.org/10.3390/microorganisms11010137

Academic Editors: Beibei Ge, Tomislav Cernava and Kyungseok Park

Received: 7 November 2022
Revised: 20 December 2022
Accepted: 24 December 2022
Published: 5 January 2023

1. Introduction

Pathogenic threats and nutritional deficiency are the main problems that affect yield and productivity in plant crops [1], increasing the need to apply new microbial agents to solve both obstacles [2]. Thus, NF have been investigated because of their dual ability as nematode controllers and as plant growth promoters that increase the bioavailabilty of nutrients [3].

Phosphorus (P) is the second most important element for plants after nitrogen. It regulates metabolism and plant development and growth, and an adequate supply is needed to keep these metabolic functions [4]. In addition, it is found in abundance in soil, although its uptake is low, which results in a limitation for the productivity of agricultural crops [5].

P can be solubilized by several soil microorganisms (saprotrophic bacteria and fungi) [6]. Phosphate-solubilizing microorganisms (PSM) play an important role in soil by mineralizing organic P, solubilizing inorganic P minerals, and storing P in biomass [7], thus improving plant growth and yield [8]. PSM can solubilize phosphorus by secreting protons and producing organic anions, such as citrate, oxalate, and gluconate [9]. However, the amount of P solubilization depends on the microbial strain and the relationship between carbon sources, P, and organic acid production [10].

Bacteria can release organic acids that solubilize P or produce acid and alkaline phosphatases that mineralize organic P [11]. There are bacterial genera that are known to be very efficient in these mechanisms, such as *Arthrobacter*, *Bacillus*, *Burkholderia*, *Natrinema*,

Pseudomonas, *Rhizobium*, and *Serratia* [12]. Bacteria are superior to fungi in colonizing plant roots, but are less tolerant to acids [13], so fungi have great potential to solubilize P in acidic conditions [14]. Fungi, in fact, have a higher capacity to release P compared with bacteria [15]. They use different mechanisms such as the secretion of fatty acids, the production of enzymes, and the discharge of metabolic substances (siderophores) that can rapidly metabolize P [16].

Despite the numerous studies on NF as biocontrol agents, the comprehensive and complete understanding of the mechanisms of action is still incipient [17]. In general, the interactions of NF have been evaluated to reduce the densities of plant pathogenic nematodes [18]. Yet, they present other benefits such as the synthesis of phytohormones, as well as favoring the absorption of P [19]. The inoculation of P solubilizing microbial agents in seeds, crops, and soil aims at improving agricultural production without affecting soil health [20]. Bioformulations based on fungal strains that increase plant yield and are also effective at solubilizing nutritional elements are needed [21]. Therefore, the objective of this review is to broaden the understanding of NF and their potential use in P solubilization, focusing on their mechanisms of solubilization and the benefits provided in the nutrition of agricultural crops.

2. Mechanisms of P solubilization by NF

Plants can harness various forms of P; nevertheless, roots take up negatively charged forms of orthophosphate ions ($HPO_4{}^{2-}$- and $HPO_4{}^{-}$) depending on the soil pH [22]. Although agriculture practices introduce considerable amounts of phosphate in soil, only 15 to 30% of P is taken up by plants [23]. It generally binds to iron and aluminum oxides and hydroxides, or to calcium in calcareous soils, becoming unavailable for roots due to chemical precipitation or physical adsorption [24]. Therefore, the application of P itself does not contribute to the improvement of agricultural production systems (Figure 1).

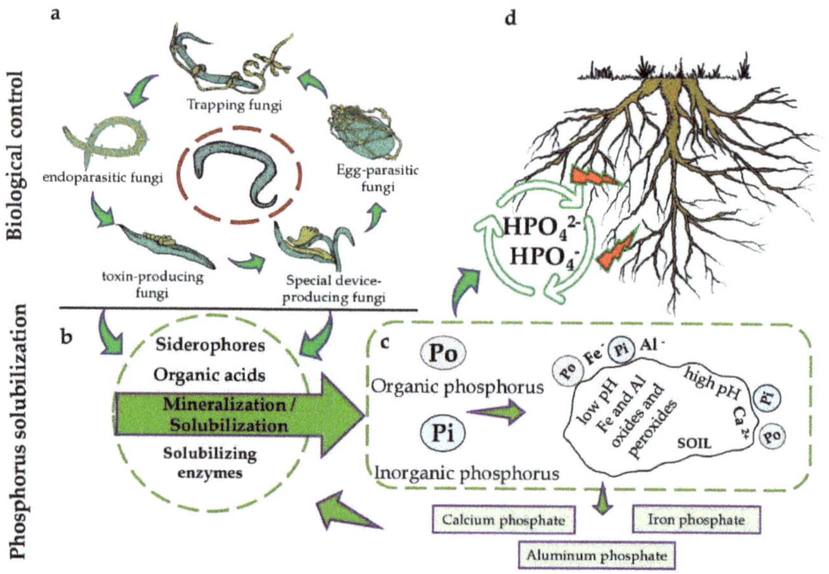

Figure 1. Systematic diagram of P solubilization by fungi used in the biological control of nematodes. (**a**) Nematodes can be parasitized and predated by (in clockwise order) trapping fungi, egg-parasite fungi, special device-producing, or toxin-producing and endoparasitic species. (**b**) Compounds produced by fungi. (**c**) Chemical reactions of P in soil, binding with elements such as calcium, iron, and aluminum that can then be mineralized or solubilized by the NF compounds. (**d**) Roots absorbing negatively charged orthophosphate ions.

Several microorganisms have been studied in relation to the solubilization of P. Fungi and bacteria use different mechanisms on soluble phosphates, especially acidification of the medium, chelation, exchange reactions, and the production of solubilizing enzymes and of various organic acids [4,8,20,25]. Therefore, some species are able to solubilize and mobilize P for plants, including, several nematode controlling fungi. An example is *Trichoderma harzianum*, which can acidify the medium, produce chelating metabolites, and produce a redox activity (capable of reducing Fe and Cu) [26].

2.1. Organic Acids

Organic acids or anions range from fatty acids to secondary metabolites [22]. They are by-products derived from bacteria and fungi that are able to mobilize P in their microenvironment. Organic acids improve P release so that it can be absorbed by plants [27]. Filamentous fungi use several mechanisms to solubilize minerals and they produce high amounts of organic acids [28]. Species of the genera *Aspergillus*, *Penicillium*, *Trichoderma*, and *Fusarium* solubilize P by producing different types of organic acids [14,29–32]. Furthermore, it is possible that some filamentous fungi such as *Aspergillus niger* solubilize P when exposed to high salt contents (4% NaCl) [33]. Even on ferric phosphate substrates, *A. niger* can secrete large amounts of citric and oxalic acids to release P [34]. Therefore, the potential of filamentous fungi in the agrobiotechnology industry is increased, especially because of their properties, for the production of organic acids [35].

2.2. Solubilizing Enzymes

Organic P from animal and plant residues can be mineralized by the action of phosphatases [36], such as phytases that hydrolyze the phospho-monoester bonds present in phytates [37]. Nematode controlling fungi can produce phytases during their development or the production of trapping structures [38]. The nematode-capturing fungus *Arthrobotrys oligospora* forms phytases while building adhesive networks, significantly increasing the release of P [39]. Among the genera of NF that mainly produce phytases and phosphatases under laboratory conditions are the filamentous fungi *Aspergillus*, *Penicillium*, and *Trichoderma* [40].

2.3. Siderophore Production

These are low molecular weight organic iron chelating compounds. They can be produced by bacteria, fungi, and plants [40]. Among the most studied fungi for siderophore production are *Aspergillus fumigatus* and *Aspergillus nidulans*, which have 55 similar types of siderophores [41,42]. In addition, *Penicillium* produces siderophores, capable of solubilizing tricalcium phosphate, even in contaminated soils, opening new opportunities in the field of phytoremediation and agrobiotechnology [43]. Therefore, the use of siderophores is increasing in use as a new alternative for agriculture, to replace pesticides and synthetic fertilizers [44].

3. Potential of NF in P Solubilization

NF can be used in agriculture through the massive production of infective spores or through formulations that improve growth, viability, and efficacy [45]. They use several mechanisms of recognition, signaling, differentiation, and penetration of the cuticle or egg shell of nematodes through mechanical and enzymatic actions [46]. For the purposes of this study, NF are classified according to their mechanism of attack, including trapping, endoparasitic, egg parasitic, toxin-producing, and special device-producing species [47]. NF interact with the environment, performing essential functions to maintain the stability of food webs and the cycling of nutrients in soil [48].

NF widely used in management plans as biological control agents include *Purpureocillium lilacinum*, *Pochonia chlamydosporia*, *Trichoderma harzianum*, *Arthrobotrys* spp., *Hirsutella* spp., etc. [49]. Some of these fungi are widely used in agriculture due to their ability to infect and kill arthropods [50]. They are recognized as biopesticides in the management of destructive pests. Some notable examples include *Beauveria bassiana* [49], *Aspergillus fijiensis* [51], *Cladospo-*

rium tenuissimum, and *Penicillium citrinum* [52]. Among NF, some have been reported as P solubilizers. *Trichoderma* is one of the most studied genera and is considered as a growth promoter, pathogen controller, and nutritional promoter of plants [53,54]. Furthermore, some NF can solubilize P, even under low temperature conditions [55].

4. Trapping Nematodes—Predatory Fungi

Fungi that capture nematodes have a double function: controlling the populations of infective nematodes and also acting as biofertilizers when applied in soil, being able to solubilizing and mobilize some important nutrients [56,57]. The adhesive networks building fungus *Duddingtonia flagrans* have been shown to solubilize P both in laboratory [56] and in greenhouse conditions [2]. Likewise, *A. oligospora* has been used successfully in the solubilization of phosphate rock in in vitro and in vivo conditions (Figure 2), showing promising results in the uptake of P by plants [58].

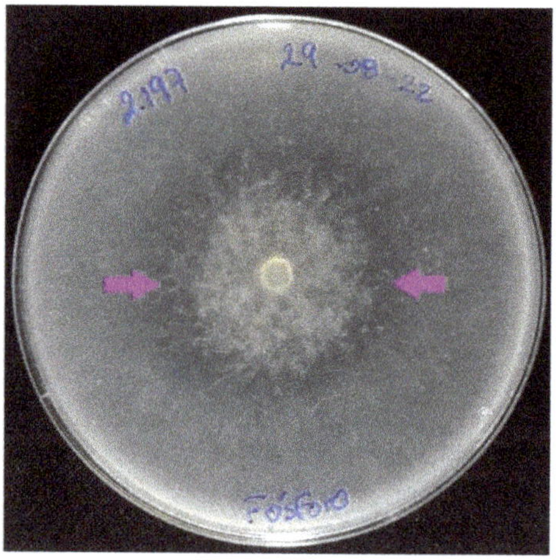

Figure 2. Nematode-capturing fungus *Arthrobotrys oligospora*. Phosphate solubilization zone on Pikovskaya agar plates, as shown by the halo (arrows).

5. Endoparasitic and Egg-parasitic Fungi

Endoparasitic fungi are obligate pathogens that infect nematodes. Conidia of the obligate endoparasitic fungus *Drechmeria coniospora* are ingested or adhere to the cuticle of its hosts [59]. The spores produce a germination tube that enters the nematode, producing a mycelium that, by degrading internal tissues, exploits the host nutrients for growth [60]. One of the most studied fungal parasites of nematode eggs is the fungus *Pochonia chlamydosporia*, able to reduce populations of several plant parasitic nematodes, including species of *Globodera*, *Heterodera*, *Meloidogyne*, *Nacobbus*, and *Rotylenchulus* [61]. *Pochonia chlamydosporia* benefits the plant, as it is able to control nematode infections, shortening the flowering and fruiting times (up to 5 and 12 days), as well as significantly increasing root growth [62]. This parasitic fungus produces phosphatases, organic acids, and propionic acid, which promote the depolymerization of phosphate compounds [63,64]. Furthermore, in association with the predatory fungus *D. flagrans*, they can significantly increase nutrient uptake, especially P, by up to 70% in tomato plants [3]. Therefore, research studies suggest the use of bioformulated *P. chlamydosporia* for the management of phytoparasitic nematodes, with the potential to support plant nutrition [65].

Purpureocillium is a further fungal genus used for the reduction of phytoparasitic nematode eggs and females. Some species attack the gall-forming nematodes *Meloidogyne* spp. [66], and show a P solubilization capability in laboratory conditions [67], as well as promoting tomato seedling growth in greenhouse conditions [68]. These fungi may secrete hydrolytic enzymes and siderophores that solubilize P and promote plant growth [69].

6. Toxin Production

Toxin-producing fungi are characterized by the release of compounds that paralyze nematodes and produce a rapid and systemic cell necrosis in multiple tissues throughout their bodies. In the order of the Agaricales, one representative is *Pleurotus ostreatus*, which is a toxin-producing fungus capable of producing necrosis after the contact of hyphae with the cilia of the sensory system of nematodes [70]. The fungus has P solubilizing properties [71] and is applied as a biofertilizer based on biomasses obtained as mycelial by-products [72].

Piriformospora indica is an endosymbiotic fungus belonging to the order Sebacinales (Basidiomycota), capable of producing metabolites that inhibit nematode populations [73]. This species has shown synergy with phosphate-solubilizing bacteria, sustaining the growth of chickpea plants [74]. Studies have indicated that the transfer of P by *P. indica* to the host plant has a significant effect on its development [75].

7. Filamentous Fungi with Nematophagous Activity

In addition to the examples mentioned above, there are a variety of fungi that do not use mechanisms based on capture, parasitism, toxins, or special devices, so it is presumed that hydrolytic enzymes are a key in the nematode infection process [76]. Among the known genera that work as both biocontrol and P solubilizers are *Trichoderma*, *Fusarium*, *Penicillium*, and *Aspergillus*. Filamentous fungi can secrete hydrolytic enzymes, organic acids, and low molecular weight natural products, which confer several functions, including P solubilization [77], and are able to solubilize calcium and iron phosphates [78]. Thus, they hold great potential for the development of biofertilizers, contributing to soil fertility and promoting plant growth [79], which are essential in sustainable agriculture (Table 1) [80].

Filamentous fungi such as *Trichoderma* spp. are often dominant in soil microbial communities and have the ability to colonize roots [81]. *Trichoderma* spp. are marketed all over the world for their potential as biocontrol and biostimulant agents for numerous agricultural crops. The application of *Trichoderma* spp. and related metabolites allegedly improves crop productivity, nutrient supply, and defense against plant pathogens [54,82], as it combines P solubilization and nematode predation.

Table 1. Fungi known as nematophagous and P solubilizers.

Type of Fungi	Fungi	Structure Involved	Substance That Is Solubilized	Solubilization Mechanism	Reference
Predatory	*Arthrobotrys oligospora*	Adhesive networks	Phosphate rock	pH of the culture medium.	[58]
	Arthrobotrys conoides and *Duddingtonia flagrans*	Adhesive networks Adhesive networks	Tricalcium, zinc, and aluminum phosphate phosphate rock	pH of the culture medium. Production of organic acids	[56]
	Duddingtonia flagrans	Adhesive networks	Phosphorus	-	[2,3]
Nematode egg-parasitic	*Pochonia chlamydosporia*	Appressoria and hyphae	Phosphorus	Phosphatases and organic acids	[3,62,63]
	Purpureocillium variotii	Toxic metabolites	Phosphorus	Siderophores and pH of the culture medium	[68]
	Purpureocillium hepiali	Toxic metabolites	Phosphate	pH of the culture medium	[67]
	Purpureocillium lilacinum	Toxic metabolites	Calcium and iron phosphate	pH of the culturemedium and organic acids	[83]
Toxin producing	*Piriformospora indica*	Metabolites	Organic and inorganic phosphorus	pH of the culture medium	[73,75,84]
	Pleurotus ostreatus	Toxins	Phosphate rock	Organic acids: tartaric, malic, citric, lactic, succinic and four unknown acids	[71,85]

Table 1. *Cont.*

Type of Fungi	Fungi	Structure Involved	Substance That Is Solubilized	Solubilization Mechanism	Reference
Filamentous fungi with indeter-mined ne-matophagous mechanism	*Trichoderma harzianum*	Hydrolytic enzymes	Calcium phosphate	pH of the culture medium, production of chelating metabolites, and redox activity	[26]
	Trichoderma asperellum	Hydrolytic enzymes	Monopotassium phosphate Phytate Phosphate rock	pH of the culture medium and organic anions	[86]
	Trichoderma spp.	Hydrolytic enzymes	Phosphate	Organic acids	[87]
	Aspergillus niger, *Penicillium canescens*, *Eupenicillium ludwigii*, and *Penicillium islandicum*	Hydrolytic enzymes	Phosphate rock	pH of the culture medium, acids: oxalic, citric, and gluconic	[31]
	Aspergillus, *Penicillium*, *Trichoderma*, *Fusarium*, *Mucor*, *Ovularopsis*, *Tritirachium*, and *Geotrichum*	Hydrolytic enzymes	Phosphate rock Tricalcium phosphate	Acids: fumaric, acetic, gluconic, lactic, and succinic	[29]
	Penicillium expansum, *Mucor ramosissimus*, and *Candida krissii*	Hydrolytic enzymes	Phosphate rock	Acids: citric, oxalic, and gluconic	[79]
	Penicillium guanacastense	Hydrolytic enzymes	Organic and inorganic phosphorus	Acids: gluconic, oxalic, lactic, and malonic	[32]
	Penicillium bilaji and *Penicillium cf.fuscum*	Hydrolytic enzymes	Phosphate rock	Organic acids	[30]
	Aspergillus niger	Hydrolytic enzymes	Various forms of soluble P calcium iron aluminum phosphate	Acids: gluconic, oxalic, tartaric	[14,33]
	Aspergillus aculeatus	Hydrolytic enzymes	Phosphate rock	pH of the culture medium	[88]
	Rhizopus stolonifer, *Aspergillus niger*, and *Alternaria alternata*	Hydrolytic enzymes	Calcium phosphate	Organic acids	[89]
	Aspergillus tubingensis	Hydrolytic enzymes	Phosphate rock	pH of the culture medium	[90]
	Penicillium oxalicum and *Aspergillus niger*	Hydrolytic enzymes	Tricalcium phosphate	Organic acids	[91]
	Penicillium bilaii	Hydrolytic enzymes	Calcium phosphate	Acids: citric and oxalic	[92]
	Aspergillus niger and *Penicillium italicum*	Hydrolytic enzymes	Tricalcium phosphate	pH of the culture medium	[93]
	Penicillium oxalicum, *Trichoderma virens*, and *Aspergillus*	Hydrolytic enzymes	Insoluble tricalcium phosphate, soluble dipotassium hydrogen phosphate	pH of the culture medium and organic acids	[94]
	Penicillium	Hydrolytic enzymes	Phosphate	pH of the culture medium	[95]
	Penicillium oxalicum	Hydrolytic enzymes	Phosphate	Organic acids	[96]
	Paecilomyces, *Trichoderma*, *Aspergillus*, *Fusarium*, and *Gongronella*	Hydrolytic enzymes	Calcium phosphate Iron phosphate		[78]

8. Agricultural Solutions: Application of NF in P Solubilization

Although P is added to the soil of agricultural crops, chemical fertilizer synthesis is an energy-consuming process with long-term impacts on the environment in terms of eutrophication, fertility decrease, and carbon footprint [97]. Nematophagous species have been used for plant growth promotion and to sustain yields, especially members of the genera *Trichoderma, Purpureocillium, Pochonia, Fusarium, Arthrobotrys,* and *Verticillium* [98].

Several NF are beneficial because of the production of phytohormones, antibiotics, or siderophores, which benefit long-cycle and short-cycle crops [99] (Table 2), as shown by the P solubilization activity in alkaline soils, which increase the yields of corn and wheat [100,101]. These NF potentially shorten the maturity period of crops, improve fruit quality, increase the availability of soluble P, and improve soil biodiversity [102]. Thus, there is a positive link between the application of these fungal strains and the P content in soil and plants [103].

Currently, several investigations suggest the use of fungi as bioinoculants to increase the yield in agricultural crops [104], as well as ornamentals. P solubilization has been in fact also reported in shrubs within ornamental greenhouses, as demonstrated by the application of *Mortierella* sp. on seedlings of *Leucaena leucocephala* [105].

Table 2. List of P-solubilizing NF and related benefits in agricultural crops.

P-Solubilizing Species	Crop	Plant Benefits	Reference
Geomyces pannorum, *Paecilomyces carneus*	*Avena sativa*	Increase availability of phosphorus in the soil and mitigate phytoparasitic nematodes	[106]
Duddingtonia flagrans *Pochonia chlamydosporia*	Soy bean and tomato	Reductions in the number of eggs and galls per gram of root and increased nutrient content in roots	[2,3]
Pochonia chlamydosporia	Tomato	Increase in secondary roots and increase in the total weight root of seedlings Reduction in flowering and fruiting times and greater weight of mature fruits	[62]
P. chlamydosporia	Maize Cowpea	Promoting root growth	[107]
Purpureocillium lilacinum *Purpureocillium lavendulum* *Metarhizium marquandii*	Maize Beans Soy bean	Plant growth promotion and availability of P and N	[104]
Purpureocillium lilacinum	Tomato	Plant growth promotion and phosphorus solubilization	[69]
Pleurotus ostreatus	*Maize*	Increase in root and shoot lengths, fresh and dry root weights, fresh and dry shoot weights, chlorophyll content, and nutrient uptake	[71]
Trichoderma, *Purpureocillium*	Banana	Active production of indole-3-acetic acid IAA and solubilize insoluble phosphate	[99]
Trichoderma	Soy bean	Plant growth promotion increase of P uptake efficiency	[82,87]
Aspergillus niger, *A. fumigatus,* *Penicillium pinophilum*	Wheat and faba bean	Yield of wheat grains and faba bean seed production	[108]
Penicillium oxalicum	Wheat and maize	Replace chemical fertilizer in alkaline soils. Improved crop production	[101]
Penicillium sp. *Penicillium oxalicum*	Maize	Increased uptake of P by plants and increased availability of P in soil	[100]

Table 2. *Cont.*

P-Solubilizing Species	Crop	Plant Benefits	Reference
Penicillium sp. *Aspergillus foetidus*	*Sorghum bicolor* L.	P uptake and increased growth	[109]
Penicillium expansum, *Mucor ramosissimus*, *Candida krissii*	Wheat	Plant growth promotion, P available in soil, and P absorption	[79]
Penicillium oxalicum	Rapeseed	Solubilize inorganic P and mineralize organic P	[110]
Mortierella capitata	Maize	Increased biomass, chlorophyll, and gibberellic acid	[111]
Mortierella sp.	Avocado	Plant growth promotion and P uptake	[112]

9. Future Research Perspectives

The main P source in the world, phosphate rock, is fundamental for food production. It is strongly threatened as it is a finite resource. P, for the most part, is limited to a few countries in the world, particularly Morocco, which holds 75% of the world reserve [113]. Therefore, in recent years, the challenge of ensuring sustainable global P management to achieve world food security has been evidenced [114]. Given the current agricultural practices dependent on the continuous supply of commercial P-based fertilizers [115], the need to search for new sustainable strategies to manage the P availability in agricultural fields has increased [116]. Therefore, biological activators must be used to accelerate the bioavailability of P for plants, making NF important not only for the biological control of nematodes, but also for P solubilization. Their applications are also in accordance with sustainable agricultural practices for developing countries [117].

Currently, microbial-based fertilizers are not only considered for productivity and economic benefits, but also for their use as environmental-friendly products [18]. The aim is to guarantee less damage to water quality, increase nutrients recycling, reduce the consumption of resources, improve soil health, and increase the biodiversity of beneficial microorganisms [118]. Therefore, there is a demand to join efforts towards the search and discovery of microfungi, especially from little-explored natural regions [119], with the potential to sustain crop nutrition [120]. In fact, the excessive use of fertilizers affects soil health, adding to the presence of pathogens and pests [121].

Filamentous fungi with a dual activity (nematicide and P solubilizers) have versatile capacities to synthesize biocompounds such as enzymes, organic acids, and metabolites [122]. Especially, for P bioavailability, they can biologically produce high concentrations of organic acids, offering new knowledge through the detection of genes related to the mechanisms of phosphate solubilization [24,120,123]. Biotechnology applications may offer sustainable solutions based on microorganisms that are tolerant to new environmental conditions, including the microbiota that favors plant nutrition [124].

10. Conclusions

Several NF have the ability to suppress nematode parasites and improve the uptake of nutritional elements in order to promote plant growth and development. They have been successfully applied as biofertilizers and biocontrol agents to establish beneficial ecological relationships within their environment. For P solubilization, they rely on different mechanisms, such as a decrease in pH and the production of siderophors, organic acids, and enzymes. The wide variety of mechanisms for P solubilization in NF can be harnessed to reduce the dependence on P-based fertilizers in agriculture. The availability of new enzymes such as phytases promotes the search for beneficial microorganisms to develop environmentally-friendly and sustainable management plans.

Author Contributions: M.V.-M.: reviewed the database, structured the sections of the article, wrote the article, and created and elaborated the figures. S.E.L.M.: wrote a section of the article, and revised and corrected text. J.N.-M.: wrote a section of the article, and revised and corrected the text. A.Q.: wrote a section of the article, and revised and corrected the text. M.F.R.: wrote a section of the article, and revised and corrected the text. All authors have read and agreed to the published version of the manuscript.

Funding: This research received no external funding.

Institutional Review Board Statement: Not applicable.

Informed Consent Statement: Not applicable.

Data Availability Statement: Not applicable.

Acknowledgments: The authors want to acknowledge Rafael F. Castañeda Ruiz and Daynet Sosa, who pioneered the studies of nematophagous fungi in Ecuador. In addition, a special acknowledgement is given to Enoy Leiva-Pantoja for his collaboration in drawings.

Conflicts of Interest: The authors declare no conflict of interest.

References

1. Hallama, M.; Pekrun, C.; Lambers, H.; Kandeler, E. Hidden Miners—The Roles of Cover Crops and Soil Microorganisms in Phosphorus Cycling through Agroecosystems. *Plant Soil* **2019**, *434*, 7–45. [CrossRef]
2. Balbino, H.M.; Monteiro, T.S.A.; Coutinho, R.R.; Pacheco, P.V.M.; de Freitas, L.G. Association of *Duddingtonia flagrans* with Microorganisms for Management of *Meloidogyne javanica* and Acquisition of Nutrients in Soybean. *Biol. Control* **2021**, *159*, 104626. [CrossRef]
3. Monteiro, T.S.A.; Valadares, S.V.; de Mello, I.N.K.; Moreira, B.C.; Kasuya, M.C.M.; de Araújo, J.V.; de Freitas, L.G. Nematophagus Fungi Increasing Phosphorus Uptake and Promoting Plant Growth. *Biol. Control* **2018**, *123*, 71–75. [CrossRef]
4. Kumar, A.; Kumar, A.; Patel, H. Role of Microbes in Phosphorus Availability and Acquisition by Plants. *Int. J. Curr. Microbiol. Appl. Sci.* **2018**, *7*, 1344–1347. [CrossRef]
5. Maharana, R.; Singh, B.S.M.; Mandal, K.; Dhal, N.K. Microbial-Mediated Mechanism to Improve Rock Phosphate Solubilization and Its Agronomic Implications. In *Enzymes for Pollutant Degradation*; Mulla, S.I., Bharagava, R.N., Eds.; Microorganisms for Sustainability; Springer Nature: Singapore, 2022; pp. 327–339. ISBN 9789811645747.
6. El Maaloum, S.; Elabed, A.; Alaoui-Talibi, Z.E.; Meddich, A.; Filali-Maltouf, A.; Douira, A.; Ibnsouda-Koraichi, S.; Amir, S.; El Modafar, C. Effect of Arbuscular Mycorrhizal Fungi and Phosphate-Solubilizing Bacteria Consortia Associated with Phospho-Compost on Phosphorus Solubilization and Growth of Tomato Seedlings (*Solanum lycopersicum* L.). *Commun. Soil Sci. Plant Anal.* **2020**, *51*, 622–634. [CrossRef]
7. Tian, J.; Ge, F.; Zhang, D.; Deng, S.; Liu, X. Roles of Phosphate Solubilizing Microorganisms from Managing Soil Phosphorus Deficiency to Mediating Biogeochemical P Cycle. *Biology* **2021**, *10*, 158. [CrossRef]
8. Kalayu, G. Phosphate Solubilizing Microorganisms: Promising Approach as Biofertilizers. *Int. J. Agron.* **2019**, *2019*, e4917256. [CrossRef]
9. Raymond, N.S.; Gómez-Muñoz, B.; van der Bom, F.J.T.; Nybroe, O.; Jensen, L.S.; Müller-Stöver, D.S.; Oberson, A.; Richardson, A.E. Phosphate-Solubilising Microorganisms for Improved Crop Productivity: A Critical Assessment. *New Phytol.* **2021**, *229*, 1268–1277. [CrossRef]
10. Marra, L.M.; de Oliveira-Longatti, S.M.; Soares, C.R.F.S.; Olivares, F.L.; Moreira, F.M.d.S. The Amount of Phosphate Solubilization Depends on the Strain, C-Source, Organic Acids and Type of Phosphate. *Geomicrobiol. J.* **2019**, *36*, 232–242. [CrossRef]
11. Billah, M.; Khan, M.; Bano, A.; Hassan, T.U.; Munir, A.; Gurmani, A.R. Phosphorus and Phosphate Solubilizing Bacteria: Keys for Sustainable Agriculture. *Geomicrobiol. J.* **2019**, *36*, 904–916. [CrossRef]
12. Kour, D.; Rana, K.L.; Kaur, T.; Yadav, N.; Yadav, A.N.; Kumar, M.; Kumar, V.; Dhaliwal, H.S.; Saxena, A.K. Biodiversity, Current Developments and Potential Biotechnological Applications of Phosphorus-Solubilizing and -Mobilizing Microbes: A Review. *Pedosphere* **2021**, *31*, 43–75. [CrossRef]
13. Puente, M.E.; Bashan, Y.; Li, C.Y.; Lebsky, V.K. Microbial Populations and Activities in the Rhizoplane of Rock-Weathering Desert Plants. I. Root Colonization and Weathering of Igneous Rocks. *Plant Biol.* **2004**, *6*, 629–642. [CrossRef] [PubMed]
14. Chuang, C.-C.; Kuo, Y.-L.; Chao, C.-C.; Chao, W.-L. Solubilization of Inorganic Phosphates and Plant Growth Promotion by *Aspergillus niger*. *Biol. Fertil. Soils* **2007**, *43*, 575–584. [CrossRef]
15. Zhang, Y.; Chen, F.-S.; Wu, X.-Q.; Luan, F.-G.; Zhang, L.-P.; Fang, X.-M.; Wan, S.-Z.; Hu, X.-F.; Ye, J.-R. Isolation and characterization of two phosphate-solubilizing fungi from rhizosphere soil of moso bamboo and their functional capacities when exposed to different phosphorus sources and pH environments. *PLoS ONE* **2018**, *13*, e0199625. [CrossRef] [PubMed]
16. Rawat, P.; Das, S.; Shankhdhar, D.; Shankhdhar, S.C. Phosphate-Solubilizing Microorganisms: Mechanism and Their Role in Phosphate Solubilization and Uptake. *J. Soil Sci. Plant Nutr.* **2021**, *21*, 49–68. [CrossRef]

17. Sun, H.; Wu, Y.; Zhou, J.; Yu, D.; Chen, Y. Microorganisms Drive Stabilization and Accumulation of Organic Phosphorus: An Incubation Experiment. *Soil Biol. Biochem.* **2022**, *172*, 108750. [CrossRef]
18. Quevedo, A.; Magdama, F.; Castro, J.; Vera-Morales, M. Interacciones ecológicas de los hongos nematófagos y su potencial uso en cultivos tropicales. *Sci. Agropecu.* **2022**, *13*, 97–108. [CrossRef]
19. Soares, F.E.d.F.; Gôlo, P.S.; Fernandes, É.K.K. Chapter 23—Nematophagous and Entomopathogenic Fungi: New Insights into the Beneficial Fungus-Plant Interaction. In *Molecular Aspects of Plant Beneficial Microbes in Agriculture*; Sharma, V., Salwan, R., Al-Ani, L.K.T., Eds.; Academic Press: Cambridge, MA, USA, 2020; pp. 295–304. ISBN 978-0-12-818469-1.
20. Alori, E.T.; Glick, B.R.; Babalola, O.O. Microbial Phosphorus Solubilization and Its Potential for Use in Sustainable Agriculture. *Front. Microbiol.* **2017**, *8*, 971. [CrossRef]
21. Bargaz, A.; Lyamlouli, K.; Chtouki, M.; Zeroual, Y.; Dhiba, D. Soil Microbial Resources for Improving Fertilizers Efficiency in an Integrated Plant Nutrient Management System. *Front. Microbiol.* **2018**, *9*, 1606. [CrossRef]
22. Wang, Y.; Lambers, H. Root-Released Organic Anions in Response to Low Phosphorus Availability: Recent Progress, Challenges and Future Perspectives. *Plant Soil* **2020**, *447*, 135–156. [CrossRef]
23. Syers, J.K.; Johnston, A.E.; Curtin, D. *Efficiency of Soil and Fertilizer Phosphorus Use. Reconciling Changing Concepts of Soil Phosphorus Behaviour with Agronomic Information*; FAO Fertilizer and Plant Nutrition Bulletin: Rome, Italy, 2008.
24. Alaylar, B.; Egamberdieva, D.; Gulluce, M.; Karadayi, M.; Arora, N.K. Integration of Molecular Tools in Microbial Phosphate Solubilization Research in Agriculture Perspective. *World J. Microbiol. Biotechnol.* **2020**, *36*, 93. [CrossRef] [PubMed]
25. Thakur, D.; Kaushal, R.; Shyam, V. Phosphate Solubilising Microorganisms: Role in Phosphorus Nutrition of Crop Plants- a Review. *Agric. Rev.* **2014**, *35*, 159–171. [CrossRef]
26. Altomare, C.; Norvell, W.; Björkman, T.; Harman, G. Solubilization of Phosphates and Micronutrients by the Plant-Growth-Promoting and Biocontrol Fungus *Trichoderma harzianum* Rifai 1295-22. *Appl. Environ. Microbiol.* **1999**, *65*, 2926–2933. [CrossRef] [PubMed]
27. Yang, X.; Kong, Y.; Guo, E.; Chen, X.; Li, L. ·Organic Acid Regulation of Inorganic Phosphorus Release from Mollisols with Different Organic Matter Contents. *Soil Use Manag.* **2022**, *38*, 576–583. [CrossRef]
28. Magnuson, J.K.; Lasure, L.L. Organic Acid Production by Filamentous Fungi. In *Advances in Fungal Biotechnology for Industry, Agriculture, and Medicine*; Tkacz, J.S., Lange, L., Eds.; Springer: Boston, MA, USA, 2004; pp. 307–340. ISBN 978-1-4419-8859-1.
29. Akintokun, A.K.; Akande, G.A.; Akintokun, P.O.; Popoola, T.O.S.; Babalola, A.O. Solubilization of Insoluble Phosphate by Organic Acid-Producing Fungi Isolated from Nigerian Soil. *Int. J. Soil Sci.* **2007**, *2*, 301–307. [CrossRef]
30. Asea, P.E.A.; Kucey, R.M.N.; Stewart, J.W.B. Inorganic Phosphate Solubilization by Two *Penicillium* Species in Solution Culture and Soil. *Soil Biol. Biochem.* **1988**, *20*, 459–464. [CrossRef]
31. De Oliveira, G.; Moreira de Freitas, A.L.; Liparini Pereira, O.; Ribeiro da Silva, I.; Bojkov Vassilev, N.; Dutra Costa, M. Mechanisms of Phosphate Solubilization by Fungal Isolates When Exposed to Different P Sources. *Ann. Microbiol.* **2014**, *64*, 239–249. [CrossRef]
32. Qiao, H.; Sun, X.-R.; Wu, X.-Q.; Li, G.-E.; Wang, Z.; Li, D.-W. The Phosphate-Solubilizing Ability of *Penicillium guanacastense* and Its Effects on the Growth of *Pinus massoniana* in Phosphate-Limiting Conditions. *Biol. Open* **2019**, *8*, bio046797. [CrossRef]
33. Li, X.; Luo, L.; Yang, J.; Li, B.; Yuan, H. Mechanisms for Solubilization of Various Insoluble Phosphates and Activation of Immobilized Phosphates in Different Soils by an Efficient and Salinity-Tolerant *Aspergillus niger* Strain An2. *Appl. Biochem. Biotechnol.* **2015**, *175*, 2755–2768. [CrossRef]
34. Tian, D.; Wang, L.; Hu, J.; Zhang, L.; Zhou, N.; Xia, J.; Xu, M.; Yusef, K.K.; Wang, S.; Li, Z.; et al. A Study of P Release from Fe-P and Ca-P via the Organic Acids Secreted by *Aspergillus niger*. *J. Microbiol.* **2021**, *59*, 819–826. [CrossRef]
35. Zhang, K.; Zhang, B.; Yang, S.-T. Production of Citric, Itaconic, Fumaric, and Malic Acids in Filamentous Fungal Fermentations. In *Bioprocessing Technologies in Biorefinery for Sustainable Production of Fuels, Chemicals, and Polymers*; John Wiley & Sons, Ltd.: Hoboken, NJ, USA, 2013; pp. 375–398. ISBN 978-1-118-64204-7.
36. Sadaf, N.; Haider, M.Z.; Iqbal, N.; Abualreesh, M.H.; Alatawi, A. Harnessing the Phytase Production Potential of Soil-Borne Fungi from Wastewater Irrigated Fields Based on Eco-Cultural Optimization under Shake Flask Method. *Agriculture* **2022**, *12*, 103. [CrossRef]
37. Azeem, M.; Riaz, A.; Chaudhary, A.N.; Hayat, R.; Hussain, Q.; Tahir, M.I.; Imran, M. Microbial Phytase Activity and Their Role in Organic P Mineralization. *Arch. Agron. Soil Sci.* **2015**, *61*, 751–766. [CrossRef]
38. Kaur, P.; Vohra, A.; Satyanarayana, T. Multifarious Applications of Fungal Phytases. In *Encyclopedia of Mycology*; Zaragoza, Ó., Casadevall, A., Eds.; Elsevier: Oxford, UK, 2021; pp. 358–369. ISBN 978-0-323-85180-0.
39. Hou, X.; Shen, Z.; Li, N.; Kong, X.; Sheng, K.; Wang, J.; Wang, Y. A Novel Fungal Beta-Propeller Phytase from Nematophagous *Arthrobotrys oligospora*: Characterization and Potential Application in Phosphorus and Mineral Release for Feed Processing. *Microb. Cell Factories* **2020**, *19*, 84. [CrossRef] [PubMed]
40. Hider, R.C.; Kong, X. Chemistry and Biology of Siderophores. *Nat. Prod. Rep.* **2010**, *27*, 637–657. [CrossRef] [PubMed]
41. Oberegger, H.; Schoeser, M.; Zadra, I.; Abt, B.; Haas, H. SREA Is Involved in Regulation of Siderophore Biosynthesis, Utilization and Uptake in *Aspergillus nidulans*. *Mol. Microbiol.* **2001**, *41*, 1077–1089. [CrossRef]
42. Schrettl, M.; Ibrahim-Granet, O.; Droin, S.; Huerre, M.; Latgé, J.-P.; Haas, H. The Crucial Role of the *Aspergillus fumigatus* Siderophore System in Interaction with Alveolar Macrophages. *Microbes Infect.* **2010**, *12*, 1035–1041. [CrossRef]

43. Tamariz-Angeles, C.; Huamán, G.D.; Palacios-Robles, E.; Olivera-Gonzales, P.; Castañeda-Barreto, A. Characterization of Siderophore-Producing Microorganisms Associated to Plants from High-Andean Heavy Metal Polluted Soil from Callejón de Huaylas (Ancash, Perú). *Microbiol. Res.* **2021**, *250*, 126811. [CrossRef]
44. Ghosh, S.K.; Bera, T.; Chakrabarty, A.M. Microbial Siderophore—A Boon to Agricultural Sciences. *Biol. Control* **2020**, *144*, 104214. [CrossRef]
45. Al-Ani, L.K.T.; Soares, F.E.d.F.; Sharma, A.; de los Santos-Villalobos, S.; Valdivia-Padilla, A.V.; Aguilar-Marcelino, L. Strategy of Nematophagous Fungi in Determining the Activity of Plant Parasitic Nematodes and Their Prospective Role in Sustainable Agriculture. *Front. Fungal Biol.* **2022**, *3*, 863198. [CrossRef]
46. López-Llorca, L.V.; Maciá-Vicente, J.G.; Jansson, H.-B. Mode of Action and Interactions of Nematophagous Fungi. In *Integrated Management and Biocontrol of Vegetable and Grain Crops Nematodes*; Ciancio, A., Mukerji, K.G., Eds.; Springer: Dordrecht, The Netherlands, 2008; pp. 51–76. ISBN 978-1-4020-6063-2.
47. Vera-Morales, M.; Castañeda-Ruiz, R.F.; Sosa, D.; Quevedo, A.; Naranjo-Morán, J.; Serrano, L.; Ratti, M.F. Mecanismos de captura, colonización y alimentación empleados por parásitos y predadores de nematodos. *Ecosistemas* **2022**, *31*, 2390. [CrossRef]
48. Zhang, Y.; Li, S.; Li, H.; Wang, R.; Zhang, K.-Q.; Xu, J. Fungi–Nematode Interactions: Diversity, Ecology, and Biocontrol Prospects in Agriculture. *J. Fungi* **2020**, *6*, 206. [CrossRef] [PubMed]
49. Islam, W.; Adnan, M.; Shabbir, A.; Naveed, H.; Abubakar, Y.S.; Qasim, M.; Tayyab, M.; Noman, A.; Nisar, M.S.; Khan, K.A.; et al. Insect-Fungal-Interactions: A Detailed Review on Entomopathogenic Fungi Pathogenicity to Combat Insect Pests. *Microb. Pathog.* **2021**, *159*, 105122. [CrossRef] [PubMed]
50. Idrees, A.; Afzal, A.; Qadir, Z.A.; Li, J. Bioassays of *Beauveria bassiana* Isolates against the Fall Armyworm, *Spodoptera frugiperda*. *J. Fungi* **2022**, *8*, 717. [CrossRef]
51. Yan, J.; Liu, H.; Idrees, A.; Chen, F.; Lu, H.; Ouyang, G.; Meng, X. First Record of *Aspergillus fijiensis* as an Entomopathogenic Fungus against Asian Citrus Psyllid, *Diaphorina citri* Kuwayama (Hemiptera: Liviidae). *J. Fungi* **2022**, *8*, 1222. [CrossRef] [PubMed]
52. Idrees, A.; Qadir, Z.A.; Akutse, K.S.; Afzal, A.; Hussain, M.; Islam, W.; Waqas, M.S.; Bamisile, B.S.; Li, J. Effectiveness of Entomopathogenic Fungi on Immature Stages and Feeding Performance of Fall Armyworm, *Spodoptera frugiperda* (Lepidoptera: Noctuidae) Larvae. *Insects* **2021**, *12*, 1044. [CrossRef]
53. França, D.V.C.; Kupper, K.C.; Magri, M.M.R.; Gomes, T.M.; Rossi, F. *Trichoderma* spp. Isolates with Potential of Phosphate Solubilization and Growth Promotion in Cherry Tomato[1]. *Pesqui. Agropecuária Trop.* **2017**, *47*, 360–368. [CrossRef]
54. Joo, J.H.; Hussein, K.A. Biological Control and Plant Growth Promotion Properties of Volatile Organic Compound-Producing Antagonistic *Trichoderma* spp. *Front. Plant Sci.* **2022**, *13*, 897668. [CrossRef]
55. Rinu, K.; Pandey, A.; Palni, L.M.S. Utilization of Psychrotolerant Phosphate Solubilizing Fungi under Low Temperature Conditions of the Mountain Ecosystem. In *Microorganisms in Sustainable Agriculture and Biotechnology*; Satyanarayana, T., Johri, B.N., Eds.; Springer: Dordrecht, The Netherlands, 2012; pp. 77–90. ISBN 978-94-007-2214-9.
56. Pandit, R.; Kunjadia, P.; Mukhopadhyaya, P.; Kunjadia, A. Inorganic Phosphate Solubilizing Potential of *Arthrobotrys conoides* and *Duddingtonia flagrans*, a Nematode Trapping Fungi a Potential Biocontrol Agent. *Int. J. Agric. Technol.* **2014**, *10*, 559–570.
57. Quevedo, A.; Vera-Morales, M.; Espinoza-Lozano, F.; Castañeda-Ruiz, R.; Sosa, D.; Magdama, F. Assessing the Predatory Activity of *Arthrobotrys oligosporus* Strain C-2197 as Biocontrol of the Root-Knot Nematode *Meloidogyne* spp. *Bionatura* **2021**, *6*, 1586–1592. [CrossRef]
58. Duponnois, R.; Kisa, M.; Plenchette, C. Phosphate-Solubilizing Potential of the Nematophagous Fungus *Arthrobotrys oligospora*. *J. Plant Nutr. Soil Sci.* **2006**, *169*, 280–282. [CrossRef]
59. Wan, J.; Dai, Z.; Zhang, K.; Li, G.; Zhao, P. Pathogenicity and Metabolites of Endoparasitic Nematophagous Fungus *Drechmeria coniospora* YMF1.01759 against Nematodes. *Microorganisms* **2021**, *9*, 1735. [CrossRef] [PubMed]
60. Lebrigand, K.; He, L.D.; Thakur, N.; Arguel, M.-J.; Polanowska, J.; Henrissat, B.; Record, E.; Magdelenat, G.; Barbe, V.; Raffaele, S.; et al. Comparative Genomic Analysis of *Drechmeria coniospora* Reveals Core and Specific Genetic Requirements for Fungal Endoparasitism of Nematodes. *PLoS Genet.* **2016**, *12*, e1006017. [CrossRef] [PubMed]
61. Manzanilla-López, R.H.; Esteves, I.; Finetti-Sialer, M.M.; Hirsch, P.R.; Ward, E.; Devonshire, J.; Hidalgo-Díaz, L. *Pochonia chlamydosporia*: Advances and Challenges to Improve Its Performance as a Biological Control Agent of Sedentary Endo-Parasitic Nematodes. *J. Nematol.* **2013**, *45*, 1–7.
62. Zavala-Gonzalez, E.A.; Escudero, N.; Lopez-Moya, F.; Aranda-Martinez, A.; Exposito, A.; Ricaño-Rodríguez, J.; Naranjo-Ortiz, M.A.; Ramírez-Lepe, M.; Lopez-Llorca, L.V. Some Isolates of the Nematophagous Fungus *Pochonia chlamydosporia* Promote Root Growth and Reduce Flowering Time of Tomato. *Ann. Appl. Biol.* **2015**, *166*, 472–483. [CrossRef]
63. Gouveia, A.D.S.; Monteiro, T.S.A.; Valadares, S.V.; Sufiate, B.L.; de Freitas, L.G.; Ramos, H.J.D.O.; de Queiroz, J.H. Understanding How *Pochonia chlamydosporia* Increases Phosphorus Availability. *Geomicrobiol. J.* **2019**, *36*, 747–751. [CrossRef]
64. Rosso, L.c.; Colagiero, M.; Salatino, N.; Ciancio, A. Observations on the Effect of Trophic Conditions on *Pochonia chlamydosporia* Gene Expression. *Ann. Appl. Biol.* **2014**, *164*, 232–243. [CrossRef]
65. Sellitto, V.M.; Curto, G.; Dallavalle, E.; Ciancio, A.; Colagiero, M.; Pietrantonio, L.; Bireescu, G.; Stoleru, V.; Storari, M. Effect of *Pochonia chlamydosporia*-Based Formulates on the Regulation of Root-Knot Nematodes and Plant Growth Response. *Front. Life Sci.* **2016**, *9*, 177–181. [CrossRef]

66. Perveen, Z.; Shahzad, S. A Comparative Study of the Efficacy of *Paecilomyces* species against Root-Knot Nematode *Meloidogyne incognita*. *Pak. J. Nematol.* **2013**, *31*, 125–135.
67. Rinu, K.; Pandey, A. Slow and Steady Phosphate Solubilization by a Psychrotolerant Strain of *Paecilomyces hepiali* (MTCC 9621). *World J. Microbiol. Biotechnol.* **2011**, *27*, 1055–1062. [CrossRef]
68. Moreno-Gavíra, A.; Diánez, F.; Sánchez-Montesinos, B.; Santos, M. *Paecilomyces variotii* as A Plant-Growth Promoter in Horticulture. *Agronomy* **2020**, *10*, 597. [CrossRef]
69. Constantin, M.; Raut, I.; Gurban, A.-M.; Doni, M.; Radu, N.; Alexandrescu, E.; Jecu, L. Exploring the Potential Applications of *Paecilomyces lilacinus* 112. *Appl. Sci.* **2022**, *12*, 7572. [CrossRef]
70. Lee, C.-H.; Chang, H.-W.; Yang, C.-T.; Wali, N.; Shie, J.-J.; Hsueh, Y.-P. Sensory Cilia as the Achilles Heel of Nematodes When Attacked by Carnivorous Mushrooms. *Proc. Natl. Acad. Sci. USA* **2020**, *117*, 6014–6022. [CrossRef] [PubMed]
71. Maharana, R.; Basu, A.; Dhal, N.K.; Adak, T. Biosolubilization of Rock Phosphate by *Pleurotus ostreatus* with Brewery Sludge and Its Effect on the Growth of Maize (*Zea mays* L.). *J. Plant Nutr.* **2021**, *44*, 395–410. [CrossRef]
72. Owaid, M.N.; Abed, I.A.; Al-Saeedi, S.S.S. Applicable Properties of the Bio-Fertilizer Spent Mushroom Substrate in Organic Systems as a Byproduct from the Cultivation of *Pleurotus* spp. *Inf. Process. Agric.* **2017**, *4*, 78–82. [CrossRef]
73. Gill, S.S.; Gill, R.; Trivedi, D.K.; Anjum, N.A.; Sharma, K.K.; Ansari, M.W.; Ansari, A.A.; Johri, A.K.; Prasad, R.; Pereira, E.; et al. *Piriformospora indica*: Potential and Significance in Plant Stress Tolerance. *Front. Microbiol.* **2016**, *7*, 332. [CrossRef]
74. Meena, K.K.; Mesapogu, S.; Kumar, M.; Yandigeri, M.S.; Singh, G.; Saxena, A.K. Co-Inoculation of the Endophytic Fungus *Piriformospora indica* with the Phosphate-Solubilising Bacterium *Pseudomonas striata* Affects Population Dynamics and Plant Growth in Chickpea. *Biol. Fertil. Soils* **2010**, *46*, 169–174. [CrossRef]
75. Kumar, M.; Yadav, V.; Kumar, H.; Sharma, R.; Singh, A.; Tuteja, N.; Johri, A.K. *Piriformospora indica* Enhances Plant Growth by Transferring Phosphate. *Plant Signal. Behav.* **2011**, *6*, 723–725. [CrossRef]
76. Soares, F.E.d.F.; Sufiate, B.L.; de Queiroz, J.H. Nematophagous Fungi: Far beyond the Endoparasite, Predator and Ovicidal Groups. *Agric. Nat. Resour.* **2018**, *52*, 1–8. [CrossRef]
77. Cairns, T.C.; Zheng, X.; Zheng, P.; Sun, J.; Meyer, V. Turning inside out: Filamentous Fungal Secretion and Its Applications in Biotechnology, Agriculture, and the Clinic. *J. Fungi* **2021**, *7*, 535. [CrossRef]
78. Vera, D.F.; Pérez, H.; Valencia, H. Aislamiento de hongos solubilizadores de fosfatos de la rizósfera de Arazá (*Eugenia stipitata*, Myrtaceae). *Acta Biológica Colomb.* **2002**, *7*, 33–40.
79. Xiao, C.; Chi, R.; He, H.; Qiu, G.; Wang, D.; Zhang, W. Isolation of Phosphate-Solubilizing Fungi from Phosphate Mines and Their Effect on Wheat Seedling Growth. *Appl. Biochem. Biotechnol.* **2009**, *159*, 330–342. [CrossRef] [PubMed]
80. Alves, G.S.; Bertini, S.C.B.; Barbosa, B.B.; Pimentel, J.P.; Ribeiro Junior, V.A.; Mendes, G.d.O.; Azevedo, L.C.B. Fungal Endophytes Inoculation Improves Soil Nutrient Availability, Arbuscular Mycorrhizal Colonization and Common Bean Growth. *Rhizosphere* **2021**, *18*, 100330. [CrossRef]
81. Tyśkiewicz, R.; Nowak, A.; Ozimek, E.; Jaroszuk-Ściseł, J. *Trichoderma*: The Current Status of its Application in Agriculture for the Biocontrol of Fungal Phytopathogens and Stimulation of Plant Growth. *Int. J. Mol. Sci.* **2022**, *23*, 2329. [CrossRef]
82. Marra, R.; Lombardi, N.; d'Errico, G.; Troisi, J.; Scala, G.; Vinale, F.; Woo, S.L.; Bonanomi, G.; Lorito, M. Application of *Trichoderma* Strains and Metabolites Enhances Soybean Productivity and Nutrient Content. *J. Agric. Food Chem.* **2019**, *67*, 1814–1822. [CrossRef] [PubMed]
83. Hernández-Leal, T.I.; Carrión, G.; Heredia, G. Solubilización in vitro de fosfatos por una cepa de *Paecilomyces lilacinus* (Thom) Samson. *Agrociencia* **2011**, *45*, 881–892.
84. Ngwene, B.; Boukail, S.; Söllner, L.; Franken, P.; Andrade-Linares, D.R. Phosphate Utilization by the Fungal Root Endophyte *Piriformospora indica*. *Plant Soil* **2016**, *405*, 231–241. [CrossRef]
85. Jayasinghearachchi, H.S.; Seneviratne, G. Fungal Solubilization of Rock Phosphate Is Enhanced by Forming Fungal–Rhizobial Biofilms. *Soil Biol. Biochem.* **2006**, *38*, 405–408. [CrossRef]
86. García-López, A.M.; Avilés, M.; Delgado, A. Plant Uptake of Phosphorus from Sparingly Available P- Sources as Affected by *Trichoderma asperellum* T34. *Agric. Food Sci.* **2015**, *24*, 249–260. [CrossRef]
87. Bononi, L.; Chiaramonte, J.B.; Pansa, C.C.; Moitinho, M.A.; Melo, I.S. Phosphorus-Solubilizing *Trichoderma* spp. from Amazon Soils Improve Soybean Plant Growth. *Sci. Rep.* **2020**, *10*, 2858. [CrossRef]
88. Narsian, V.; Patel, H.H. *Aspergillus aculeatus* as a Rock Phosphate Solubilizer. *Soil Biol. Biochem.* **2000**, *32*, 559–565. [CrossRef]
89. Ceci, A.; Pinzari, F.; Russo, F.; Maggi, O.; Persiani, A.M. Saprotrophic Soil Fungi to Improve Phosphorus Solubilisation and Release: In Vitro Abilities of Several Species. *Ambio* **2018**, *47*, 30–40. [CrossRef]
90. Jamshidi, R.; Jalili, B.; Bahmanyar, M.A.; Salek-Gilani, S. Isolation and Identification of a Phosphate Solubilising Fungus from Soil of a Phosphate Mine in Chaluse, Iran. *Mycology* **2016**, *7*, 134–142. [CrossRef] [PubMed]
91. Li, Z.; Bai, T.; Dai, L.; Wang, F.; Tao, J.; Meng, S.; Hu, Y.; Wang, S.; Hu, S. A Study of Organic Acid Production in Contrasts between Two Phosphate Solubilizing Fungi: *Penicillium oxalicum* and *Aspergillus niger*. *Sci. Rep.* **2016**, *6*, 25313. [CrossRef] [PubMed]
92. Cunningham, J.; Kuiack, C. Production of Citric and Oxalic Acids and Solubilization of Calcium Phosphate by *Penicillium bilaii*. *Appl. Environ. Microbiol.* **1992**, *58*, 1451–1458. [CrossRef] [PubMed]
93. El-Azouni, I. Effect of Phosphate Solubilizing Fungi on Growth and Nutrient Uptake of Soybean (*Glycine max* L.) Plants. *J. Appl. Sci. Res.* **2008**, *4*, 592–598.

94. Kumari, P.D.S.U.; Nanayakkara, C.M. Phosphate-Solubilizing Fungi for Efficient Soil Phosphorus Management. *Sri Lanka J. Food Agric.* **2017**, *3*, 1–9. [CrossRef]
95. Pandey, A.; Das, N.; Kumar, B.; Rinu, K.; Trivedi, P. Phosphate Solubilization by *Penicillium* spp. Isolated from Soil Samples of Indian Himalayan Region. *World J. Microbiol. Biotechnol.* **2008**, *24*, 97–102. [CrossRef]
96. Yang, T.; Li, L.; Wang, B.; Tian, J.; Shi, F.; Zhang, S.; Wu, Z. Isolation, Mutagenesis, and Organic Acid Secretion of a Highly Efficient Phosphate-Solubilizing Fungus. *Front. Microbiol.* **2022**, *13*, 793122. [CrossRef]
97. Sharma, S.B.; Sayyed, R.Z.; Trivedi, M.H.; Gobi, T.A. Phosphate Solubilizing Microbes: Sustainable Approach for Managing Phosphorus Deficiency in Agricultural Soils. *SpringerPlus* **2013**, *2*, 587. [CrossRef]
98. Peiris, P.U.S.; Li, Y.; Brown, P.; Xu, C. Fungal Biocontrol against *Meloidogyne* spp. in Agricultural Crops: A Systematic Review and Meta-Analysis. *Biol. Control* **2020**, *144*, 104235. [CrossRef]
99. Napitupulu, T.P.; Ramadhani, I.; Kanti, A.; Sudiana, I.M. Diversity, Phosphate Solubilizing, and IAA Production of Culturable Fungi Associated with Healthy and Wilt Banana. *Arch. Phytopathol. Plant Prot.* **2021**, *54*, 2306–2332. [CrossRef]
100. Qarni, A.; Billah, M.; Hussain, K.; Shah, S.H.; Ahmed, W.; Alam, S.; Sheikh, A.A.; Jafri, L.; Munir, A.; Malik, K.M.; et al. Isolation and Characterization of Phosphate Solubilizing Microbes from Rock Phosphate Mines and Their Potential Effect for Sustainable Agriculture. *Sustainability* **2021**, *13*, 2151. [CrossRef]
101. Singh, H.; Reddy, M.S. Effect of Inoculation with Phosphate Solubilizing Fungus on Growth and Nutrient Uptake of Wheat and Maize Plants Fertilized with Rock Phosphate in Alkaline Soils. *Eur. J. Soil Biol.* **2011**, *47*, 30–34. [CrossRef]
102. Chang, C.-H.; Yang, S.-S. Thermo-Tolerant Phosphate-Solubilizing Microbes for Multi-Functional Biofertilizer Preparation. *Bioresour. Technol.* **2009**, *100*, 1648–1658. [CrossRef]
103. Djuuna, I.A.F.; Prabawardani, S.; Massora, M. Population Distribution of Phosphate-Solubilizing Microorganisms in Agricultural Soil. *Microbes Environ.* **2022**, *37*, ME21041. [CrossRef]
104. Baron, N.C.; Pollo, A.d.S.; Rigobelo, E.C. *Purpureocillium lilacinum* and *Metarhizium marquandii* as Plant Growth-Promoting Fungi. *PeerJ* **2020**, *8*, e9005. [CrossRef]
105. Osorio, N.W.; Habte, M. Effect of a Phosphate-Solubilizing Fungus and an Arbuscular Mycorrhizal Fungus on *Leucaena* Seedlings in Tropical Soils with Contrasting Phosphate Sorption Capacity. *Plant Soil* **2015**, *389*, 375–385. [CrossRef]
106. Lima-Rivera, D.L.; Lopez-Lima, D.; Desgarennes, D.; Velázquez-Rodríguez, A.S.; Carrion, G. Phosphate Solubilization by Fungi with Nematicidal Potential. *J. Soil Sci. Plant Nutr.* **2016**, *16*, 507–524. [CrossRef]
107. Gomezjurado, M.E.; de Abreu, L.M.; Marra, L.M.; Pfenning, L.H.; de S. Moreira, F.M. Phosphate Solubilization by Several Genera of Saprophytic Fungi and Its Influence on Corn and Cowpea Growth. *J. Plant Nutr.* **2015**, *38*, 675–686. [CrossRef]
108. Abdul, O.A.; Mehana, T.A. Impact of Phosphate-Solubilizing Fungi on the Yield and Phosphorus-Uptake by Wheat and Faba Bean Plants. *Microbiol. Res.* **2000**, *155*, 221–227. [CrossRef]
109. Salih, H.M.; Yahya, A.I.; Abdul-Rahem, A.M.; Munam, B.H. Availability of Phosphorus in a Calcareous Soil Treated with Rock Phosphate or Superphosphate as Affected by Phosphate-Dissolving Fungi. *Plant Soil* **1989**, *120*, 181–185. [CrossRef]
110. Wang, J.; Zhao, Y.-G.; Maqbool, F. Capability of *Penicillium oxalicum* Y2 to Release Phosphate from Different Insoluble Phosphorus Sources and Soil. *Folia Microbiol. (Praha)* **2021**, *66*, 69–77. [CrossRef] [PubMed]
111. Li, F.; Zhang, S.; Wang, Y.; Li, Y.; Li, P.; Chen, L.; Jie, X.; Hu, D.; Feng, B.; Yue, K.; et al. Rare Fungus, *Mortierella capitata*, Promotes Crop Growth by Stimulating Primary Metabolisms Related Genes and Reshaping Rhizosphere Bacterial Community. *Soil Biol. Biochem.* **2020**, *151*, 108017. [CrossRef]
112. Tamayo-Velez, A.; Osorio, N.W. Co-Inoculation with an Arbuscular Mycorrhizal Fungus and a Phosphate-Solubilizing Fungus Promotes the Plant Growth and Phosphate Uptake of Avocado Plantlets in a Nursery. *Botany* **2017**, *95*, 539–545. [CrossRef]
113. Cordell, D.; White, S. Tracking Phosphorus Security: Indicators of Phosphorus Vulnerability in the Global Food System. *Food Secur.* **2015**, *7*, 337–350. [CrossRef]
114. Chowdhury, R.B.; Moore, G.A.; Weatherley, A.J.; Arora, M. Key Sustainability Challenges for the Global Phosphorus Resource, Their Implications for Global Food Security, and Options for Mitigation. *J. Clean. Prod.* **2017**, *140*, 945–963. [CrossRef]
115. Condron, L.M.; Spears, B.M.; Haygarth, P.M.; Turner, B.L.; Richardson, A.E. Role of Legacy Phosphorus in Improving Global Phosphorus-Use Efficiency. *Environ. Dev.* **2013**, *8*, 147–148. [CrossRef]
116. Zhu, J.; Li, M.; Whelan, M. Phosphorus Activators Contribute to Legacy Phosphorus Availability in Agricultural Soils: A Review. *Sci. Total Environ.* **2018**, *612*, 522–537. [CrossRef]
117. Chittora, P.; Sharma, D.; Aseri, G.K.; Sohal, J.S.; Singh, D.; Khare, N.; Jain, N. Chapter 9—Fungi as Phosphate-Solubilizing Microorganisms in Arid Soil: Current Perspective. In *New and Future Developments in Microbial Biotechnology and Bioengineering*; Singh, J., Gehlot, P., Eds.; Elsevier: Amsterdam, The Netherlands, 2020; pp. 99–108. ISBN 978-0-12-821007-9.
118. Bruulsema, T.W.; Peterson, H.M.; Prochnow, L.I. The Science of 4R Nutrient Stewardship for Phosphorus ManAgement across Latitudes. *J. Environ. Qual.* **2019**, *48*, 1295–1299. [CrossRef]
119. Landínez-Torres, A.Y.; Becerra Abril, J.L.; Tosi, S.; Nicola, L. Soil Microfungi of the Colombian Natural Regions. *Int. J. Environ. Res. Public. Health* **2020**, *17*, E8311. [CrossRef]
120. Hyde, K.D.; Xu, J.; Rapior, S.; Jeewon, R.; Lumyong, S.; Niego, A.G.T.; Abeywickrama, P.D.; Aluthmuhandiram, J.V.S.; Brahamanage, R.S.; Brooks, S.; et al. The Amazing Potential of Fungi: 50 Ways We Can Exploit Fungi Industrially. *Fungal Div.* **2019**, *97*, 1–136. [CrossRef]

121. Vassileva, M.; Mocali, S.; Canfora, L.; Malusá, E.; García Del Moral, L.F.; Martos, V.; Flor-Peregrin, E.; Vassilev, N. Safety Level of Microorganism-Bearing Products Applied in Soil-Plant Systems. *Front. Plant Sci.* **2022**, *13*, 862875. [CrossRef] [PubMed]
122. Madhavan, A.; Arun, K.; Sindhu, R.; Alphonsa Jose, A.; Pugazhendhi, A.; Binod, P.; Sirohi, R.; Reshmy, R.; Kumar Awasthi, M. Engineering Interventions in Industrial Filamentous Fungal Cell Factories for Biomass Valorization. *Bioresour. Technol.* **2022**, *344*, 126209. [CrossRef] [PubMed]
123. Dörsam, S.; Fesseler, J.; Gorte, O.; Hahn, T.; Zibek, S.; Syldatk, C.; Ochsenreither, K. Sustainable Carbon Sources for Microbial Organic Acid Production with Filamentous Fungi. *Biotechnol. Biofuels* **2017**, *10*, 242. [CrossRef]
124. Vassilev, N.; Eichler-Löbermann, B.; Vassileva, M. Stress-Tolerant P-Solubilizing Microorganisms. *Appl. Microbiol. Biotechnol.* **2012**, *95*, 851–859. [CrossRef]

microorganisms

MDPI

Article

Membrane Protein Bcest Is Involved in Hyphal Growth, Virulence and Stress Tolerance of *Botrytis cinerea*

Wei Zhang [1], Bei-Bei Ge [1], Zhao-Yang Lv [1], Kyung Seok Park [2], Li-Ming Shi [1,*] and Ke-Cheng Zhang [1,*]

[1] State Key Laboratory of Biology of Plant Diseases and Insect Pests, Institute of Plant Protection, Chinese Academy of Agricultural Sciences, 2 Yuanmingyuan West Road, Beijing 100193, China
[2] International Agricultural Technology Information Institute, Hankyong National University, 327 Jungang Road, Anseong 17579, Republic of Korea
* Correspondence: shiliming@caas.cn (L.-M.S.); kczhang@ippcaas.cn (K.-C.Z.)

Abstract: *Botrytis cinerea* is a necrotrophic model fungal plant pathogen that causes grey mould, a devastating disease responsible for large losses in the agriculture sector. As important targets of fungicides, membrane proteins are hot spots in the research and development of fungicide products. We previously found that membrane protein Bcest may be closely related to the pathogenicity of *Botrytis cinerea*. Herein, we further explored its function. We generated and characterised Δ*Bcest* deletion mutants of *B. cinerea* and constructed complemented strains. The Δ*Bcest* deletion mutants exhibited reduced conidia germination and germ tube elongation. The functional activity of Δ*Bcest* deletion mutants was investigated by reduced necrotic colonisation of *B. cinerea* on grapevine fruits and leaves. Targeted deletion of Bcest also blocked several phenotypic defects in aspects of mycelial growth, conidiation and virulence. All phenotypic defects were restored by targeted-gene complementation. The role of Bcest in pathogenicity was also supported by reverse-transcriptase real-time quantitative PCR results indicating that melanin synthesis gene *Bcpks*13 and virulence factor *Bccdc*14 were significantly downregulated in the early infection stage of the Δ*Bcest* strain. Taken together, these results suggest that Bcest plays important roles in the regulation of various cellular processes in *B. cinerea*.

Keywords: Bcest; *Botrytis cinerea*; germination; hyphal growth; pathogenicity; stress tolerance

Citation: Zhang, W.; Ge, B.-B.; Lv, Z.-Y.; Park, K.S.; Shi, L.-M.; Zhang, K.-C. Membrane Protein Bcest Is Involved in Hyphal Growth, Virulence and Stress Tolerance of *Botrytis cinerea*. *Microorganisms* **2023**, *11*, 1225. https://doi.org/10.3390/microorganisms11051225

Academic Editor: James F. White

Received: 28 March 2023
Revised: 27 April 2023
Accepted: 4 May 2023
Published: 6 May 2023

1. Introduction

Botrytis cinerea Pers.: Fr. (teleomorph: *Botryotinia fuckeliana Whetzel*) is a typical necrotrophic ascomycete and worldwide plant pathogen that infects crop hosts during both pre- and post-harvesting phases [1,2]. Moreover, *B. cinerea* infects over 1400 plant species including many economically important crops, and leads to tremendous economic losses [3]. Due to the lack of resistant varieties, chemical control remains the most effective strategy for grey mould management [4]. Nevertheless, through genetic plasticity, *B. cinerea* has developed resistance to many types of fungicides. Researchers use *B. cinerea* as a model fungus in molecular studies; therefore, exploring the molecular mechanisms underlying the development and virulence of *B. cinerea* will contribute to establishing more effective disease control strategies.

Membrane proteins play key roles in the physiological processes of microorganisms, including transportation, intercellular communication and drug targets. Membrane proteins are recognised and inserted into the lipid bilayer by exquisite cellular machineries, such as the GlpG rhomboid protease, which is thought to allow docking of a transmembrane substrate [5]. However, transporters are integral membrane proteins with central roles in the efficient movement of molecules across biological membranes. The nucleobase ascorbate transporter UapA from *Aspergillus nidulans* must dimerise for correct trafficking to the membrane [6]. Many membrane proteins are primary drug targets, especially those involved in converting extracellular signals into intracellular processes. Among them, G protein-coupled receptors (GPCRs) are crucial for cellular responses to a range of bioactive molecules, and they play a key role in signalling, including increasing the basal activity of the cannabinoid 2 receptor [7]. Interestingly, cell membrane proteins are important targets of fungi in the prevention and control of fungal diseases by fungicides [8].For example, natamycin inhibits the growth of yeast and fungi by inhibiting plasma membrane transporters that regulate amino acid and glucose transport [9].

In our previous study, we identified and characterised membrane protein Bcest, which was associated with the pathogenicity of *B. cinerea* [10]. Through homologous protein searches, we found that homologs of Bcest included adhesins (Ata/Bmac), peptidoglycan DL_endopeptidase, ATP-binding cassette (ABC) transporters, EkdA and lipasesproteins, and others. Among them, adhesin Ata is an important virulence factor that can promote the formation of bacterial biofilms [11]. Adhesin BmaC facilitates attachment between bacteria and hosts [12]. Peptidoglycan DL_endopeptidase SadA causes cell aggregation and promotes biofilm formation [13]. The absence of transporter PltI affects the inability of strains to produce antibiotic Plt [14]. In the present study, we investigated the role of membrane protein Bcest in fungal growth and pathogenesis to bridge this gap in knowledge about *B. cinerea*.

In order to determine the role of Bcest in *B. cinerea*, we constructed and characterised Δ*Bcest* deletion mutants. Deletion of the *Bcest* gene not only led to reduced pathogenicity and lower conidiation, but also increased sensitivity to H_2O_2. These results indicate that Bcest is involved in several processes in *B. cinerea*, including vegetative differentiation, adaptation to oxidative stress and triadimefon, and virulence.

2. Materials and Methods

2.1. Fungal Strains and Culture Conditions

Strain B05.10 of *B. cinerea* Pers.: Fr. (*B. fuckeliana* (de Bary) Whetzel) was isolated from *Vitis vinifera* and has been widely used as a standard reference strain [15]. *B. cinerea* was grown on potato dextrose agar (PDA; 200 g potato, 20 g dextrose, 20 g agar, 1 L water), potato dextrose broth (PDB; 200 g potato, 20 g dextrose, 1 L water), a complete medium (CM; 1 g yeast extract, 0.5 g casein acid hydrolysate, 0.5 g hydrolysed casein, 10 g glucose, 4 mM $Ca(NO_3)_2 \cdot 4H_2O$, 1.5 mM KH_2PO_4, 1 mM $MgSO_4 \cdot 7H_2O$, 2.5 mM NaCl, 20 g agar, 1 L water) and a minimal medium (MM; 10 mM K_2HPO_4, 10 mM KH_2PO_4, 4 mM $(NH_4)_2SO_4$, 2.5 mM NaCl, 2 mM $MgSO_4$, 0.45 mM $CaCl_2$, 9 μM $FeSO_4$, 10 mM glucose, 20 g agar, 1 L water, pH 6.9) [16,17].

Conidia were quantified after 10 days of incubation on the PDA medium, washed from plates, diluted to 5 mL with ddH_2O, and counted with a hemocytometer under a light microscope (×40). Growth tests under different stress conditions were performed on PDA plates supplemented with different agents including H_2O_2 (10 mM), KCl (1 M) and sodium dodecyl sulphate (SDS; 0.02%) [18]. The percentage of mycelial radial growth inhibition (RGI) was calculated using the formula $RGI = ([C - N]/[C - 5]) \times 100$, where C and N indicate the colony diameter of the control and treatments, respectively [18]. Each experiment was repeated three times.

2.2. Gene Deletion and Complementation

To replace the *Bcest* gene in wild-type (WT) strain B05.10, 1000 bp upstream and 1000 bp downstream flanking sequences of *Bcest* were amplified by PCR from the genomic DNA of B05.10. The resulting amplicons were fused with the HPH hygromycin resistance gene using double-joint PCR [19]. Protoplast preparation and transformation were performed as previously described [20]. The resulting hygromycin-resistant transformants were preliminarily screened by PCR with primers (Supplementary Table S1), and further confirmed by Southern blotting analysis. The upstream fragment of *Bcest* was used as a probe and labelled with digoxigenin (DIG) using High Prime DNA Labelling and Detection Starter Kit I (Roche Diagnostics, Mannheim, Germany) in accordance with the manufacturer's protocol. Genomic DNA was digested with *EcoRI* endonuclease. For complementation assays, a *Bcest* green fluorescent protein (GFP) cassette was generated by amplifying the entire open reading frame (ORF) of *Bcest* (without a stop codon) and cloning it into the pNAN–OGG vector containing the GFP gene and the nourseothricin resistance gene [21]. The resulting construct was confirmed by DNA sequencing and transformed into the *Bcest* deletion mutant.

2.3. Transcriptome Analyses

The mycelia of WT B05.10 and *Bcest* gene deletion mutant Δ*Bcest* were harvested after growth on the PDA medium at 22 °C under 12 h of light and 12 h of darkness for 3 days (with three biological replicates). Total RNA was extracted using a fungal RNA kit (R6840-01; Omega Bio-Tek, Norcross, GA, USA), using Nanodrop (Thermo Scientific, Waltham, MA, USA) for quality testing and using Agilent 2100 Bioanalyzer (Agilgent, Santa Clara, CA, USA) for obtaining the RNA integrity number (RIN) [22]. RNA samples with a RIN of > 7.0, a 260/280 ratio of > 1.8, and a 260/230 ratio of > 1.9 were analysed by Allwegene Technology Co., Ltd. (Beijing, China). Briefly, Trimmomatic (v0.33) software was used to filter the sequencing data [23]. A reference genome index was built and filtered reads were mapped to the reference genome using STAR (v2.5.2b). Mapping statistics are shown in Supplementary Table S2. HTSeq (v0.5.4) was used to compare the read count values for each gene with the original gene expression level, and the value of fragments per kilobase of exon per million mapped reads (FPKM) was used to standardise the expression. DESeq (v1.10.1) was used to analyse differentially expressed genes (DEGs) with an absolute log2 value of > 1 and a p value of < 0.05 as the cutoff criteria. All DEGs are listed in Supplementary Table S3. Gene ontology (GO) categories of upregulated and downregulated genes were identified using the g:Profiler toolset [24].

2.4. Nucleic Acid Manipulation and qRT-PCR

Fungal genomic DNA was extracted as described previously [25]. Plasmid DNA was isolated using RapidLyse Plasmid Mini Kit (DC211; Vazyme, Nanjing, China).

Real-time quantitative reverse-transcription PCR (qRT-PCR) was used to measure the expression of disease-related genes in the *Bcest* disruption mutant Δ*Bcest* and the WT B05.10 strain. The total RNA remaining from the transcriptome experiment was used. RNA was reverse-transcribed using HiScript III 1st Strand cDNA Synthesis Kit (R312; Vazyme, Nanjing, China). qRT-PCR was performed using Taq Pro Universal SYBR qPCR Master Mix (Q712; Vazyme, Nanjing, China) and using specific primers and a thermal program to determine the expression of *Bcactin* (reference gene), Bcin14g02370, Bcin10g01030 and other genes. The expression level of each transcript was calculated using the $\Delta\Delta C_t$ method [26]. For the normalisation of the data, the transcription level of each gene in the hyphae of B05.10 was given a value of 1.0, and the scale was used to calibrate the transcript levels of genes in the hyphae of Δ*Bcest*. qRT-PCR was repeated three times. All genes and primers used for qRT-PCR are listed in Supplementary Table S4.

2.5. Pathogenicity and Infection-Related Morphogenesis Assay

Infection tests were performed on grape fruits and leaves. Briefly, the tested plant tissues were point-inoculated with 5 mm diameter mycelial plugs of 3-day-old cultures. Before inoculation, the cuticles of hosts were wounded with a sterilised needle tip to facilitate the penetration of the fungus into plant tissues. The inoculated samples were placed under high relative humidity conditions (~95%) at 25 °C with 16 h of daylight. These experiments were repeated three times and each included 10 samples. Infection-related morphogenesis was observed on an onion epidermis using a published method [27].

2.6. Morphology and Ultrastructure of Fungal Hyphae

To investigate the role of Bcest on hypha morphology and ultrastructure in Δ*Bcest* and WT B05.10 strains, scanning electron microscopy (SEM) and transmission electron microscopy (TEM) were performed. The mycelial morphology and ultrastructure were observed by SEM/TEM in accordance with a modified method [28]. The hyphae on coverslips were immersed in 4 °C glutaraldehyde (4%) and incubated in darkness at 4 °C for 16 h. Mycelia were washed three times with phosphate-buffered saline (PBS), dehydrated, and dried in a vacuum freeze-dryer. Samples were sprayed with gold powder and examined with an SU8000 SEM instrument (Hitachi, Tokyo, Japan). One millilitre of spore suspension (1×10^5 spores mL^{-1}) was added to 100 mL of PDB and incubated at 25 °C with shaking a 150 rpm for 72 h. Mycelia were centrifuged, washed three times with PBS, and postfixed with 1% osmium tetroxide for 2 h. Samples were washed three more times with PBS, further dehydrated in a graded ethanol series (30%, 50%, 60%, 70%, 80%, 90%, 95% and 100%), then embedded in Spurr's low-viscosity resin. Sections were observed using an H-7500 TEM instrument (Hitachi, Tokyo, Japan).

2.7. Abiotic Stress and Pathogenic Factor Assay

Mycelia-responsive trials were carried out to determine the responses of B05.10, Δ*Bcest*, and the complemented strain Δ*Bcest*-C to abiotic stresses, including osmotic pressure, H_2O_2, SDS, protease, polygalacturonase and cellulase, and their ability to produce an infectious agent. Specifically, mycelial agar plugs were removed from the margin area of a 2-day-old PDA culture of an isolate and inoculated in Petri dishes containing PDA with KCl (1 M), H_2O_2 (10 mM) and 20 mg/L of SDS (w/v). Cultures were incubated at 20 °C for 2 days. Secretions of proteases, polygalacturonases and cellulases were assessed using nutrient agar (NA; 50 g skimmed milk powder, 5 g NaCl, 10 g sucrose, 3 g beef extract, 3 g yeast extract, 20 g agar, 1 L water, pH 7.0), polygalacuronic acid agar (PGAA; 10 g polygalacuronic acid, 20 g sucrose, 2 g $(NH_4)_2SO_4$, 20 g agar, 1 L water) and carboxymethyl cellulose sodium agar (CMC-Na; 10 g carboxymethyl cellulose sodium salt, 10 g yeast extract, 1 g tryptone, 4 g $(NH_4)_2SO_4$, 2 g K_2HPO_4, 0.5 g $MgSO_4{\cdot}7H_2O$, 20 g agar, 1 L water) mediums, respectively. Cultures were incubated at 22 °C for 3 days. Experiments included one mycelial agar plug per dish and three dishes (replicates) for each treatment. The diameter of each colony was measured, and the mycelial growth inhibition rate (MGIR) was calculated using the following formula [29]:

$$\text{MGIR} = (\text{AD}_{\text{CK}} - \text{Ds})/\text{AD}_{\text{CK}} \times 100\%.$$

where AD_{CK} is the average colony diameter of an investigated isolate in the control treatment, and Ds is the diameter of that isolate in the presence of a stress generation chemical (KCl, H_2O_2 or SDS). Each assay was repeated three times.

2.8. Statistical Analyses

All assays were conducted in triplicate, unless otherwise indicated. Conidia number, colony diameter and lesion diameter analyses were performed using IBM SPSS statistics 20.0 software (IBM Corp., Armonk, NY, USA). The significance of the effects of different treatments on various indices was evaluated with an analysis of variance (ANOVA) with multiple least-significant-difference comparisons at the $p \leq 0.05$ level. After analysis, average angular values were individually back-transformed to numerical values.

3. Results

3.1. Identification of Bcest in B. cinerea

The *Bcest* gene (Bcin15g00520) of B. cinerea was identified via transcriptome data analysis. Bioinformatic analysis showed that this 874 bp gene with three exons and two introns encodes an 81-amino acid protein. Homologs of Bcest were identified via BLASTp and phylogenetic trees of Bcest proteins were constructed vis MEGA 10.0.5 (Figure 1). Evolutionary history was inferred using the neighbour-joining method with 1000 bootstrap replications [30]. It can be seen from Figure 1 that the Bcest protein A0A384K4A6 is highly homologous to putative EkdA protein M7TXI3 from the *B. cinerea* BcDW1 strain.

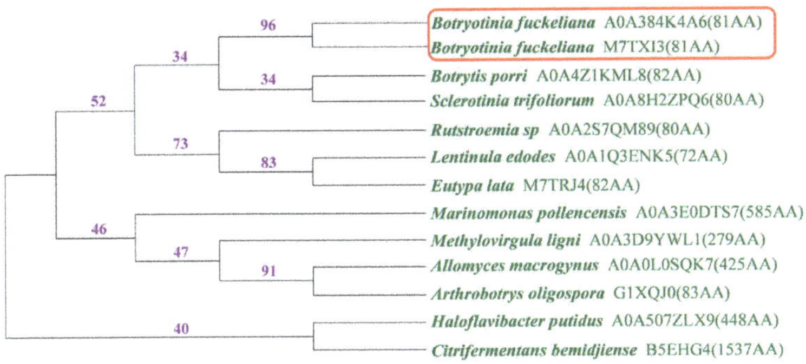

Figure 1. Phylogenetic tree of Bcest proteins based on a neighbour-joining analysis using MEGA. Numbers represent boostrap values.

3.2. Deletion and Complementation of Bcest in *B. cinerea*

To investigate the functions of the Bcest protein in *B. cinerea*, we generated single-gene deletion mutants of Δ*Bcest* using homologous recombination (Figure 2A). The left and right arms (1000 bp) of the *Bcest* gene and the hygromycin gene (2145 bp) of plasmid pUCHYG were amplified. The recombinant *Bcest* gene containing the above fragments was obtained via fusion PCR.

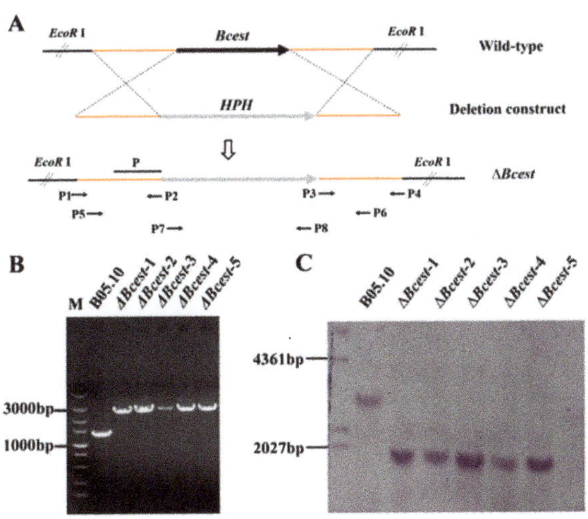

Figure 2. Target gene deletion. (**A**) Schematic diagram of the *Bcest* homologous replacement strategy. (**B**) Amplification of *Bcest* recombinant fragments in B05.10 and *Bcest* gene deletion mutants. (**C**) Southern blotting of *Bcest* gene deletion mutant strains.

We obtained independent transformants via screening a selection medium supplemented with hygromycin B and PCR verification. After single0spore isolation, transformants were verified as homozygous via PCR and further confirmed to be single-copy insertions via Southern blotting analysis (Figure 2B,C). To confirm that the phenotypic changes of the mutants were due to gene deletion, Δ*Bcest* mutants were complemented with the full-length *Bcest* gene to generate complemented strains of Δ*Bcest*-C.

3.3. *Bcest Is Involved in Hyphal Growth and Conidiation*

The mycelial growth rate and the conidium of Δ*Bcest* were significantly different from the those of the WT parent B05.10. The Δ*Bcest* strain had a slower growth rate than the

Microorganisms 2023, 11, 1225

3.3. *Bcest* Is Involved in Hyphal Growth and Conidiation

The mycelial growth rate and the conidium of Δ*Bcest* were significantly different from the those of the WT parent B05.10. The Δ*Bcest* strain had a slower growth rate than the Δ*Bcest*-C complemented strains and the B05.10 WT strain did on PDA, CM and MM, but especially on PDA (Figure 3A,B). Conidia are the primary inoculum for the disease cycle of *B. cinerea* [31]. After incubating it on PDA for 10 days, the abundance of conidia for Δ*Bcest* was significantly less than that for B05.10 and Δ*Bcest*-C (Figure 3C). Interestingly, the conidia of Δ*Bcest* showed morphological and size abnormalities, and some spores produced by Δ*Bcest* had folds and cracks on the surface (Figure 3D). In addition, when incubated on the PDA medium at 22 °C for 10 h, all spores of B05.10 germinated, whereas the average germination rate of Δ*Bcest* was only 78.03%.

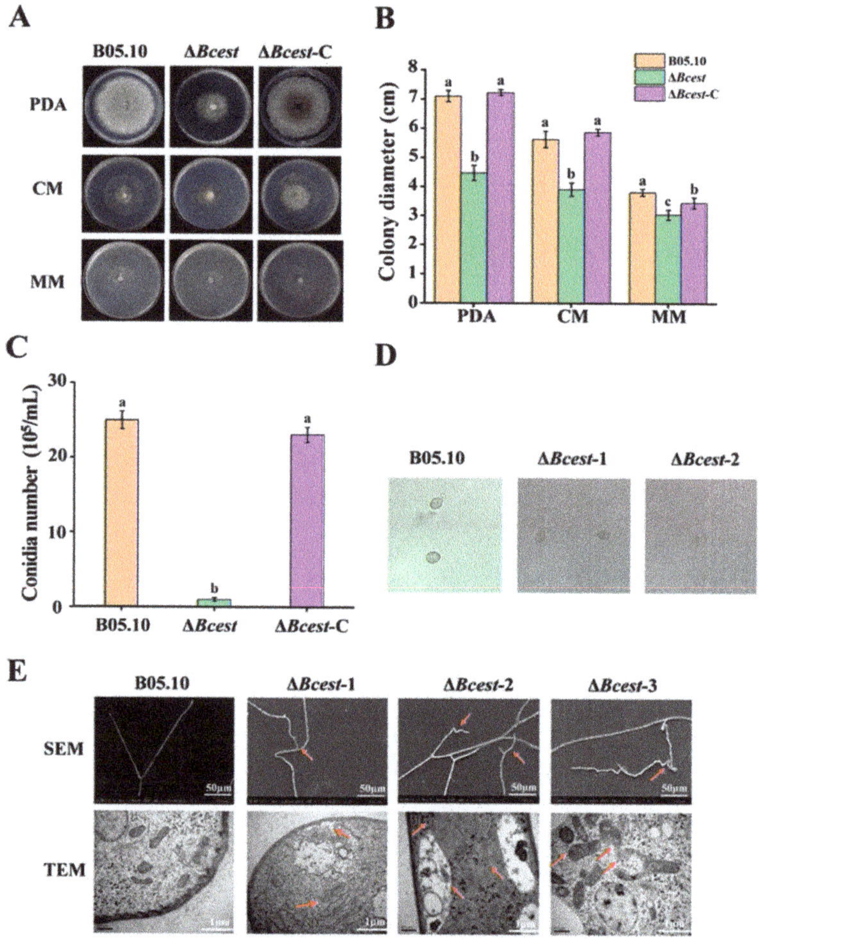

Figure 3. Effects of Bcest deletion on mycelial growth, sporulation and conidial germination. Bars represent standard errors from three replicates. Values on bars followed by different letters indicate significant differences at $p = 0.05$. (**A**) Mycelial growth of Δ*Bcest*, B05.10 and Δ*Bcest*-C strains on PDA plates after 3 days of cultivation. (**B**) Quantification of colony diameter of the indicated strains grown on PDA plates for 3 days. (**C**) Quantification of conidia produced by the indicated strains. (**D**) Conidia morphology of different strains. (**E**) SEM and TEM observations of hyphae of *B. cinerea* and Δ*Bcest* strains grown on PDA plates (diameter: 4 cm). Note: The red arrows in TEM highlight abnormal hyphal growth and uneven branching, and in TEM highlight cell membrane invaginations, mitochondrial swelling, and mitochondrial membrane disappearance.

SEM results showed that control strain B05.10 hyphae grew upright, with uniform thickness, a smooth surface, and no distortion or deformities, and apical hyphae branches were relatively uniform; by contrast, those of Δ*Bcest* showed distortion and deformities, with surface folds and depressions, differences in thickness, and uneven apical branches. TEM results showed that compared with control strain B05.10, the amount of endoplasmic reticulum and the number of mitochondria were increased in Δ*Bcest*, the mitochondrial membrane had disappeared at one end, and mitochondria were larger, looser and more folded. Furthermore, cell membranes were relatively intact, but tended to sink inwards. The above features are indicated by red arrows in Figure 3E.

These results also indicate that Bcest is important for vegetative growth and the conidiation of *B. cinerea*.

3.4. Bcest Participates in Regulating the Pathogenicity of B. cinerea

To determine and visually observe whether or not Bcest is involved in regulating the infection capacity of *B. cinerea*, an onion epidermis was inoculated with the Δ*Bcest* mutant spore suspension. At 12 h, unlike WT spores, Δ*Bcest* spores failed to form an infection structure after germination, and the WT strain successfully infected the onion epidermis by generating attachment cells after spore germination (Figure 4A).

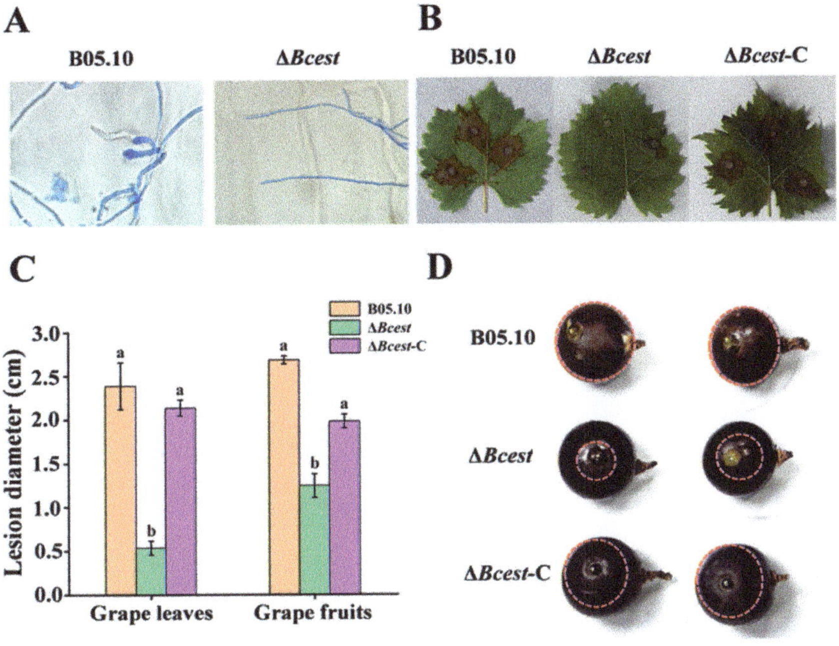

Figure 4. Effects of Δ*Bcest* deletion on mycelial infection and pathogenicity. Bars represent standard errors from three replicates. Values on bars followed by different letters indicate significant differences at $p = 0.05$. (**A**) Onion epidermis penetration by B05.10 and Δ*Bcest*. Pictures were taken after 12 h of inoculation of onion epidermis with conidia from B05.10 and Δ*Bcest*. (**B**) Disease symptoms caused by each strain on wounded grape leaves and fruits. Images were captured at 96 h after inoculation. (**C**) Pathogenicity on grape leaves after 96 h of incubation. (**D**) Pathogenicity on grape fruits after 96 h of incubation on PDA plates (diameter 4 cm). Note: we used Δ*Bcest*-1, Δ*Bcest*-2 and Δ*Bcest*-3 strains for the pathogenicity test of the Δ*Bcest* strain. The red circle represents the diameter of the lesion.

To determine whether or not Bcest is involved in regulating pathogenicity in *B. cinerea*, grapevine leaves and fruits were inoculated with Δ*Bcest* mutants. Compared with the

3 strains for the pathogenicity test of the Δ*Bcest* strain. The red circle represents the diameter of the lesion.

To determine whether or not Bcest is involved in regulating pathogenicity in *B. cinerea*, grapevine leaves and fruits were inoculated with Δ*Bcest* mutants. Compared with the WT strain, Δ*Bcest* mutants exhibited reduced virulence in different hosts (Figure 4B). At 96 h, grape leaves inoculated with Δ*Bcest* mutants showed no or small lesions, while WT-inoculated leaves showed larger lesions and the average lesion size reached 0.53 and 2.13 cm, respectively (Figure 4C). Similarly, lesion size was considerably decreased on Δ*Bcest* mutant-inoculated grape fruits compared with WT-inoculated fruits (Figure 4D). The complemented strain Δ*Bcest*-C exhibited almost the same level of virulence as the WT strain did. These results suggest that Bcest plays a crucial role in virulence of *B. cinerea*.

3.5. Effects of ΔBcest Deletion on Sensitivity to Abiotic Stresses and Pathogenicity Factors

3.5. Effects of ΔBcest Deletion on Sensitivity to Abiotic Stresses and Pathogenicity Factors

The results of the mycelial responsive assays showed that compared with Δ*Bcest*-C and B05.10, Δ*Bcest* mutants exhibited suppressed mycelial growth in the presence of KCl and H_2O_2, and the transparent zone in the protease and cellulase test medium was significantly reduced. However, the growth diameter under the SDS treatment and the polygalacturonase production capacity of Δ*Bcest* were significantly higher than those of B05.10 and Δ*Bcest*-C, respectively (Figure 5A). The values were not significantly different (p = 0.05) from the inhibition rates for B05.10 and Δ*Bcest*-C in response to these chemicals, and the enzyme-producing capacity was not significantly different either (Figure 5B,C). These results suggest that disruption of Bcest may have marginal effects on mycelial growth in response to abiotic stresses, and that pathogenicity factor production capacity may also be affected.

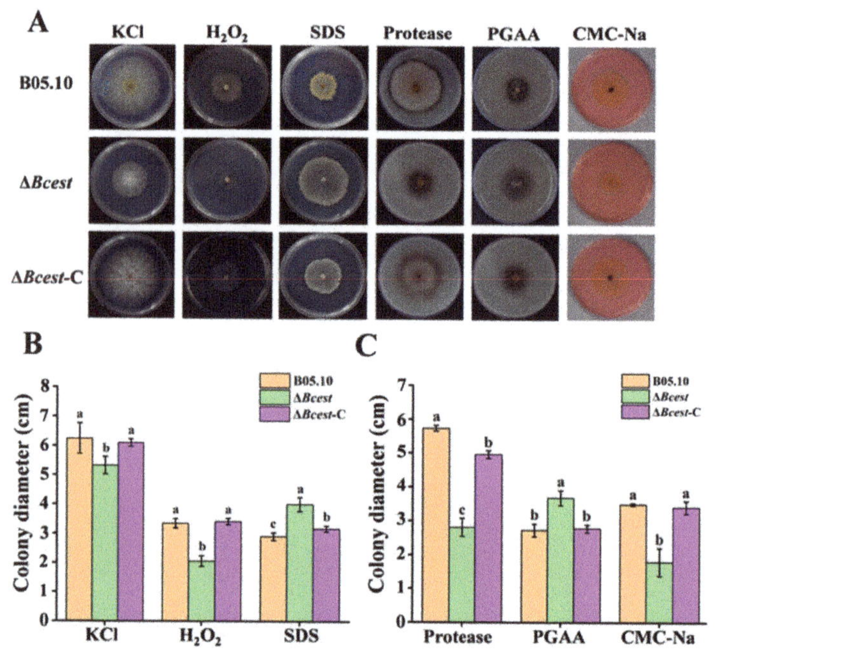

Figure 5. Detection of sensitivity to abiotic stress and the ability to secrete virulence factors in the Δ*Bcest*, B05.10, and Δ*Bcest*-C strains. (**A**) All strains were grown on PDA plates amended with NaCl, KCl, or SDS at the indicated concentrations at 20 °C for 2 days, and with skimmed milk powder, polygalacuronic acid or carboxymethyl cellulose at 20 °C for 3 days. (**B**) Sensitivity of Δ*Bcest*, B05.10 and Δ*Bcest*-C to KCl, H_2O_2 and SDS. (**C**) Ability of Δ*Bcest*, B05.10 and Δ*Bcest*-C to produce proteases, polygalacturonase and cellulases. Bars represent standard errors from three replicates. Values on bars followed by different letters indicate significant differences at p = 0.05.

polygalacuronic acid or carboxymethyl cellulose at 20 °C for 3 days. (**B**) Sensitivity of Δ*Bcest*, B05.10 and Δ*Bcest*-C to KCl, H_2O_2 and SDS. (**C**) Ability of Δ*Bcest*, B05.10 and Δ*Bcest*-C to produce proteases, polygalacturonase and cellulases. Bars represent standard errors from three replicates. Values on bars followed by different letters indicate significant differences at $p = 0.05$.

3.6. *Bcest Deletion Affects Transcription and Pathogenicity-Related Genes*

We performed a RNA-seq analysis to identify genes that might exhibit changes in regulation affected by Bcest in *B. cinerea*. Three biological replicates with mRNA isolated from WT B05.10 and Δ*Bcest* strains were performed, and 162 downregulated and 189 upregulated (fold change ≥ 2, $p \leq 0.05$) genes were identified in Δ*Bcest* and compared with those of B05.10 (Figure 6A).

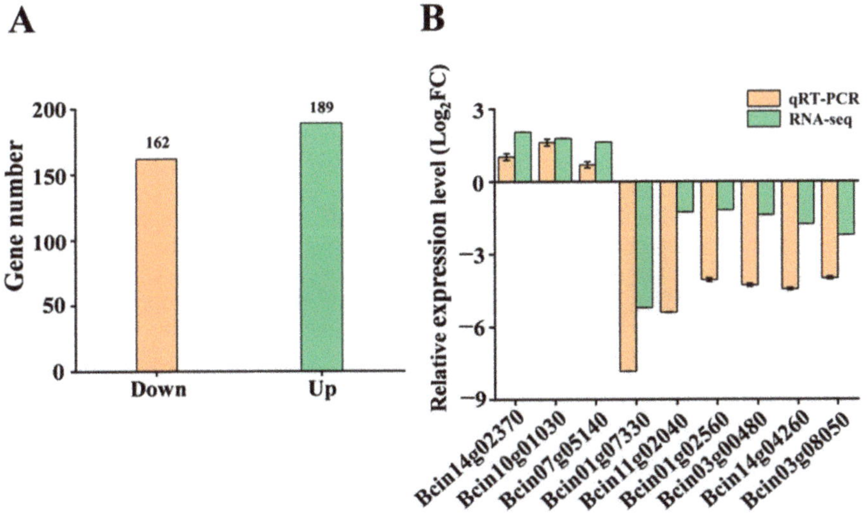

Figure 6. RNA-seq analysis of Δ*Bcest* deletion strains. (**A**) Numbers of upregulated and downregulated genes ($p < 0.05$, >2-fold change) in Δ*Bcest* strains compared with those in WT B05.10. (**B**) qRT-PCR test of transcriptome DEGs.

Functional annotation of DEGs via GO analysis was performed to identify genes belonging to molecular function, cellular component and biological process categories (Figure S1). Among them, catalytic activity/oxidoreductase activity, organelle envelope/mitochondrial outer membrane, the oxidation–reduction process and the monocarboxylic acid metabolic process were the main molecular function, cellular composition, and biological process subcategories, respectively. These results showed that growth and pathogenicity defects caused by the absence of Bcest in the B05.10 strain may be closely related to these six terms.

In addition, enriched DEGs can be functionally classified into metabolism, cellular processes and genetic information processing categories. Among the 45 metabolic pathways belonging to these three categories, the top 10 metabolic pathways of enriched DEGs were metabolic pathways, biosynthesis of secondary metabolites, biosynthesis of antibiotics, carbon metabolism, biosynthesis of amino acids, oxidative phosphorylation, glutathione metabolism, tricarboxylic acid (TCA)/citrate cycle, the pentose phosphate pathway, and ubiquitin–mediated proteolysis (Figure S2). These metabolic pathways may be closely related to the growth and pathogenicity defects of Δ*Bcest*.

To verify the reliability of DEGs identified from transcriptome sequencing, qRT-PCR was performed on the remaining Δ*Bcest* RNA samples used for transcriptome analysis. qRT-PCR validation was performed by randomly selecting nine growth- and pathogenicity-associated genes. These genes encode proteins that participate in the synthesis and metabolism of substances (Bcin14g02370, Bcin07g05140, Bcin03g08050, and Bcin11g02040), the catalytic reactions of enzymes (Bcin03g00480, Bcin01g07330, and Bcin10g01030) and virulence and stress factors (Bcin14g04260 and Bcin01g02560). The qRT-PCR results were consistent with the RNA-seq results (Figure 6B), indicating the reliability of the RNA-seq data in this study.

4. Discussion

To explore the functions of membrane protein Bcest, we first disrupted the *Bcest* gene and characterised the resulting mutant, which showed severe defects in hyphal growth and pathogenicity. These results are consistent with those of experiments on *BcHBF1* and *BcATG6* genes of *B. cinerea* [32,33]. We thus hypothesised that the Bcest protein might be involved in regulating hyphal growth and pathogenicity-related genes in this fungal species. Bcest is a vital virulence determinant, since deletion of the *Bcest* gene also compromised the penetration ability of *B. cinerea*, indicating that the reduced virulence of the Δ*Bcest* mutant was likely due, as least in part, to defects in the penetration of host cells. The Δ*Bcest* mutant exhibited increased sensitivity to H_2O_2 and osmotic stress, which may be due to the absence of membrane protein Bcest, which destroys mitochondrial membrane integrity, reducing tolerance to osmotic pressure and reactive oxygen species (ROS) [34]. However, under SDS-mediated stress, the growth rate of the Δ*Bcest* mutant strain was higher than that of WT strain B05.10. However, other studies have shown that SDS can inhibit the growth of *B. cinerea* [33]; hence, the specific mode of action will be explored in our future research. In addition, the ability of the Δ*Bcest* mutant strain to produce cellulases and proteases was significantly reduced, which is another important factor of pathogenicity [35,36]. Based on the above results, the Bcest protein appears to be an important virulence factor of *B. cinerea*. However, the regulatory mechanisms involving Bcest remain poorly understood, and further research such as a comparative analysis of transcription profiles could provide valuable information.

We used transcriptome sequencing technology to compare and analyse transcriptional regulation differences and DEGs of Δ*Bcest* mutant strains in culture for 72 h, and 351 DEGs were screened. GO functional analysis showed that in biological process and molecular function categories, DEGs resulting from the loss of the *Bcest* gene were mostly involved in metabolic and cellular processes, and catalytic activity and binding, respectively. Among the cellular component terms, the most enriched ones were linked to cellular structure and membrane, which is consistent with the fact that the deletion of the *Bcest* gene affected the mitochondrial membrane in the TEM experiment. Our RNA-seq analysis results suggest that global changes in genes involved in metabolic pathways, the biosynthesis of secondary metabolites, the pentose phosphate pathway and the TCA cycle are likely to underlie this defect.

The reliability of the transcriptome data was verified by via qRT-PCR of nine growth- and pathogenicity-related genes. Sclerotia are an important virulence factor in *B. cinerea*, and this study found that the melanin synthesis gene *Bcpks13* (Bcin03g08050) was significantly downregulated in the Δ*Bcest* mutant strain, which may explain why this strain failed to form sclerotia [37]. Deletion of *CDC14* in several plant pathogen species severely impairs virulence, demonstrating that *CDC14* function is important for host infection [38]. *Fusarium graminearum*, *Magnaporthe oryzae* and *Aspergillus flavus* lacking *CDC14* gene expression can lead to conidia formation defects and reduced pathogenicity [39–41]. Polygalacturonase and β-1,4-endoxylanase *BcXyn11A* are important virulence factors of *B. cinerea*, and they mainly promote the virulence of *B. cinerea* through necrotic activity rather than enzymatic activity [42]. We found that *BcXyn11a* (Bcin03g00480), *Bccdc14* (Bcin01g02560) and endopolygalacturonase-encoding gene *Bcpg5* (Bcin01g07330) were significantly downregulated, which suggests that the decrease in the number of conidia and weakening of the pathogenicity of Δ*Bcest* mutant strains were the result of a combination of polygenes. *Bcgas2* (Bcin14g04260) is required for *B. cinerea* to cope with stress [43,44], and a decrease in the stress capacity of the Δ*Bcest* mutants was closely related to the downregulation of this gene. In addition, both the mitochondrial membrane and the mitochondrial electron transport chain can induce a burst of ROS, which in turn affects the growth and development of the strain. In this study, the expression of the cytochrome C synthesis gene

Bccyc1 (Bcin14g02370), which is involved in the mitochondrial electron transport chain, was significantly upregulated in Δ*Bcest* mutant strains [45]. Interestingly, we found that the membrane protein Bcest had little effect on cell membranes, but TEM showed that the mitochondrial membrane of the Δ*Bcest* strain was damaged, enlarged, wrinkled and relaxed. Thus, we hypothesised that loss of Bcest may destroy the integrity of the mitochondrial membrane, affect the transmission of the mitochondrial electron chain, and influence metabolic pathways in the strain. Therefore, we concluded that an absence of Bcest can reduce the growth rate and pathogenicity of *B. cinerea*, while increasing sensitivity to environmental stress.

5. Conclusions

In conclusion, membrane protein Bcest can inhibit the growth of *B. cinerea*, reduce the spore germination rate and pathogenicity, disrupt the integrity of the mitochondrial structure, and increase sensitivity to oxidative and osmotic stress.

Supplementary Materials: The following supporting information can be downloaded at https://www.mdpi.com/article/10.3390/microorganisms11051225/s1. Figure S1: GO functional classification of Δ*Bcest* transcriptome DEGs. Figure S2: KEGG enrichment analysis bubble diagram of the Δ*Bcest* transcriptome. Table S1: Mapping statistics in transcriptome analyses. Table S2: Oligonucleotide primers used in this study. Table S3: List of all DEGs in transcriptome analyses. Table S4: Genes and primers used for qRT-PCR.

Author Contributions: Conceptualisation, K.-C.Z. and L.-M.S.; methodology, L.-M.S. and W.Z.; validation, W.Z. and Z.-Y.L.; formal analysis, L.-M.S., W.Z. and B.-B.G.; investigation, L.-M.S., W.Z., B.-B.G. and K.S.P.; supervision, K.-C.Z., L.-M.S. and B.-B.G.; writing—original draft preparation, W.Z.; writing—review and editing, W.Z. and L.-M.S. All authors have read and agreed to the published version of the manuscript.

Funding: This work was supported by the National Key Research and Development Program of China (no. 2019YFD1002000), the Agricultural Science and Technology Innovation Program of CAAS, and the Major Research Project of Guangxi for Science and Technology (no. AA18242026).

Institutional Review Board Statement: Not applicable.

Informed Consent Statement: Not applicable.

Data Availability Statement: The data used in the study analyses can be made available by the corresponding author upon reasonable request.

Conflicts of Interest: The authors declare no conflict of interest.

References

1. Fan, L.; Wei, Y.Y.; Chen, Y.; Jiang, S.; Xu, F.; Zhang, C.D.; Wang, H.F.; Shao, X.F. Epinecidin-1, a marine antifungal peptide, inhibits *Botrytis cinerea* and delays gray mold in postharvest peaches. *Food Chem.* **2023**, *403*, 134419. [CrossRef] [PubMed]
2. Bu, S.W.; Munir, S.; He, P.F.; Li, Y.M.; Wu, Y.X.; Li, X.Y.; Kong, B.H.; He, P.B.; He, Y.Q. *Bacillus subtilis* L1-21 as a biocontrol agent for postharvest gray mold of tomato caused by *Botrytis cinerea*. *Biol. Control* **2021**, *157*, 104568. [CrossRef]
3. Elad, Y.; Pertot, I.; Cotes Prado, A.M.; Stewart, A. Plant Hosts of *Botrytis* spp. In *Botrytis—The Fungus, the Pathogen and Its Management in Agricultural Systems*; Fillinger, S., Elad, Y., Eds.; Springer International Publishing: Cham, Switzerland, 2016; pp. 413–486.
4. Wang, C.C.; Yuan, S.S.; Zhang, W.W.; Ng, T.; Ye, X.J. Buckwheat antifungal protein with biocontrol potential to Inhibit fungal (*Botrytis cinerea*) infection of cherry tomato. *J. Agric. Food Chem.* **2019**, *67*, 6748–6756. [CrossRef]
5. Engberg, O.; Ulbricht, D.; Döbel, V.; Siebert, V.; Frie, C.; Penk, A.; Lemberg, M.K.; Huster, D. Rhomboid-catalyzed intramembrane proteolysis requires hydrophobic matching with the surrounding lipid bilayer. *Sci. Adv.* **2022**, *8*, eabq8303. [CrossRef]
6. Alguel, Y.; Cameron, A.D.; Diallinas, G.; Byrne, B. Transporter oligomerization: Form and function. *Biochem. Soc. Trans.* **2016**, *44*, 1737–1744. [CrossRef] [PubMed]
7. Yeliseev, A.; Iyer, M.R.; Joseph, T.T.; Coffey, N.J.; Cinar, R.; Zoubak, L.; Kunos, G.; Gawrisch, K. Cholesterol as a modulator of cannabinoid receptor CB_2 signaling. *Sci. Rep.* **2021**, *11*, 3706. [CrossRef]
8. Sant, D.G.; Tupe, S.G.; Ramana, C.V.; Deshpande, M.V. Fungal cell membrane—Promising drug target for antifungal therapy. *J. Appl. Microbiol.* **2016**, *121*, 1498–1510. [CrossRef]

9. Te Welscher, Y.M.; van Leeuwen, M.R.; de Kruijff, B.; Dijksterhuis, J.; Breukink, E. Polyene antibiotic that inhibits membrane transport proteins. *Proc. Natl. Acad. Sci. USA* **2012**, *109*, 11156–11159. [CrossRef]
10. Shi, L.M.; Ge, B.B.; Wang, J.Z.; Liu, B.H.; Ma, J.J.; Wei, Q.H.; Zhang, K.C. iTRAQ-based proteomic analysis reveals the mechanisms of *Botrytis cinerea* controlled with Wuyiencin. *BMC Microbiol.* **2019**, *19*, 280. [CrossRef]
11. Weidensdorfer, M.; Ishikawa, M.; Hori, K.; Linke, D.; Djahanschiri, B.; Iruegas, R.; Ebersberger, I.; Riedel-Christ, S.; Enders, G.; Leukert, L.; et al. The *Acinetobacter* trimeric autotransporter adhesin Ata controls key virulence traits of *Acinetobacter baumannii*. *Virulence* **2019**, *10*, 68–81. [CrossRef]
12. Bialer, M.G.; Ferrero, M.C.; Delpino, M.V.; Ruiz-Ranwez, V.; Posadas, D.M.; Baldi, P.C.; Zorreguieta, A. Adhesive functions or pseudogenization of type va autotransporters in *Brucella Species*. *Front. Cell. Infect. Microbiol.* **2021**, *11*, 607610. [CrossRef]
13. Raghunathan, D.; Wells, T.J.; Morris, F.C.; Shaw, R.K.; Bobat, S.; Peters, S.E.; Paterson, G.K.; Jensen, K.T.; Leyton, D.L.; Blair, J.M.; et al. SadA, a trimeric autotransporter from *Salmonella enterica serovar* Typhimurium, can promote biofilm formation and provides limited protection against infection. *Infect. Immun.* **2011**, *79*, 4342–4352. [CrossRef] [PubMed]
14. Huang, X.Q.; Yan, A.; Zhang, X.H.; Xu, Y.Q. Identification and characterization of a putative ABC transporter PltHIJKN required for pyoluteorin production in *Pseudomonas* sp. M18. *Gene* **2006**, *376*, 68–78. [CrossRef]
15. Quidde, T.; Büttner, P.; Tudzynski, P. Evidence for three different specific saponin-detoxifying activities in *Botrytis cinerea* and cloning and functional analysis of a gene coding for a putative avenacinase. *Eur. J. Plant Pathol.* **1999**, *105*, 273–283. [CrossRef]
16. Zhang, N.; Yang, Z.Z.; Zhang, Z.H.; Liang, W.X. BcRPD3-Mediated histone deacetylation is involved in growth and pathogenicity of *Botrytis cinerea*. *Front. Microbiol.* **2020**, *11*, 1832. [CrossRef]
17. Ren, W.C.; Liu, N.; Sang, C.W.; Shi, D.Y.; Zhou, M.G.; Chen, C.J.; Qin, Q.M.; Chen, W.C. The autophagy gene *BcATG8* regulates the vegetative differentiation and pathogenicity of *Botrytis cinerea*. *Appl. Environ. Microbiol.* **2018**, *84*, e02455-17. [CrossRef] [PubMed]
18. Yan, L.Y.; Yang, Q.Q.; Sundin, G.W.; Li, H.Y.; Ma, Z.H. The mitogen-activated protein kinase kinase BOS5 is involved in regulating vegetative differentiation and virulence in *Botrytis cinerea*. *Fungal Genet. Biol.* **2010**, *47*, 753–760. [CrossRef]
19. Yu, J.H.; Hamari, Z.; Han, K.H.; Seo, J.A.; Reyes-Domínguez, Y.; Scazzocchio, C. Double-joint PCR: A PCR-based molecular tool for gene manipulations in filamentous fungi. *Fungal Genet. Biol.* **2004**, *41*, 973–981. [CrossRef]
20. Gronover, C.S.; Kasulke, D.; Tudzynski, P.; Tudzynski, B. The role of G protein alpha subunits in the infection process of the gray mold fungus *Botrytis cinerea*. *Mol. Plant-Microbe Interact.* **2001**, *14*, 1293–1302. [CrossRef]
21. Ren, W.C.; Sang, C.W.; Shi, D.Y.; Song, X.S.; Zhou, M.G.; Chen, C.J. Ubiquitin-like activating enzymes BcAtg3 and BcAtg7 participate in development and pathogenesis of *Botrytis cinerea*. *Curr. Genet.* **2018**, *64*, 919–930. [CrossRef]
22. Mueller, O.M.; Lightfoot, S.; Schroeder, A. *RNA Integrity Number (RIN)—Standardization of RNA Quality Control*; Agilent Technology Application: Santa Clara, CA, USA, 2004; pp. 1–8.
23. Bolger, A.M.; Lohse, M.; Usadel, B. Trimmomatic: A flexible trimmer for Illumina sequence data. *Bioinformatics* **2014**, *30*, 2114–2120. [CrossRef] [PubMed]
24. Raudvere, U.; Kolberg, L.; Kuzmin, I.; Arak, T.; Adler, P.; Peterson, H.; Vilo, J. g: Profiler: A web server for functional enrichment analysis and conversions of gene lists (2019 update). *Nucleic Acids Res.* **2019**, *47*, W191–W198. [CrossRef] [PubMed]
25. McDonald, B.A.; Martinez, J.P. Restriction fragment length polymorphisms in *Septoria tritici* occur at a high frequency. *Curr. Genet.* **1990**, *17*, 133–138. [CrossRef]
26. Livak, K.J.; Schmittgen, T.D. Analysis of relative gene expression data using real-time quantitative PCR and the 2(-Delta Delta C(T)) Method. *Methods* **2001**, *25*, 402–408. [CrossRef]
27. Yang, Q.Q.; Zhang, J.N.; Hu, J.C.; Wang, X.; Lv, B.N.; Liang, W.X. Involvement of *BcYak1* in the regulation of vegetative differentiation and adaptation to oxidative stress of *Botrytis cinerea*. *Front. Microbiol.* **2018**, *9*, 281. [CrossRef]
28. Zou, X.R.; Wei, Y.Y.; Jiang, S.; Xu, F.; Wang, H.F.; Zhan, P.P.; Shao, X.F. ROS stress and cell membrane disruption are the main antifungal mechanisms of 2-phenylethanol against *Botrytis cinerea*. *J. Agric. Food Chem.* **2022**, *70*, 14468–14479. [CrossRef]
29. Zhou, Y.J.; Song, J.J.; Wang, Y.C.; Yang, L.; Wu, M.D.; Li, G.Q.; Zhang, J. Biological characterization of the melanin biosynthesis gene *Bcscd1* in the plant pathogenic fungus *Botrytis cinerea*. *Fungal Genet. Biol.* **2022**, *160*, 103693. [CrossRef]
30. Russo, C.A.d.M.; Selvatti, A.P. Bootstrap and rogue identification tests for phylogenetic analyses. *Mol. Biol. Evol.* **2018**, *35*, 2327–2333. [CrossRef]
31. Abbey, J.A.; Percival, D.; Abbey, L.; Asiedu, S.K.; Prithiviraj, B.; Schilder, A. Biofungicides as alternative to synthetic fungicide control of grey mould (*Botrytis cinerea*)—Prospects and challenges. *Biocontrol Sci. Technol.* **2019**, *29*, 207–228. [CrossRef]
32. Liu, Y.; Liu, J.K.; Li, G.H.; Zhang, M.Z.; Zhang, Y.Y.; Wang, Y.Y.; Hou, J.; Yang, S.; Sun, J.; Qin, Q.M. A novel *Botrytis cinerea*-specific gene *BcHBF1* enhances virulence of the grey mould fungus via promoting host penetration and invasive hyphal development. *Mol. Plant Pathol.* **2019**, *20*, 731–747. [CrossRef]
33. Liu, N.; Zhou, S.Y.; Li, B.H.; Ren, W.C. Involvement of the autophagy protein Atg6 in development and virulence in the gray mold fungus *Botrytis cinerea*. *Front. Microbiol.* **2021**, *12*, 3943. [CrossRef] [PubMed]
34. He, Z.J.; Zhang, S.H.; Keyhani, N.O.; Song, Y.L.; Huang, S.S.; Pei, Y.; Zhang, Y.J. A novel mitochondrial membrane protein, Ohmm, limits fungal oxidative stress resistance and virulence in the insect fungal pathogen *Beauveria bassiana*. *Environ. Microbiol.* **2015**, *17*, 4213–4238. [CrossRef] [PubMed]
35. Yuan, X.M. Molecular Mechanism of *BcPDR1* Gene in Regulating Pathogenicity of *Botrytis cinerea*. Mater's Dissertation, Hebei Agricultural University, Baoding, China, 2019.

36. He, J.Y. Functional Analysis of Phytochrome *Bcphy3* in *Botrytis cinerea*. Mater's Dissertation, East China Normal University, Shanghai, China, 2014.
37. Zhang, C.H.; He, Y.F.; Zhu, P.K.; Chen, L.; Wang, Y.W.; Ni, B.; Xu, L. Loss of bcbrn1 and *bcpks13* in *Botrytis cinerea* not only blocks melanization but also increases vegetative growth and virulence. *Mol. Plant-Microbe Interact.* **2015**, *28*, 1091–1101. [CrossRef] [PubMed]
38. DeMarco, A.G.; Milholland, K.L.; Pendleton, A.L.; Whitney, J.J.; Zhu, P.; Wesenberg, D.T.; Nambiar, M.; Pepe, A.; Paula, S.; Chmielewski, J.; et al. Conservation of Cdc14 phosphatase specificity in plant fungal pathogens: Implications for antifungal development. *Sci. Rep.* **2020**, *10*, 12073. [CrossRef] [PubMed]
39. Li, C.H.; Melesse, M.; Zhang, S.J.; Hao, C.F.; Wang, C.F.; Zhang, H.C.; Hall, M.C.; Xu, J.R. *FgCDC14* regulates cytokinesis, morphogenesis, and pathogenesis in *Fusarium graminearum*. *Mol. Microbiol.* **2015**, *98*, 770–786. [CrossRef]
40. Li, C.H.; Cao, S.L.; Zhang, C.K.; Zhang, Y.H.; Zhang, Q.; Xu, J.R.; Wang, C.F. *MoCDC14* is important for septation during conidiation and appressorium formation in *Magnaporthe oryzae*. *Mol. Plant Pathol.* **2018**, *19*, 328–340. [CrossRef]
41. Yang, G.; Hu, Y.; Fasoyin, O.E.; Yue, Y.W.; Chen, L.J.; Qiu, Y.; Wang, X.N.; Zhuang, Z.H.; Wang, S.H. The *Aspergillus flavus* phosphatase CDC14 regulates development, aflatoxin biosynthesis and pathogenicity. *Front. Cell. Infect. Microbiol.* **2018**, *8*, 141. [CrossRef]
42. Zhuang, W. Functional Analysis of Two Putative Secreted Proteins in *Botrytis cinerea*. Mater's Dissertation, Fujian Agriculture and Forestry University, Fuzhou, China, 2019.
43. Schamber, A.; Leroch, M.; Diwo, J.; Mendgen, K.; Hahn, M. The role of mitogen-activated protein (MAP) kinase signalling components and the *Ste12* transcription factor in germination and pathogenicity of *Botrytis cinerea*. *Mol. Plant Pathol.* **2010**, *11*, 105–119. [CrossRef]
44. Tundo, S.; Paccanaro, M.C.; Elmaghraby, I.; Moscetti, I.; D'Ovidio, R.; Favaron, F.; Sella, L. The xylanase inhibitor TAXI-I increases plant resistance to *Botrytis cinerea* by inhibiting the BcXyn11a xylanase necrotizing activity. *Plants* **2020**, *9*, 601. [CrossRef]
45. Schumacher, J.; Simon, A.; Cohrs, K.C.; Viaud, M.; Tudzynski, P. The transcription factor *BcLTF1* regulates virulence and light responses in the necrotrophic plant pathogen *Botrytis cinerea*. *PLoS Genet.* **2014**, *10*, e1004040. [CrossRef]

microorganisms

MDPI

Article

The *Epichloë festucae* Antifungal Protein *Efe*-AfpA Has Activity against Numerous Plant Pathogens

Patrick A. Fardella, Bruce B. Clarke and Faith C. Belanger *

Department of Plant Biology, Rutgers University, New Brunswick, NJ 08901, USA
* Correspondence: belanger@sebs.rutgers.edu

Abstract: Fungal plant pathogens can present major problems for most crop species. Currently, control of fungal diseases relies heavily on the use of fungicides. However, there are problems associated with fungicide use, including potential toxicity to non-target organisms and the development of resistance in the target fungus. New strategies are being sought to reduce fungicide use. One area of active research is the potential use of antifungal proteins from various fungal species as alternatives or complements to traditional fungicides. An antifungal protein, *Efe*-AfpA, from the fungal endophyte *Epichloë festucae* was previously found to protect plants from the pathogen *Clarireedia jacksonii*, the causal agent of dollar spot disease. Here we report that *Efe*-AfpA also has inhibitory activity against other important plant pathogens. These results suggest that it may be possible to develop *Efe*-AfpA as a biofungicide to target a broad range of destructive plant pathogens.

Keywords: antifungal protein; fungal endophyte; fungal plant pathogens

Citation: Fardella, P.A.; Clarke, B.B.; Belanger, F.C. The *Epichloë festucae* Antifungal Protein *Efe*-AfpA Has Activity against Numerous Plant Pathogens. *Microorganisms* **2023**, *11*, 828. https://doi.org/10.3390/microorganisms11040828

Academic Editors: Tomislav Cernava, Beibei Ge and Kyungseok Park

Received: 23 February 2023
Revised: 17 March 2023
Accepted: 22 March 2023
Published: 24 March 2023

1. Introduction

Management of fungal plant pathogens is challenging for many crops. Current management strategies rely heavily on the use of fungicides, which are considered critical for global food security [1,2]. Fungicides are also used heavily on amenity crops such as turfgrasses [3]. Although deemed critical for disease management, there are several problems associated with fungicide use, including potential toxicity to non-target organisms and the development of resistance in the target fungus [1]. Strategies to reduce fungicide usage include the breeding of more disease-tolerant cultivars [4,5] as well as genetic modification of plants to enhance disease resistance [6]. Improved cultural management strategies can also help to improve disease control and reduce fungicide usage [7].

Another strategy is the development of alternatives or complements to synthetic fungicides that have reduced potential for causing toxicities to the environment or non-target organisms. This approach can involve the application of biological control organisms or products derived from living organisms to plants to reduce disease severity [8]. One area of active research is the potential use of antifungal proteins from various fungal species as biofungicides. The most extensively researched antifungal proteins are PAF and PAFB from *Penicillium chrysogenum*, and AFP and NFAP from *Aspergillus* spp. [9–12]. These proteins have been tested against a variety of fungi, yeasts, and bacteria to determine their efficacy in inhibiting growth, showing both stark differences and striking similarities in their effectiveness [10,13–15]. This underscores the importance of continuing to test these and additional antifungal proteins to determine their repertoire of sensitive microorganisms. This information would allow for the selection of appropriate antifungal proteins to help control specific plant pathogens [16].

We are interested in the potential development of an antifungal protein from the endophytic fungus *Epichloë festucae*, which infects the grass strong creeping red fescue (*Festuca rubra* subsp. *rubra*), as a possible alternative or complement to fungicides for the control of important plant pathogens. Strong creeping red fescue is a commercially important turfgrass species [17]. *Epichloë* spp. are common endophytic symbionts of many cool-season grass species, often conferring insect resistance to the hosts due to the production of toxic alkaloids [18]. In addition to insect resistance, *E. festucae* infection is known to confer field-level disease resistance in strong creeping red fescue against both dollar spot disease (*Clarireedia jacksonii*) and red thread disease (*Laetisaria fuciformis*) [19,20]. This endophyte-mediated disease resistance is unique to the *E. festucae*/fine fescue symbiosis and has not been seen in other *Epichloë* spp./commercial grass species interactions [20–22]. The genome of *E. festucae* isolates that infect strong creeping red fescue contains a gene encoding an antifungal protein [23]. The *E. festucae* antifungal protein gene is designated *Efe-afpA* and the protein is designated *Efe*-AfpA [24]. Such a gene is not present in the genomes of most *Epichloë* spp. for which whole genome sequencing is available, including other strains of *E. festucae* [21,23] This suggested that the *E. festucae* antifungal protein is a likely factor in the endophyte-mediated disease resistance observed in strong creeping red fescue. To validate the presumption that *Efe*-AfpA was a major component of this endophyte-mediated disease resistance, *Efe*-AfpA was produced in a *Pe. chrysogenum* expression system and tested against *Cl. jacksonii*. Purified *Efe*-AfpA inhibited the growth of *Cl. jacksonii* in culture and direct application to *Cl. jacksonii*-inoculated plants reduced the severity of disease symptoms on endophyte-free strong creeping red fescue and creeping bentgrass (*Agrostis stolonifera*) [25].

These results support the original hypothesis that *Efe*-AfpA is a major contributor to the dollar spot resistance seen in endophyte-infected strong creeping red fescue plants and suggested that *Efe*-AfpA may have potential as a biofungicide. It was therefore of interest to determine if *Efe*-AfpA also had activity against other plant pathogens. The objective of this work was to assay *Efe*-AfpA activity on important plant fungal pathogens in culture. We also assayed the *Pe. chrysogenum* antifungal protein PAF since we previously found differences between the activities of *Efe*-AfpA and PAF against *Cl. jacksonii* [25]. In particular, we focused on historically and economically important plant pathogens and major turfgrass pathogens, as *Efe*-AfpA is from a turfgrass endophyte. *Efe*-AfpA did have inhibitory activity against several other plant pathogens, expanding its potential usefulness beyond *Cl. jacksonii*.

2. Materials and Methods

2.1. Fungi and Culture Conditions

Pe. chrysogenum paf, an isolate over-expressing *paf*, and *Pe. chrysogenum*$^{Efe\text{-}AfpA}$, an isolate expressing *Efe-afpA*, were maintained on PcMM (*Penicillium chrysogenum* Minimal Media) [25,26] supplemented with 200 μg mL^{-1} nourseothricin and 0.6 μg mL^{-1} pyrithiamine. All other fungi were maintained on potato dextrose agar (PDA) (HiMedia Laboratories, Mumbai, India). To generate conidia for *Pyricularia oryze*, a fungal plug grown on PDA was subcultured onto ryegrass-amended plates and grown for at least 2 weeks. Ryegrass-amended plates were made by autoclaving ryegrass clippings, approximately 20 g in 500 mL of distilled water twice. The clippings were removed by filtering through cheesecloth. The volume of the filtered solution was brought back up to 500 mL with distilled water, agar was added to a final concentration of 1.5%, and the solution was autoclaved.

2.2. Purification of Efe-AfpA from Pe. chrysogenum$^{Efe\text{-}AfpA}$ and PAF from Pe. chrysogenum paf

The protein purification method was previously described in detail [25]. Briefly, for purification of *Efe*-AfpA 2 × 10^8 conidia of *Pe. chrysogenum*$^{Efe\text{-}AfpA}$ were added to 200 mL *Aspergillus nidulans* Complete Media and shaken at 200 rpm for 48 h. The mycelium was then harvested by filtering the culture through cheesecloth, which was then resuspended in 200 mL PcMM and incubated for 72 h while shaking at 200 rpm. The culture supernatant was collected by filtering through cheesecloth to remove mycelia, and any excess debris was pelleted by centrifugation at 10,000 rpm for 10 min. The cleared culture supernatant was applied to a carboxymethyl cellulose (CMC52) (Biophoretics, Sparks, NV, USA) column pre-equilibrated with 10 mM $NaPO_4$, 25 mM NaCl, 0.15 mM EDTA, pH 6.6. The column was then washed with excess buffer and *Efe*-AfpA was eluted with increasing salt concentrations from 0.1 to 0.5 M NaCl. Similarly, for purification of PAF, 2×10^8 conidia of *Pe. chrysogenum paf* was added to 200 mL PcMM and shaken for 72 h at 200 rpm. The culture supernatant was then processed as described above for *Efe*-AfpA.

Eluted fractions from both purifications were evaluated for the presence of the protein utilizing their respective molecular weights and extinction coefficients at A_{280} measured by using a Nanodrop ND-1000 spectrophotometer (Thermo Fisher Scientific, Waltham, MA, USA). *Efe*-AfpA molecular weight: 6.278 kDa, extinction coefficient: 5220 M^{-1} cm^{-1}. PAF molecular weight: 6.25 kDa, extinction coefficient: 4845 M^{-1} cm^{-1}. Protein-containing fractions were filtered through a 30 kDa Amicon® Ultra-15 Centrifugal Filter (Millipore-Sigma, Burlington, MA, USA) to remove high molecular weight proteins. The proteins were then concentrated and desalted on a 3 kDa Amicon® Ultra-15 Centrifugal Filter with sterile distilled water. The protein samples were finally sterilized by filtering through a 0.2 µm polyethersulfone syringe filter (Corning Inc., Corning, NY, USA).

2.3. Antifungal Activity Assays

For fungi that easily produced spores (*B. cinerea*, *Co. graminicola*, *F. graminearum* PH1, *Py. oyzae*), antifungal activities of *Efe*-AfpA and PAF were assayed in 24 and 96-well plates. For the 24-well plate activity assays, spores were harvested in Spore Buffer (0.9% NaCl, 0.01% Tween 20), washed twice in Spore Buffer, washed once in sterile distilled water, and resuspended in sterile distilled water. Spores were counted using a hemocytometer and diluted to 2×10^5 conidia mL^{-1}. Five µL of spores was plated onto 500 µL of PDA amended with increasing concentrations of *Efe*-AfpA or PAF (0, 0.75, 1.5, 3, 6, 12, 25, 500, 100 µg mL^{-1}) in a 24-well plate. Plates were incubated for 72 to 96 h, depending on the growth rate of the fungus. Assays were performed in duplicate and the experiment was completed twice.

For 96-well activity assays, spores were harvested in Spore Buffer, washed twice in Spore Buffer, washed once in $2 \times$ low cation media (LCM) ($1 \times$ LCM is 2 g L^{-1} glucose, 0.1 g L^{-1} yeast extract, 0.05 g L^{-1} peptone), and resuspended in fresh $2 \times$ LCM. Spores were counted on a hemocytometer and diluted to 2×10^4 conidia mL^{-1} with $2 \times$ LCM. One hundred µL of spores and 100 µL of water containing increasing concentrations of *Efe*-AfpA or PAF were incubated in each well to final concentrations of 0, 0.3, 0.6, 1.2, 5, 10, 20, 30, 40, 50, and 100 µg mL^{-1} antifungal protein. Plates were incubated at room temperature for 24 to 48 h, depending on how fast the fungi grew. Growth was monitored at A_{620} using a microtiter plate reader (Absorbance 96, Byonoy GmbH, Hamburg, Germany). Optical density at A_{620} was measured at 0 h and subtracted from subsequent readings to correct for background absorbance. The corrected absorbance of untreated control conidia was considered 100% growth. The percent inhibition by each protein was calculated by comparing the conidia growth of the treated samples to the untreated samples. Wells were visualized microscopically (EVOS M5000, Invitrogen, Waltham, MA, USA). Each treatment had 3 replicates and the experiment was completed twice.

Fungi that did not produce spores easily (*Cr. parasitica* Ep155, *L. fuciformis*, *R. solani*) had their sensitivity to *Efe*-AfpA and PAF determined using fungal plugs. These fungi were grown from 4 to 10 days on PDA, depending on the growth rate of the fungus, prior to being subcultured. A 5 mm plug of each fungus was placed in the center of 8 mL PDA plates amended with increasing concentrations of *Efe*-AfpA or PAF (0, 0.5, 1, 10, 20, 30, 40, 50, 100 µg mL^{-1}). Cross-sectional fungal diameter was measured, and plates were photographed daily. Each treatment had three replicates and the experiment was completed twice.

3. Results

3.1. *Botrytis cinerea*

Botrytis cinerea is a necrotrophic ascomycete responsible for gray mold on over 200 crop species worldwide. Although it causes disease issues in the field and in greenhouses, it twice.

3. Results

3.1. *Botrytis cinerea*

Botrytis cinerea is a necrotrophic ascomycete responsible for gray mold on over 200 crop species worldwide. Although it causes disease issues in the field and in greenhouses, it is also a considerable post-harvest problem because the fungus can remain quiescent for extended periods of time before becoming active [27]. *B. cinerea* was listed as the second most important fungal plant pathogen and causes between $10 to $100 billion in losses annually worldwide [28,29]. Previous work reported variable activity of PAF against *B. cinerea*, which could be due to differences in experimental assays used [30–32]. In the current study, *Efe*-AfpA had a minimal inhibitory concentration (MIC) against *B. cinerea* of 0.6 μg mL^{-1} (Figures 1 and S1, Table 1). MIC is defined as the minimal concentration resulting in at least 90% inhibition of the growth of the target organism. PAF did have activity against *B. cinerea* but its activity did not result in 90% inhibition at any of the concentrations tested (Figure 1, Table 1).

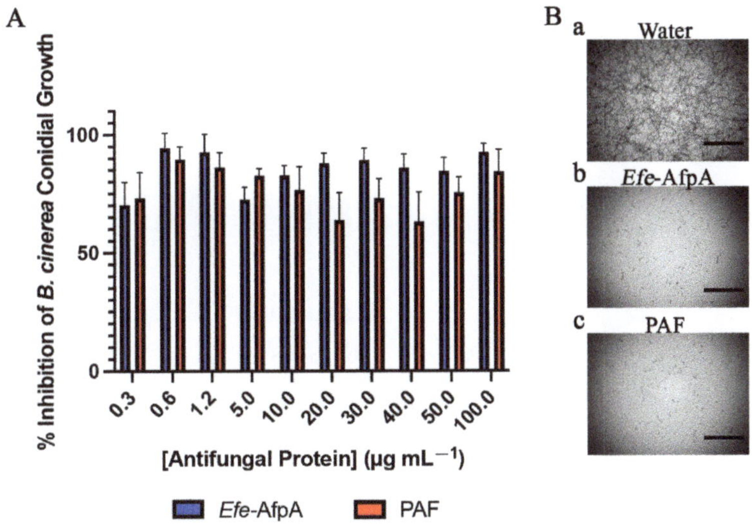

Figure 1. Activity of *Efe*-AfpA and PAF against *B. cinerea* conidial growth. (**A**) Growth inhibition of *B. cinerea* conidia treated with increasing concentrations of either *Efe*-AfpA or PAF incubated at room temperature for 30 h. The data presented are the means and standard deviations of three replicates. (**B**) Microscopy of *B. cinerea* conidia treated with water or 0.6 μg mL^{-1} of either *Efe*-AfpA or PAF. Bars are 750 μm.

Table 1. Minimal and maximum inhibitory concentrations of *Efe*-AfpA and PAF against fungal plant pathogens.

Organism	Minimal Inhibitory Concentration [1]				Maximum Inhibitory Concentration [2]			
	Efe-AfpA		PAF		*Efe*-AfpA		PAF	
	μg mL^{-1}	% Inhibition	μg mL^{-1}	% Inhibition	μg mL^{-1}	% Inhibition	μg mL^{-1}	% Inhibition
Ascomycetes								
B. cinerea	0.6	94.4	–	–	0.6	94.4	0.6	89.6
Co. cereale	1.2	92.9	40	95.6	40	96.6	100	96.4
Cr. parasitica	0.5	98.8	0.5	100	0.5	98.8	0.5	100
F. graminearum	10	91	–	–	100	98.7	100	75.7
P. oryzae	–	–	–	–	100	87.1	20	75.3
Basidiomycetes								
L. fuciformis	–	–	–	–	100	46.5	–	–
R. solani	–	–	–	–	–	–	–	–

[1] The minimal inhibitory concentration is defined as the concentration needed for 90% inhibition. "–" indicates activity did not result in 90% inhibition at any of the concentrations tested. [2] The maximum inhibitory concentration is defined as the highest inhibition obtained at any concentration tested. "–" indicates activity did not result in inhibition at any of the concentrations tested.

Basidiomycetes								
L. fuciformis	–	–	–	–	100	46.5	–	–
R. solani	–	–	–	–	–	–	–	–

The minimal inhibitory concentration is defined as the concentration needed for 90% inhibition. [1] "–" indicates activity did not result in 90% inhibition at any of the concentrations tested. The maximum inhibitory concentration is defined as the highest inhibition obtained at any concentration tested. [2] "–" indicates activity did not result in inhibition at any of the concentrations tested.

3.2. *Colletotrichum cereale*

Colletotrichum spp. are common and destructive pathogens on many plant species and are particularly destructive on most agricultural crops. They are ranked the eighth most important fungal pathogens of plants [28]. *Co. cereale* is responsible for anthracnose on turfgrasses such as *Poa annua* and *Agrostis* species, where the disease occurs as either a foliar blight or basal rot [33]. Both *Efe*-AfpA and PAF had activity against *Co. cereale* that resulted in greater than 90% inhibition, with MIC concentrations of 1.2 and 100 μg mL^{-1}, respectively (Figures 2 and S2, Table 1).

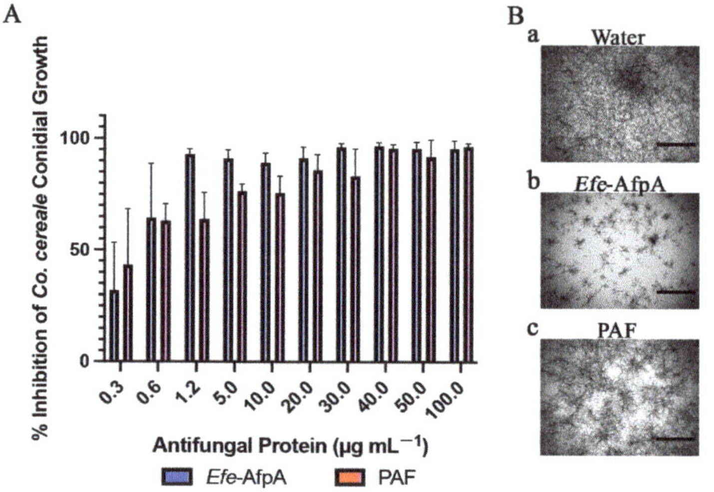

Figure 2. Activity of *Efe*-AfpA and PAF against *Co. cereale* conidial growth. (**A**) Growth inhibition of *Co. cereale* conidia treated with increasing concentrations of either *Efe*-AfpA or PAF incubated at room temperature for 48 h. The data presented are the means and standard deviations of three replicates. (**B**) Microscopy of *Co. cereale* conidia treated with water or 1.2 μg mL^{-1} of either *Efe*-AfpA or PAF. Bars are 750 μm.

3.3. *Cryphonectria parasitica*

Cryphonectria parasitica EP155 is the causal agent of chestnut blight, which nearly eliminated the American chestnut tree (*Castanea dentata*) worldwide [34]. Current methods of alleviating the disease and restoring the American chestnut to its native range include the development of interspecific hybrids, backcross breeding, and genetic engineering [35]. Since *Cr. parasitica* produces spores slowly, taking 3 to 4 weeks [36], the inhibitory activity of the antifungal proteins was determined in 8 mL agar plate assays using fungal mycelium plugs. Both *Efe*-AfpA and PAF were active against *Cr. parasitica*, with the same MIC value of 0.5 μg mL^{-1} (Figures 3, S3 and S4, Table 1). Since there was nearly complete inhibition at the lowest concentration tested, 0.5 μg mL^{-1}, higher concentrations are not shown in the graphs in Figure 3.

3.4. *Fusarium graminearum*

Fusarium graminearum PH1 causes both Fusarium head blight on wheat as well as barley and Fusarium stalk and ear rot on maize and was considered the fourth most important fungal plant pathogen [28,37]. In addition to its destructive impact on wheat and barley, *F. graminearum* produces mycotoxins detrimental to human and animal health, such as deoxynivalenol (DON) [38]. Both *Efe*-AfpA and PAF had activity against *F. graminearum*, but only *Efe*-AfpA had an activity that resulted in greater than 90% inhibition, with a MIC of 10 μg mL^{-1} (Figures 4 and S5, Table 1).

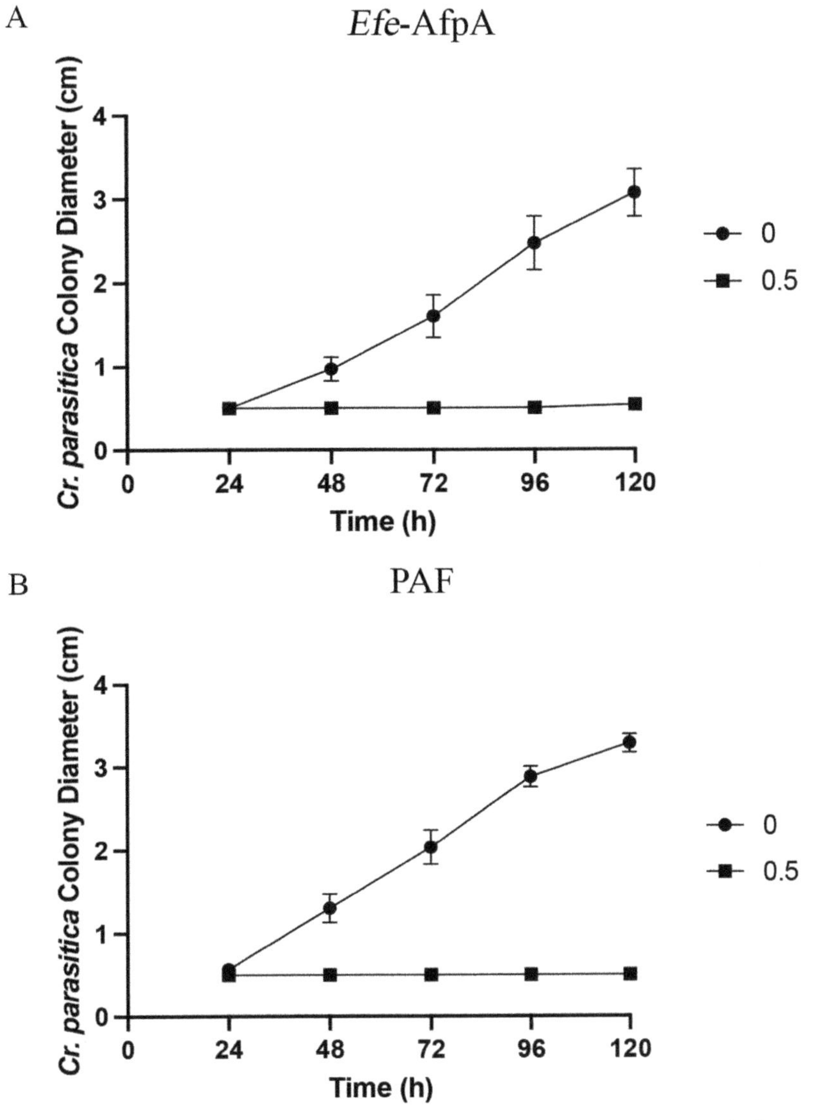

Figure 3. Activity of (**A**) *Efe*-AfpA and (**B**) PAF against *Cr. parasitica* mycelial growth. *Cr. parasitica* mycelial plugs were subcultured onto PDA plates amended with 0.5 μg mL^{-1} of *Efe*-AfpA or PAF. The colony diameters were measured daily. The data presented are the means and standard deviations of three replicates.

3.5. *Pyricularia oryzae*

Pyricularia oryzae (previously *Magnaporthe oryzae*) is the causal agent of rice blast and was voted the number one fungal plant pathogen from a scientific and economic perspective [28,39]. Rice (*Oryza sativa*) is an economically and agriculturally important crop that feeds about half the world's population, and from 10 to 30% of yield can be lost annually due to rice blast disease [40]. *Py. oryzae* also causes gray leaf spot of many genera of turfgrass including *Cynodon*, *Eremochloa*, *Festuca*, *Lolium*, *Paspalum*, *Pennisetum*, and *Stenotaphrum* [33]. Gray leaf spot was also recently identified on the turfgrass *Festuca brevipila* (hard fescue), which had not previously been reported to be sensitive to *Py. oryzae* [41]. As the pathogen presents problems on both rice and turf, new mechanisms of plant protection are needed. The well-studied antifungal proteins AFP from *A. niger* and PAF from *Pe. chrysogenum* have been tested against *Py. oryzae*. AFP was highly effective at completely inhibiting growth at the low concentration of 4 μM but PAF was less effective, requiring a concentration greater than 200 μg mL^{-1} (approximately 32 μM) to completely inhibit growth [32,42]. Here, both *Efe*-AfpA and PAF did have activity against *Py. oryzae*, but their activities did not result in 90% inhibition at any of the concentrations tested (Figures 5 and S6, Table 1).

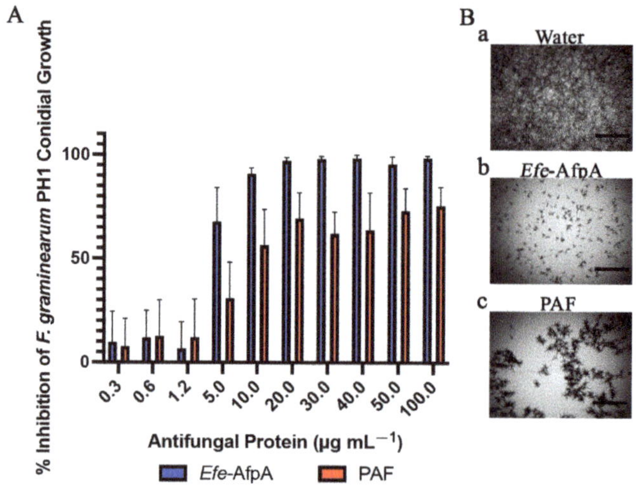

Figure 4. Activity of *Efe*-AfpA and PAF against *F. graminearum* PH1 conidial growth. (**A**) Growth inhibition of *F. graminearum* conidia treated with increasing concentrations of either *Efe*-AfpA or PAF incubated at room temperature for 30 h. The data presented are the means and standard deviations of three replicates. (**B**) Microscopy of *F. graminearum* conidia treated with water or 10 μg mL^{-1} of either *Efe*-AfpA or PAF. Bars are 750 μm.

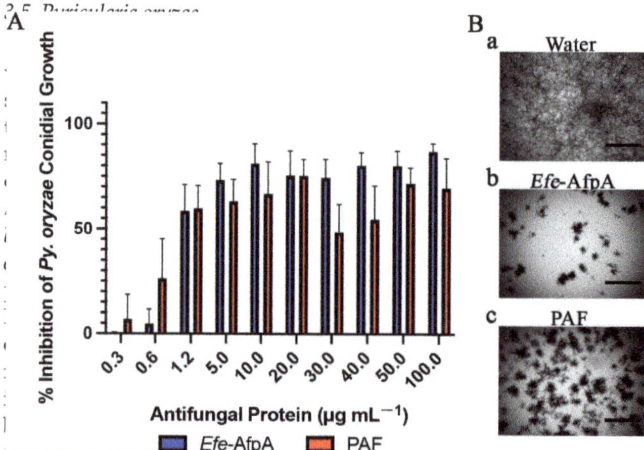

Figure 5. Activity of *Efe*-AfpA and PAF against *Py. oryzae* conidial growth. (**A**) Growth inhibition of *Py. oryzae* conidia treated with increasing concentrations of either *Efe*-AfpA or PAF incubated at room temperature for 48 h. The data presented are the means and standard deviations of three replicates. (**B**) Microscopy of *Py. oryzae* conidia treated with water or 100 μg mL^{-1} of either *Efe*-AfpA or PAF. Bars are 750 μm.

3.6. *Laetisaria fuciformis and Rhizoctonia solani*

Laetisaria fuciformis causes red thread of turfgrasses, and is particularly damaging to *Lolium perenne* (perennial ryegrass) and *Festuca rubra* [33]. *F. rubra* has been shown to have enhanced tolerance to red thread when infected by the fungal endophyte *E. festucae* [19]. *Rhizoctonia* spp. cause several diseases in different types of turfgrass species. *R. solani*

3.6. *Laetisaria fuciformis* and *Rhizoctonia solani*

Laetisaria fuciformis causes red thread of turfgrasses, and is particularly damaging to *Lolium perenne* (perennial ryegrass) and *Festuca rubra* [33]. *F. rubra* has been shown to have enhanced tolerance to red thread when infected by the fungal endophyte *E. festucae* [19]. *Rhizoctonia* spp. cause several diseases in different types of turfgrass species. *R. solani* causes brown patch in cool-season turf and large patch in warm-season turf [33]. *Efe*-AfpA had some activity against *L. fuciformis* but none against *R. solani* (Figures 6 and 7, Table 1, Figures S7–S10). PAF was ineffective against both *L. fuciformis* and *R. solani*. Only the two highest concentrations tested are shown in Figures 6 and 7 since there was negligible inhibitory activity at the lower concentrations.

A

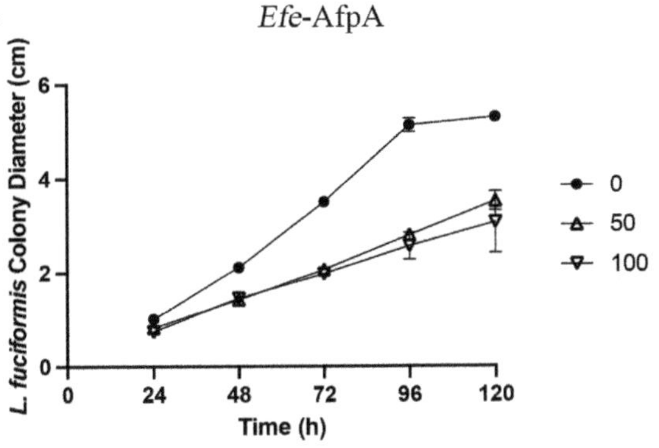

B

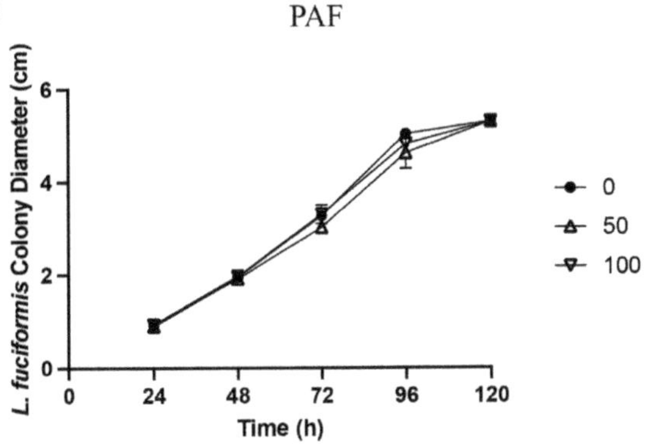

Figure 6. Activity of (**A**) *Efe*-AfpA and (**B**) PAF against *L. fuciformis* mycelial growth. *L. fuciformis* mycelial plugs were subcultured onto PDA plates amended with increasing concentrations of *Efe*-AfpA or PAF. The colony diameters were measured daily. The data presented are the means and standard deviations of three replicates.

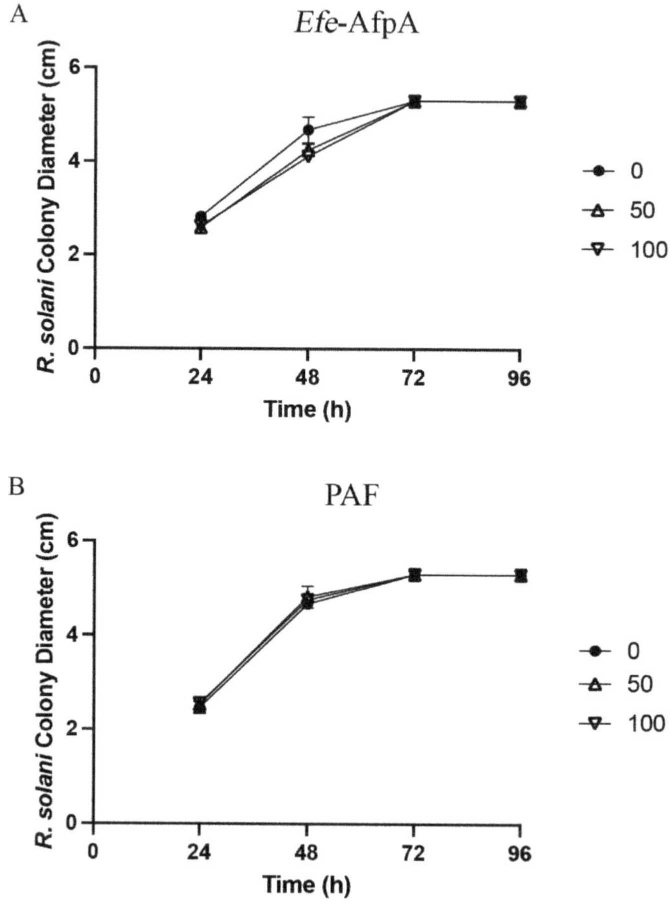

Figure 7. Activity of (**A**) *Efe*-AfpA and (**B**) PAF against *R. solani* mycelial growth. *R. solani* mycelial plugs were subcultured onto PDA plates amended with increasing concentrations of *Efe*-AfpA or PAF. The colony diameters were measured daily. The data presented are the means and standard deviations of three replicates.

4. Discussion

Here, we evaluated the antifungal activity of *Efe*-AfpA and PAF against some important fungal plant pathogens in culture. Both proteins were effective at inhibiting the growth of some of the pathogens, but there were some clear differences in their activity levels despite the sequence similarity of the two proteins. The amino acid sequences of the two proteins are 65% identical. *Efe*-AfpA and PAF were previously found to differ in activity against *Cl. jacksonii*, the causal agent of dollar spot disease on turfgrasses [25]. *Efe*-AfpA inhibited the growth of *Cl. jacksonii* but PAF had no inhibitory activity against the pathogen. Antifungal proteins have been identified from numerous fungal species and differences in the activity of other similar antifungal proteins have previously been reported [43,44]. Here, both *Efe*-AfpA and PAF had activity against all the Ascomycete fungi tested but the level of inhibition varied. *Efe*-AfpA had some activity against the Basidiomycete *L. fuciformis* whereas PAF had no activity. Neither protein had activity against the Basidiomycete *R. solani*.

The differences in activities of *Efe*-AfpA and PAF against the pathogens tested may reflect the different environments in which *E. festucae* and *Pe. chrysogenum* exist and the roles of these antifungal proteins in the biology of two fungi. *E. festucae* is a fungal endophyte of grasses and does not exist in nature independently of its grass host. PAF was of these antifungal proteins in the biology of two fungi. *E. festucae* is a fungal endophyte of grasses and does not exist in nature independently of its grass host. PAF was originally reported from *Pe. chrysogenum* [9], which is commonly found in indoor environments and is a food spoilage fungus [45]. It is unknown if the antifungal activities of the two proteins are their main functions. If so, their differences in activity could be the result of evolutionary changes resulting from different competing fungi present in their environments. However, additional functions for both *Efe*-AfpA and PAF have been suggested, which could also contribute to the sequence differences that result in activity differences. *Efe*-AfpA was suggested to be critical for the interaction of the fungus with its host grass [46]. PAF was proposed to have roles in conidiation and autophagy [47,48]. AfpB from *Pe. digitatum* is another similar antifungal protein, but is also active against its parent strain. AfpB may play a role in regulating the *Pe. digitatum* population, as well as other fungal populations [49].

New strategies are needed to combat plant pathogens and the development of antifungal proteins could provide some alternatives or complements to traditional fungicides. The data presented here indicate that specific antifungal proteins could be developed to target particular plant pathogens. Since *Efe*-AfpA had activity against numerous other important plant pathogens, in addition to its activity against *Cl. jacksonii*, it could be developed to target a broader range of destructive pathogens in both turfgrass and agronomic systems.

Supplementary Materials: The following supporting information can be downloaded at: https://www.mdpi.com/article/10.3390/microorganisms11040828/s1, Figure S1: Effect of *Efe*-AfpA and PAF on *Botrytis cinerea* growth; Figure S2: Effect of *Efe*-AfpA and PAF on *Colletotrichum cereale* growth; Figure S3: Effect of *Efe*-AfpA on *Cryphonectria parasitica* EP155 growth; Figure S4: Effect of PAF on *Cryphonectria parasitica* EP155 growth; Figure S5: Effect of *Efe*-AfpA and PAF on *Fusarium graminearum* Ph1 growth; Figure S6: Effect of *Efe*-AfpA and PAF on *Pyricularia oryzae* growth; Figure S7: Effect of *Efe*-AfpA on *Laetisaria fuciformis* growth; Figure S8: Effect of PAF on *Laetisaria fuciformis* growth; Figure S9: Effect of *Efe*-AfpA on *Rhizoctonia solani* growth; Figure S10: Effect of PAF on *Rhizoctonia solani* growth.

Author Contributions: Conceptualization, P.A.F., B.B.C. and F.C.B.; methodology, P.A.F., B.B.C. and F.C.B.; investigation, P.A.F.; data curation, P.A.F.; writing—original draft preparation, P.A.F. and F.C.B.; writing—review and editing, P.A.F., B.B.C. and F.C.B.; visualization, P.A.F.; supervision, F.C.B.; project administration, F.C.B.; funding acquisition, B.B.C. and F.C.B. All authors have read and agreed to the published version of the manuscript.

Funding: This work was supported with funds provided by the United States Golf Association Davis Program, the Rutgers Center for Turfgrass Science, and the USDA National Institute of Food and Agriculture Hatch project accession number 1024790 through the New Jersey Agricultural Experiment Station, Hatch project NJ12108.

Data Availability Statement: All data supporting the findings of this study are available within the paper and within its Supplementary Materials published online.

Acknowledgments: We thank Florentine Marx for providing the *Penicillium chrysogenum paf* knockout and overexpressing isolates and the pSK275*paf* plasmid. We would like to thank the following labs and researchers for providing the fungal pathogens for our experiments: Bruce Clarke (*Colletotrichum cereale, Pyricularia oryzae, Laetisaria fuciformis, Rhizoctonia solani*), John McLaughlin (*Fusarium graminearum* PH1), Bradley Hillman (*Cryphonectria parasitica*), and Thomas Gianfagna (*Botrytis cinerea*). We also thank the Rutgers University Core Facility in Foran Hall for the variety of instruments accessible for use.

Conflicts of Interest: Rutgers University has filed a provisional patent concerning the purification and use of *Efe*-AfpA for fungal pathogen disease control. The funders had no role in the design of the study; in the collection, analyses, or interpretation of data; in the writing of the manuscript, or in the decision to publish the results.

References

1. Zubrod, J.P.; Bundschuh, M.; Arts, G.; Bruhl, C.A.; Imfeld, G.; Knabel, A.; Payraudeau, S.; Rasmussen, J.J.; Rohr, J.; Scharmuller, A.; et al. Fungicides: An overlooked pesticide class? *Environ. Sci. Technol.* **2019**, *53*, 3347–3365. [CrossRef]
2. Steinberg, G.; Gurr, S.J. Fungi, fungicide discovery and global food security. *Fungal Genet. Biol.* **2020**, *144*, 103476. [CrossRef] [PubMed]
3. Morton, V.; Staub, T.A. Short history of fungicides. *APSnet Features* **2008**, *308*, 1–12. [CrossRef]
4. Nelson, R.; Wiesner-Hanks, T.; Wisser, R.; Ballint-Kurti, P. Navigating complexity to breed disease-resistant crops. *Nat. Rev. Genet.* **2018**, *19*, 21–33. [CrossRef]
5. Sanchez-Martin, J.; Keller, B. Contribution of recent technological advances to future resistance breeding. *Theor. Appl. Genet.* **2019**, *132*, 713–732. [CrossRef] [PubMed]
6. van Esse, H.P.; Reuber, T.L.; van der Does, D. Genetic modification to improve disease resistance in crops. *New Phytol.* **2020**, *225*, 70–86. [CrossRef] [PubMed]
7. Green, K.K.; Stenberg, J.A.; Lankinen, A. Making sense of integrated pest management (IPM) in the light of evolution. *Evol. Appl.* **2020**, *13*, 1791–1805. [CrossRef] [PubMed]
8. Ons, L.; Bylemans, D.; Thevissen, K.; Cammue, P.A. Combining biocontrol agents with chemical fungicides for integrated plant fungal disease control. *Microorganisms* **2020**, *8*, 1930. [CrossRef]
9. Marx, F.; Haas, H.; Reindl, M.; Stoffler, G.; Lottspeich, F.; Redl, B. Cloning, structural organization and regulation of expression of the *Penicillium chrysogenum* paf gene encoding an abundantly secreted protein with antifungal activity. *Gene* **1995**, *167*, 167–171. [CrossRef]
10. Huber, A.; Galgoczy, L.; Varadi, G.; Holzknecht, J.; Kakar, A.; Malanovic, N.; Leber, R.; Koch, J.; Keller, M.A.; Batta, G.; et al. Two small, cysteine-rich and cationic antifungal proteins from *Penicillium chrysogenum*: A comparative study of PAF and PAFB. *Biochim. Biophys. Acta Biomembr.* **2020**, *1862*, 183246. [CrossRef]
11. Wnendt, S.; Ulbrich, N.; Stahl, U. Molecular cloning, sequence analysis and expression of the gene encoding an antifungal-protein from *Aspergillus giganteus*. *Curr. Genet.* **1994**, *25*, 519–523. [CrossRef] [PubMed]
12. Kovacs, L.; Viragh, M.; Tako, M.; Papp, T.; Vagvolgyi, C.; Galgoczy, L. Isolation and characterization of *Neosartorya fischeri* antifungal protein (NFAP). *Peptides* **2011**, *32*, 1724–1731. [CrossRef] [PubMed]
13. Theis, T.; Wedde, M.; Meyer, V.; Stahl, U. The antifungal protein from *Aspergillus giganteus* causes membrane permeabilization. *Antimicrob. Agents Chemother.* **2003**, *47*, 588–593. [CrossRef] [PubMed]
14. Marx, F. Small, basic antifungal proteins secreted from filamentous ascomycetes: A comparative study regarding expression, structure, function and potential application. *App. Microbiol. Biotechnol.* **2004**, *65*, 133–142. [CrossRef]
15. Toth, L.; Boros, E.; Poor, P.; Ordog, A.; Kele, Z.; Varadi, G.; Holzknecht, J.; Bratschun-Khan, D.; Nagy, I.; Toth, G.K.; et al. The potential use of the *Penicillium chrysogenum* antifungal protein PAF, the designed variant PAF^{opt} and its γ-core peptide $P\gamma^{opt}$ in plant protection. *Microb. Biotechnol.* **2020**, *13*, 1403–1414. [CrossRef]
16. Leiter, E.; Gall, T.; Csernoch, L.; Pocsi, I. Biofungicide utilizations of antifungal proteins of filamentous ascomycetes: Current and foreseeable future developments. *BioControl* **2017**, *62*, 125–138. [CrossRef]
17. Braun, R.C.; Patton, A.J.; Watkins, E.; Koch, P.L.; Anderson, N.P.; Bonos, S.A.; Brilman, L.A. Fine fescues: A review of the species, their improvement, production, establishment, and management. *Crop Sci.* **2020**, *60*, 1142–1187. [CrossRef]
18. Schardl, C.L.; Young, C.A.; Hesse, U.; Amyotte, S.G.; Andreeva, K.; Calie, P.J.; Fleetwood, D.J.; Haws, D.C.; Moore, N.; Oeser, B.; et al. Plant-symbiotic fungi as chemical engineers: Multi-genome analysis of the Clavicipitaceae reveals dynamics of alkaloid loci. *PLoS Genet.* **2013**, *9*, e1003323. [CrossRef]
19. Bonos, S.A.; Wilson, M.M.; Meyer, W.A.; Funk, C.R. Suppression of red thread in fine fescues through endophyte-mediated resistance. *Appl. Turfgrass Sci.* **2005**, *10*, 1094. [CrossRef]
20. Clarke, B.B.; White, J.F., Jr.; Hurley, R.H.; Torres, M.S.; Sun, S.; Huff, D.R. Endophyte-mediated suppression of dollar spot disease in fine fescues. *Plant Dis.* **2006**, *90*, 994–998. [CrossRef]
21. Tian, Z.; Wang, R.; Ambrose, K.V.; Clarke, B.B.; Belanger, F.C. The *Epichloë festucae* antifungal protein has activity against the plant pathogen *Sclerotinia homoeocarpa*, the causal agent of dollar spot disease. *Sci. Rep.* **2017**, *7*, 5643. [CrossRef] [PubMed]
22. Heineck, G.C.; Qiu, Y.; Ehlke, N.J.; Watkins, E. The fungal endophyte *Epichloë festucae* var. *lolii* plays a limited role in mediating crown rust severity in perennial ryegrass. *Crop Sci.* **2020**, *60*, 1090–1104. [CrossRef]
23. Ambrose, K.V.; Belanger, F.C. SOLiD-SAGE of endophyte-infected red fescue reveals numerous effects on host transcriptome and an abundance of highly expressed fungal secreted proteins. *PLoS ONE* **2012**, *7*, e53214. [CrossRef] [PubMed]
24. Schardl, C.L.; Scott, B. Recommendations for gene nomenclature for *Epichloë* species and related Clavicipitaceae. In *Epichloae, Endophytes of Cool Season Grasses: Implications, Utilization and Biology*; Young, C.A., Aiken, G.E., McCulley, R.L., Strickland, J.R., Schardl, C.L., Eds.; The Samuel Roberts Noble Foundation: Ardmore, OK, USA, 2012; pp. 84–87.
25. Fardella, P.A.; Tian, Z.; Clarke, B.B.; Belanger, F.C. The *Epichloë festucae* antifungal protein *Efe*-AfpA protects creeping bentgrass (*Agrostis stolonifera*) from the plant pathogen *Clarireedia jacksonii*, the causal agent of dollar spot disease. *J. Fungi* **2022**, *8*, 1097. [CrossRef] [PubMed]
26. Sonderegger, C.; Galgoczy, L.; Garrigues, S.; Fizil, A.; Borics, A.; Manzanares, P.; Hededus, N.; Huber, A.; Marcos, J.F.; Batta, G.; et al. A *Penicillium chrysogenum*-based expression system for the production of small, cysteine-rich antifungal proteins for structural and functional analyses. *Microb. Cell Factories* **2016**, *15*, 192. [CrossRef]

27. Williamson, B.; Tudzynski, B.; Tudzynski, P.; Van Kan, J.A.L. *Botrytis cinerea*: The cause of grey mould disease. *Mol. Plant Pathol.* **2007**, *8*, 561–580. [CrossRef]
28. Dean, R.; Van Kan, J.A.L.; Pretorius, Z.A.; Hammond-Kosack, K.E.; Di Pietro, A.; Spanu, P.D.; Rudd, J.J.; Dickman, M.; Kahmann, R.; Ellis, J.; et al. The top 10 fungal pathogens in molecular plant pathology. *Mol. Plant Pathol.* **2012**, *13*, 414–430. [CrossRef]
29. Weiberg, A.; Wang, M.; Lin, F.M.; Zhao, H.; Zhang, Z.; Kaloshian, I.; Huang, H.D.; Jin, H. Fungal small RNAs suppress plant immunity by hijacking host RNA interference pathways. *Science* **2013**, *342*, 118–123. [CrossRef]
30. Kaiserer, L.; Oberparleiter, C.; Weiler-Görz, R.; Burgstaller, W.; Leiter, E.; Marx, F. Characterization of the *Penicillium chrysogenum* antifungal protein PAF. *Arch. Microbiol.* **2003**, *180*, 204–210. [CrossRef]
31. Marx, F.; Binder, U.; Leiter, É.; Pócsi, I. The *Penicillium chrysogenum* antifungal protein PAF, a promising tool for the development of new antifungal therapies and fungal cell biology studies. *Cell. Mol. Life Sci.* **2008**, *65*, 445–454. [CrossRef]
32. Garrigues, S.; Gandía, M.; Popa, C.; Borics, A.; Marx, F.; Coca, M.; Marcos, J.F.; Manzanares, P. Efficient production and characterization of the novel and highly active antifungal protein AfpB from *Penicillium digitatum*. *Sci. Rep.* **2017**, *7*, 14663. [CrossRef] [PubMed]
33. Tredway, L.P.; Tomaso-Peterson, M.; Kerns, J.P.; Clarke, B.B. *Compendium of Turfgrass Diseases*, 4th ed.; APS Press: St. Paul, MN, USA, 2022.
34. Crouch, J.A.; Dawe, A.; Aerts, A.; Barry, K.; Churchill, A.C.L.; Grimwood, J.; Hillman, B.I.; Milgroom, M.G.; Pangilinan, J.; Smith, M.; et al. Genome sequence of the chestnut blight fungus *Cryphonectria parasitica* EP155: A fundamental resource for an archetypical invasive plant pathogen. *Phytopathology* **2020**, *110*, 1180–1188. [CrossRef] [PubMed]
35. Newhouse, A.; Powell, W. Intentional introgression of a blight tolerance transgene to rescue the remnant population of American chestnut. *Conserv. Sci. Pract.* **2020**, *3*, e348. [CrossRef]
36. Yue, Q.; Miller, C.J.; White, J.P.; Richardson, M.D. Isolation and characterization of fungal inhibitors from *Epichloë festucae*. *J. Agric. Food Chem.* **2000**, *48*, 4687–4692. [CrossRef] [PubMed]
37. King, R.; Urban, M.; Hammond-Kosack, M.C.U.; Hassani-Pak, K.; Hammond-Kosack, K.E. The completed genome sequence of the pathogenic ascomycete fungus *Fusarium graminearum*. *BMC Genom.* **2015**, *16*, 544. [CrossRef]
38. Leplat, J.; Friberg, H.; Abid, M.; Steinberg, C. Survival of *Fusarium graminearum*, the causal agent of Fusarium Head Blight. A review. *Agron. Sustain. Dev.* **2013**, *33*, 97–111. [CrossRef]
39. Luo, J.; Zhang, N. Magnaporthiopsis, a new genus in Magnaporthaceae (Ascomycota). *Mycologia* **2013**, *105*, 1019–1029. [CrossRef]
40. Skamnioti, P.; Gurr, S.J. Against the grain: Safeguarding rice from rice blast disease. *Trends Biotechnol.* **2009**, *27*, 141–150. [CrossRef]
41. Vines, P.L.; Daddio, R.M.; Luo, J.; Wang, R.; Murphy, J.A.; Zhang, N.; Clarke, B.B.; Meyer, W.A.; Bonos, S.A. *Pyricularia oryzae* incites gray leaf spot disease on hard fescue (*Festuca brevipila*). *Int. Turfgrass Soc. Res. J.* **2022**, *14*, 997–1002. [CrossRef]
42. Vila, L.; Lacadena, V.; Fontanet, P.; Martinez del Pozo, A.; San Segundo, B. A Protein from the mold *Aspergillus giganteus* is a potent inhibitor of fungal plant pathogens. *Mol. Plant-Microbe Interact.* **2001**, *14*, 1327–1331. [CrossRef]
43. Gandia, M.; Kakar, A.; Giner-Llorca, M.; Holzknecht, J.; Martinez-Culebras, P.; Galgoczy, L.; Marx, F.; Marcos, J.F.; Manzanares, P. Potential of antifungal proteins (AFPs) to control *Penicillium* postharvest fruit decay. *J. Fungi* **2021**, *7*, 449. [CrossRef] [PubMed]
44. Martinez-Culebras, P.V.; Gandia, M.; Boronat, A.; Marcos, J.F.; Manzanares, P. Differenial susceptibility of mycotoxin-producing fungi to distinct antifungal proteins (AFPs). *Food Microbiol.* **2021**, *97*, 103760. [CrossRef] [PubMed]
45. Houbraken, J.; Frisvad, J.C.; Samson, R.A. Fleming's penicillin producing strain is not *Penicillium chrysogenum* but *P. rubens*. *IMA Fungus* **2011**, *2*, 87–95. [CrossRef] [PubMed]
46. Wang, R.; Luo, S.; Clarke, B.B.; Belanger, F.C. The *Epichloë festucae* antifungal protein *Efe*-AfpA is also a possible effector protein required for the interaction of the fungus with its host grass *Festuca rubra* subsp. *rubra*. *Microorganisms* **2021**, *9*, 140. [CrossRef]
47. Hegedus, N.; Sigl, C.; Zadra, I.; Pocsi, I.; Marx, F. The *paf* gene product modulates asexual development in *Penicillium chrysogenum*. *J. Basic Microbiol.* **2011**, *51*, 253–262. [CrossRef]
48. Kovacs, B.; Hegedus, N.; Balint, M.; Szabo, Z.; Emri, T.; Kiss, G.; Antal, M.; Pocsi, I.; Leiter, E. *Penicillium* antifungal protein (PAF) is involved in the apoptotic and autophagic processes of the producer *Penicillium chrysogenum*. *Acta Microbiol. Immunol. Hung.* **2014**, *61*, 379–388. [CrossRef]
49. Bugeda, A.; Garrigues, S.; Gandia, M.; Manzanares, P.; Marcos, J.F.; Coca, M. The antifungal protein AfpB induces regulated cell death in its parental fungus *Penicillium digitatum*. *MSphere* **2020**, *5*, e00595-20. [CrossRef]

microorganisms

Article

Response Surface Methodology (RSM) Mediated Optimization of Medium Components for Mycelial Growth and Metabolites Production of *Streptomyces alfalfae* XN-04

Jing Chen, Xingjie Lan, Ruimin Jia, Lifang Hu and Yang Wang *

College of Plant Protection, Northwest A & F University, Xianyang 712100, China
* Correspondence: wangyang2006@nwafu.edu.cn; Tel.: +86-187-9280-9011

Abstract: *Streptomyces alfalfae* XN-04 has been reported for the production of antifungal metabolites effectively to control Fusarium wilt of cotton, caused by *Fusarium oxysporum* f. sp. *vasinfectum* (*Fov*). In this study, we used integrated statistical experimental design methods to investigate the optimized liquid fermentation medium components of XN-04, which can significantly increase the antifungal activity and biomass of XN-04. Seven variables, including soluble starch, KNO_3, soybean cake powder, K_2HPO_4, $MgSO_4{\cdot}7H_2O$, $CaCO_3$ and $FeSO_4{\cdot}7H_2O$, were identified as the best ingredients based on one-factor-at-a-time (OFAT) method. The results of Plackett–Burman Design (PBD) showed that soluble starch, soybean cake powder and K_2HPO_4 were the most significant variables among the seven variables. The steepest climbing experiment and response surface methodology (RSM) were performed to determine the interactions among these three variables and fine-tune the concentrations. The optimal compositions of medium were as follows: soluble starch (26.26 g/L), KNO_3 (1.00 g/L), soybean cake powder (23.54 g/L), K_2HPO_4 (0.27 g/L), $MgSO_4{\cdot}7H_2O$ (0.50 g/L), $CaCO_3$ (1.00 g/L) and $FeSO_4{\cdot}7H_2O$ (0.10 g/L). A verification experiment was then carried out under the optimized conditions, and the results revealed the mycelial dry weight of *S. alfalfae* XN-04 reaching 6.61 g/L. Compared with the initial medium, a 7.47-fold increase in the biomass was achieved using the optimized medium. Moreover, the active ingredient was purified from the methanol extract of *S. alfalfae* XN-04 mycelium and then identified as roflamycoin (a polyene macrolide antibiotic). The results may provide new insights into the development of *S. alfalfae* XN-04 fermentation process and the control of the Fusarium wilt of cotton and other plant diseases.

Keywords: *Streptomyces alfalfae*; antifungal activity; biomass; optimization; response surface methodology

Citation: Chen, J.; Lan, X.; Jia, R.; Hu, L.; Wang, Y. Response Surface Methodology (RSM) Mediated Optimization of Medium Components for Mycelial Growth and Metabolites Production of *Streptomyces alfalfae* XN-04. *Microorganisms* **2022**, *10*, 1854. https://doi.org/10.3390/microorganisms10091854

Academic Editor: Fred O. Asiegbu

Received: 11 August 2022
Accepted: 13 September 2022
Published: 16 September 2022

Publisher's Note: MDPI stays neutral with regard to jurisdictional claims in published maps and institutional affiliations.

1. Introduction

The soil-inhabiting fungus *Fusarium oxysporum* has been reported to infect more than a hundred different crops, causing vascular wilt and even death of the plants [1]. The widely use of chemical pesticides has caused several environmental concerns such as pesticide residues and soil contamination, although it can also control the Fusarium wilt [2]. Therefore, the biocontrol of Fusarium wilt has become an environmentally friendly strategy of the management of plant diseases during the recent years [2]. Among various potential biocontrol methods, the most widely studied are beneficial microorganisms and their metabolites. As one of the most biotechnological and economic value prokaryotes, actinomycetes are responsible for the production of nearly half of the discovered secondary metabolites with economic importance [3]. Currently, more than 75% of bioactive compounds are reported from the members of *Streptomyces* [4–10]. Some bioactive compounds showed great activity against phytopathogenic fungi, such as kasugamycin produced by *Streptomyces kasugaensis* [11], validamycin produced by *Streptomyces hygroscopicus* [12], ningnanmycin produced by *Streptomyces noursei* [13] and wuyiencin produced by *Streptomyces albulus* var. *wuyiensis* [14].

The optimization of productive conditions is an extremely important step in the fermentation process due to their high complexity. The fermentation processes of microorganisms are influenced by many variables, including physical parameters and the nutritional compositions of culture medium [15]. The yield and metabolic profile of microorganisms are influenced by minor variations in the composition of fermentation medium [16]. The techniques used for the medium optimization of various fermentation parameters including both classical and statistical tools. Classical studies of fermentation optimization are usually performed "one-factor-at-a-time (OFAT)", in which the level of one factor is changed while keeping the other factors constant [17]. It has the advantage of being simple and easy. However, the conventional OFAT method is tedious, time-consuming and uneconomical. In addition, when plenty of variables are involved, it ignores the combined interactions among various nutritional and physical parameters [18]. Therefore, statistical-based measures have been employed to achieve medium optimization by changing more than one variable at a time. Some of the famous statistics-based experimental designs are used in fermentation optimization, including full factorial design, fractional factorial design (FFD), Plackett–Burman design (PBD), Box–Behnken design (BBD) and central composite design (CCD) [19,20].

Response surface methodology (RSM) is a compilation of mathematical statistical techniques for designing experiments, plotting models and evaluating the effects of various variables, and establishing optimum conditions of a multivariable system to achieve significant responses [18]. This method has been extensively used for optimizing factors of fermentation medium and discovering the interactions among a multitude of fermentation parameters using a minimum of experiments [21]. In many cases, RSM uses statistically based experimental designs such as BBD and CCD to develop empirical models, which mathematically describe the relationships existing among the independent and dependent variables [22]. Further, RSM provides three-dimensional (3D) graphs and two-dimensional (2D) contour plots to clarify the shape of a response surface. Currently, RSM has been successfully applied to optimize variables for fermentation culture across a series of microorganisms including fungi, bacteria and actinomycetes in the production of industrially important metabolites [23–25]. For example, *Streptomyces* sp. WP-1 culture under the optimal medium designed by RSM increased fungichromin yield to 5741.7 mg/L [26]. Medium optimization using RSM contributed to a 45% (220.7 $\pm$ 5.7 mg/L) increase in rapamycin production for the *Streptomyces hygroscopicus*, compared with the unoptimized production medium (151.9 $\pm$ 22.6 mg/L) [27].

Streptomyces alfalfae XN-04 is a plant growth-promoting rhizobacterium (PGPR) and an excellent biocontrol strain. A previous study indicated that *S. alfalfae* XN-04 was able to produce metabolites with significant activity against some plant fungal pathogens [28]. Due to the extremely low production of biomass and long fermentation time of *S. alfalfae* XN-04, it is necessary to improve *S. alfalfae* XN-04 yield in submerged fermentation. Moreover, the main antifungal ingredient in the metabolites of *S. alfalfae* XN-04 remains unclear. In this study, we assessed the following objectives: (1) optimize the fermentation medium of *S. alfalfae* XN-04 for the maximum biomass at the level of flask fermentation; (2) purify and identify the main antifungal ingredient of *S. alfalfae* XN-04 using various chromatographic and spectroscopic methods.

2. Materials and Methods

2.1. Microorganisms and Culture Conditions

S. alfalfae XN-04 was originally isolated from the rhizosphere soil sample collected from Qinghai Province, China [28]. The strain XN-04 was cultured on mannitol soybean agar (MS) and incubated at 28 °C for 14 days to produce spores. The phytopathogenic fungus *Fusarium oxysporum* f. sp. *vasinfectum* (*Fov*) was preserved in the Biological Control of Plant Disease Laboratory of Northwest Agriculture and Forestry University. The fungus was cultured on potato dextrose agar (PDA) plates and incubated at 28 °C for 7 days.

2.2. Inoculum Preparation

Inoculum was prepared by inoculating 5 spore cakes (6 mm) of *S. alfalfae* XN-04 into a 250 mL Erlenmeyer flasks containing 100 mL Gauze's synthetic No. 1 (GS) medium. Then, the flasks were incubated at 28 °C on a shaker at 180 rpm for 3 days. The inoculum quantity was controlled at 5% (v/v) in all the fermentation experiments.

2.3. Antifungal Activity Assay

Fermentation culture was centrifuged at 5000× g rpm at 4 °C for 10 min, and the precipitate was collected. Three volumes of methanol (MeOH) were added, and the sample was then processed with ultrasonication for 10 min. After filtration using filter paper, the MeOH extract was evaporated to dryness using a rotary vacuum evaporator under reduced pressure at 50 °C. The MeOH extract was redissolved in MeOH to prepare a stock solution with a final concentration of 50 mg/mL. The antifungal activity was estimated with the method of mycelium growth in sealed plates described by previous study [29]. Briefly, 50 µL of the MeOH extract and 50 mL of sterile PDA medium were uniformly mixed and poured into three culture dishes (9 cm in diameter). Inverted mycelial plugs (6 mm in diameter) taken from the active periphery of 7-day-old fungus colonies were placed on the center of these plates. Plates with an appropriate volume of MeOH were used as control. Plates were incubated for 7 days in a growth chamber at 28 °C, and the colony diameter was measured. Each treatment had three plates, and the experiment was repeated three times. The levels of inhibition were calculated by using the equation as follows:

$$\text{Inhibition rate (\%)} = (\text{the colony diameter of control} - \text{the colony diameter of treatment})/(\text{the colony diameter of control}) \times 100 \quad (1)$$

2.4. Single-Factor Optimization of Fermentation Medium Composition

Single-factor experiments were used to screen the most significant medium substrate. GS medium was used as the initial medium. A total of 9 different carbon sources and 9 different nitrogen sources to replace corresponding carbon and nitrogen source in the initial medium while other compositions were kept constant at their original concentration. Glucose, sucrose, xylose, galactose, fructose, millet flour, rice flour, corn steep liquor and soluble starch (at 20 g/L) were investigated as carbon sources. The influence of KNO_3, $(NH_4)_2SO_4$, NH_4NO_3, $NaNO_3$, urea, yeast extract, beef extract, soybean cake powder and peptone (at 1 g/L) as nitrogen sources were simultaneously investigated. Similarly, K_2HPO_4, $MgSO_4{\cdot}7H_2O$, $CaCO_3$, KCl, KH_2PO_4 and NaCl (at 0.1 g/L) were selected to optimize mineral salts. Additionally, $CuSO_4{\cdot}5H_2O$, $CoCl_2{\cdot}6H_2O$, $MnCl_2{\cdot}4H_2O$, $FeSO_4{\cdot}7H_2O$ and $ZnSO_4{\cdot}7H_2O$ (at 0.01 g/L) were chosen as trace elements. At the end of fermentation, samples obtained from fermentation culture were used to calculate mycelium dry weight and antifungal activity.

2.5. Single-Factor Concentration Screening Test

In order to select the most significant concentration by using the single variable procedure, soluble starch (0, 20, 40, 60, 80 and 100 g/L), KNO_3 (0, 1, 2, 3, 4 and 5 g/L), soybean cake powder (0, 20, 40, 60, 80 and 100 g/L), K_2HPO_4 (0, 0.2, 0.4, 0.6, 0.8 and 1.0 g/L), $MgSO_4{\cdot}7H_2O$ (0, 0.2, 0.4, 0.6, 0.8 and 1.0 g/L), $CaCO_3$ (0, 0.2, 0.4, 0.6, 0.8 and 1.0 g/L) and $FeSO_4{\cdot}7H_2O$ (0, 0.02, 0.04, 0.06, 0.08 and 0.1 g/L) were added to the initial medium, respectively. The mycelium dry weight and antifungal activity of samples were determined as described above.

2.6. Plackett–Burman Design (PBD)

The Plackett–Burman design (PBD) is an effective method to investigate the effect of medium composition, and is very helpful for screening the most important variables with respect to their main effects. The results of PBD do not describe the interaction among these variables but it is used to screen and evaluate the variables that have a significant impact on response. The total number of experiments performed according to PBD is $n + 1$, where n is the number of variables. In this study, a range of 12 experiments were constructed using the Design-Expert software version 8.0.6.1 (Stat-Ease, Inc., Minneapolis, MN, USA) for 7 different independent variables including soluble starch, KNO_3, soybean cake powder, K_2HPO_4, $MgSO_4 \cdot 7H_2O$, $CaCO_3$ and $FeSO_4 \cdot 7H_2O$. Each independent variable was tested at 2 levels, high and low, denoted by (+) and (−), respectively (Table 1). A total of 4 dummy variables were designed in these 12 experiments to calculate the standard error. All tests were performed in triplicate, and the average of mycelium dry weight was treated as responses. The effect of medium components on mycelium dry weight was determined by p-values obtained by analysis of variance (ANOVA). A p-value (Prob > F) of less than 0.05 to indicate when factors are mathematically significant.

Table 1. The level and code of variables chosen for PBD.

Name	Factors	Code Value (g/L)	
		−1	+1
X_1	Soluble starch	20	25
X_2	KNO_3	1.00	1.25
X_3	Soybean cake powder	20	25
X_4	K_2HPO_4	0.20	0.25
X_5	$MgSO_4 \cdot 7H_2O$	0.40	0.50
X_6	$CaCO_3$	1.00	1.25
X_7	$FeSO_4 \cdot 7H_2O$	0.100	0.125

2.7. The Steepest Ascent Experiment

The steepest ascent experiment was designed to determine a suitable direction according to the results of PBD, and the change step was determined according to the effect value of each factor. The direction of the steepest ascent experiment and the step of change were determined by the three main influencing factors of soluble starch, soybean cake powder and K_2HPO_4 according to the results of PBD, and the area of mycelial dry weight maximum production could be approached rapidly.

2.8. Box–Behnken Design (BBD)

Response surface methodology (RSM) based on Box–Behnken design (BBD) was used to optimize optimal levels of medium compositions. Based on the trajectory of the steepest ascent experiments, 3 variables were selected and their concentrations were arranged at 17 levels, with 5 replicates at the center point to study their interaction effect. According to the design, 3 variables with the highest confidence levels were prescribed at 3 levels, coded −1, 0 and +1 for low, middle and high concentrations respectively (Table 2). In order to predicting the maximum value of response, a ternary quadratic equation was established to correlate the relationship between variables and response. The validity of this model was determined based on Student's t test. Data obtained from the BBD were subjected to first and second order multiple regression analysis using the method of least squares to obtain the parameters of the mathematical models. Model coefficients, R_2 values, F values and significance probabilities generated by the Design-Expert software version 8.0.6.1 (Stat-Ease, Inc., Minneapolis, MN, USA) provided a confirmation of the significance of each experimental variable. The optimal medium composition for improving the biomass of *S. alfalfae* XN-04 were obtained by solving this ternary quadratic equation and by analyzing the response surface contour plots. Combined with the regression equation, the 3D response surfaces and 2D contour plots were plotted by the Design-Expert software to understand the interaction effects of medium components and optimum concentration of each variable required for maximum biomass production. The 2D contour plot is used to determine the interaction strength between the two variables according to the radius of the curved surface of the arc [30]. The elliptic order of contour indicates that the interactions between corresponding variables were significant, while the circular order reveals non-significant interactions [30].

Table 2. The level and code of variables chosen for BBD.

Name	Factors	Code Value (g/L)		
		−1	0	+1
X_1	Soluble starch	25.00	26.50	28.00
X_3	Soybean cake powder	22.00	23.50	25.00
X_4	K_2HPO_4	0.250	0.265	0.280

In order to determine the accuracy of this model and verify the results, an experiment under the optimal conditions obtained from BBD was performed, followed by comparing the response value with the predicted data.

2.9. Purification and Identification of Antifungal Compounds

S. alfalfae XN-04 was inoculated into the optimized medium and incubated at 28 °C with shaking at 180 rpm for 14 days. After incubation, a total of 7.5 L fermentation broth was centrifuged at 5000× *g* rpm at 4 °C for 10 min, and the precipitate was collected. Three volumes of MeOH were added, and the sample was then processed with ultrasonication for 10 min. After filtration using filter paper, the MeOH extract was evaporated to dryness using a rotary vacuum evaporator under reduced pressure at 50 °C to yield a brown crude extract (13.2 g). The crude extract was purified on silica gel column chromatography (300 mesh, Qingdao Marine Chemical Inc., Qingdao, China) and subjected to gradient elution with a mixture solvent consisting of dichloromethane/MeOH from 10:1 to 2:1 (*v*/*v*) to obtain 10 fractions (Frs. 1–10). The paper disk (6 mm in diameter) diffusion assay was used to evaluate the antifungal activities of each fraction [31]. Further separation of Fr. 5 (672 mg) by semi-preparative HPLC on a C18 column using acetonitrile/water (45/55, *v*/*v*) as the mobile phase with a flow rate of 10 mL/min, collected the peak at 10.08 min (UV detector 363 nm) to yield compound 1 (220 mg, yellow amorphous solid).

The antifungal compound 1 was analyzed by high performance liquid chromatography-tandem mass spectrometry (HPLC-MS/MS; LC-30A + TripleTOF5600+, AB SCIEX, Framingham, MA, USA). Mass spectrometry (MS) and tandem mass spectrometry (MS/MS) were performed using electrospray ionization (ESI) detection in the positive mode. Nuclear magnetic resonance (NMR) spectra were recorded on a Bruker Avance III instrument at 600 MHz to elucidate the structure of compound **1**.

2.10. Statistical Analysis

All experiments were repeated three times independently and the summary statistics are expressed as mean ± standard deviation (SD). Data were analyzed by one-way ANOVA with Duncan's post hoc pairwise multiple range test. Differences between sample mean values of $p < 0.05$ were considered to be significant. Data obtained from the PBD and BBD were analyzed with a statistical software package Design-Expert software (Version 8.0.6.1, Stat-Ease Inc., Minneapolis, MN, USA).

3. Results

3.1. Effect of Different Nutrient Sources on Antifungal Metabolites Production and Biomass

The results showed that different nutrient sources had significant effects on the cell growth and antifungal activity of *S. alfalfae* XN-04 (Figure 1).

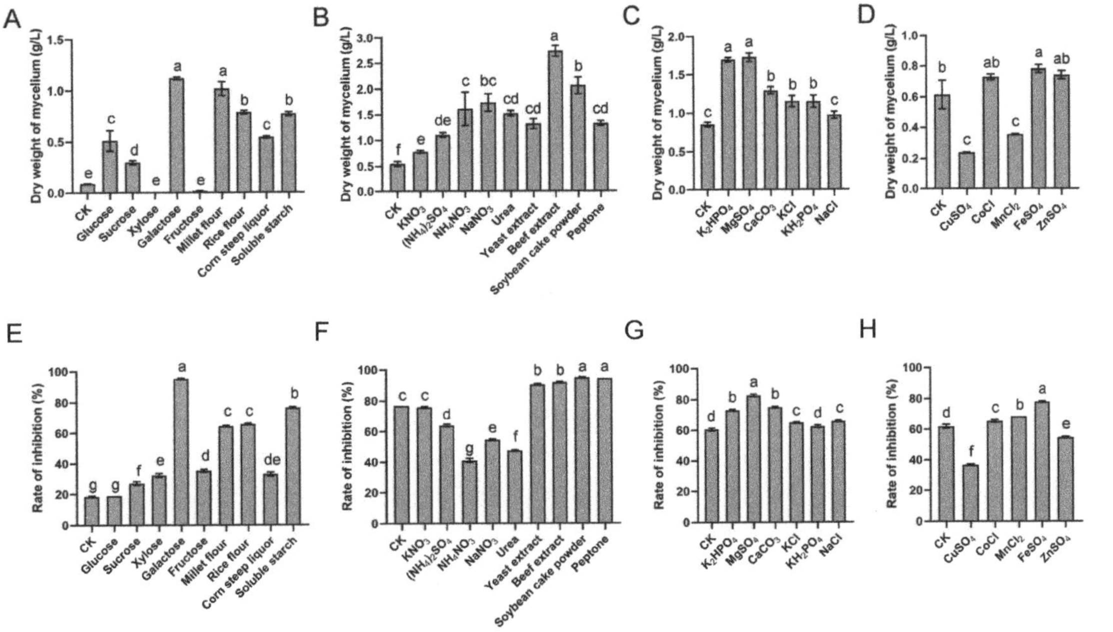

Figure 1. Effect of different nutrient sources on the biomass yield and antifungal metabolites production of *S. alfalfae* XN-04. (**A**,**E**) Effect of carbon sources. (**B**,**F**) Effect of nitrogen sources. (**C**,**G**) Effect of mineral salts. (**D**,**H**) Effect of trace elements. The dry weight of mycelium was used as indication of biomass, and the rate of inhibition was used as indication of antifungal metabolites production. Bars represent the SD of three replicates. Different lowercase letters indicate a significant difference at $p < 0.05$ level by Duncan' s new multiple range test.

3.1.1. Carbon Sources

The fermentation medium was supplemented with 20 g/L of different carbon sources to determine their individual influence. The results revealed that the maximum biomass was obtained with galactose (1.12 g/L), followed by millet flour (1.02 g/L) (Figure 1A). Rate of inhibition (%) was used to indicate the production of antifungal metabolites. Maximum inhibition rate was achieved with galactose (95.5%), followed by soluble starch (76.33%) (Figure 1E). Notably, the results showed that the fermentation medium containing millet flour and rice flour increased the levels of biomass but these agents were less efficient in the production of antifungal metabolites (Figure 1A,E). Although maximum biomass and antifungal metabolites was produced by galactose, soluble starch was selected for further studies due to cost consideration. The results of concentration screening test showed that the maximum biomass was produced at the 20 g/L soluble starch level (Figure 2A). However, the antifungal activity remained unchanged between the concentrations of 20 g/L and 100 g/L (Figure 2A). This means the highest antifungal metabolites production was also obtained at the 20 g/L soluble starch level. Therefore, 20 g/L soluble starch was chosen as the carbon source for further experiments.

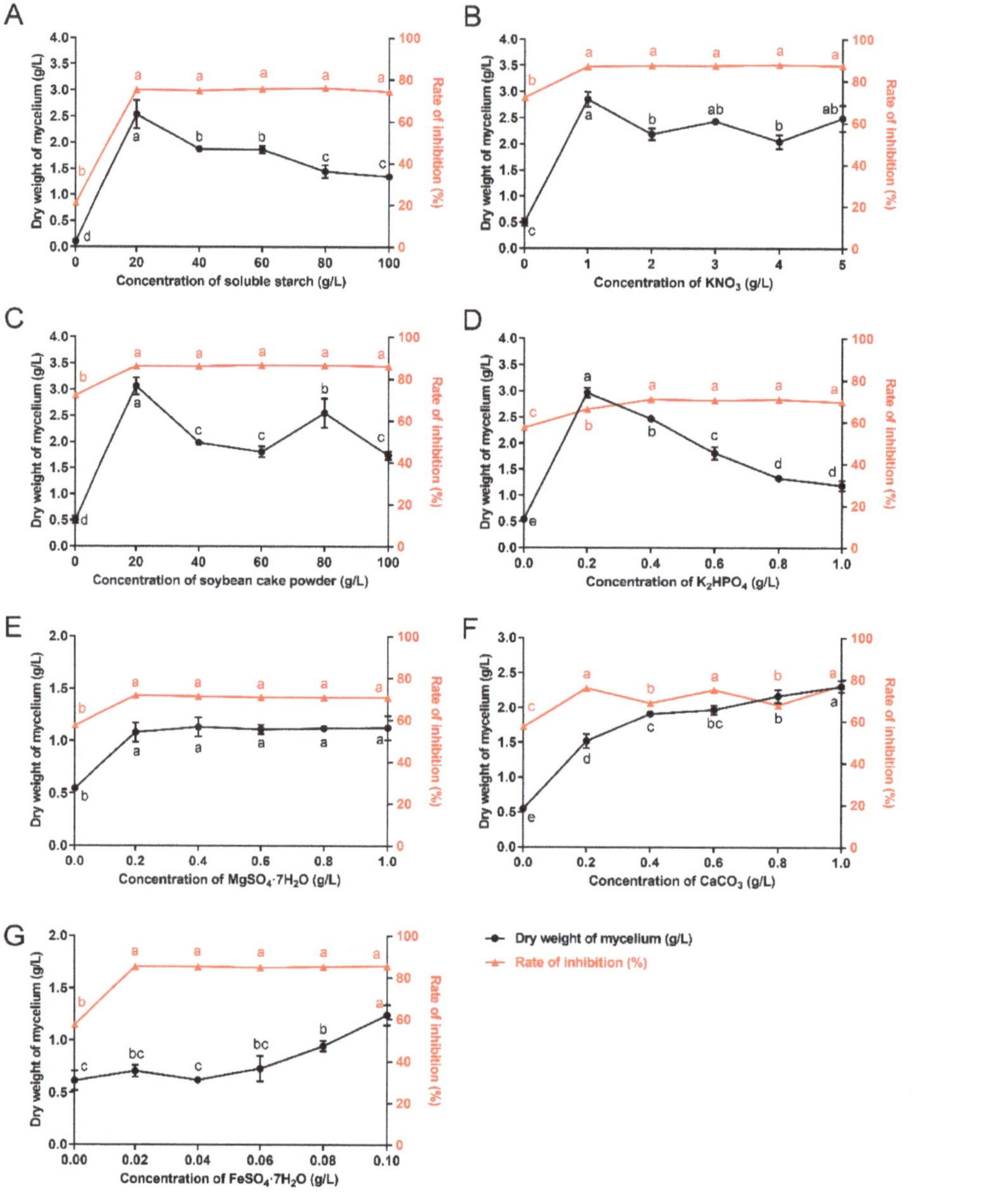

Figure 2. Effect of nutrient sources at different concentrations on the biomass yield and antifungal metabolites production of *S. alfalfae* XN-04. (**A**) Soluble starch. (**B**) KNO_3. (**C**) Soybean cake powder. (**D**) K_2HPO_4. (**E**) $MgSO_4 \cdot 7H_2O$. (**F**) $CaCO_3$. (**G**) $FeSO_4 \cdot 7H_2O$. The dry weight of mycelium was used as indication of biomass (corresponds to the left X axis), and the rate of inhibition was used as indication of antifungal metabolites production (corresponds to the right Y axis). Bars represent the SD of three replicates. Different lowercase letters indicate a significant difference at $p < 0.05$ level by Duncan' s new multiple range test.

3.1.2. Nitrogen Sources

Each organic and inorganic nitrogen source supported growth (Figure 1B). The results showed that the maximum biomass was obtained with beef extract (2.74 g/L), followed by soybean cake powder (2.06 g/L) (Figure 1B). Maximum inhibition rate was achieved with soybean cake powder (94.68%), followed by peptone (94.2%) (Figure 1F). Generally, a higher mycelial growth inhibition rate was achieved when the complex organic nitrogen source was added in fermentation medium (Figure 1F). Notably, the production of antifungal metabolites was negatively affected by nitrogen sources favorable for growth, such as NH_4NO_3, $NaNO_3$ and urea (Figure 1F). Although the maximum biomass was obtained by beef extract, soybean cake powder was selected for further studies due to cost considerations. Additionally, considering the ratio of organic nitrogen sources to inorganic nitrogen sources, KNO_3 was selected as the inorganic nitrogen source. Therefore, KNO_3 at 1 g/L and soybean cake powder at 20 g/L were the most suitable nitrogen sources for maximum biomass (Figure 2B,C).

3.1.3. Mineral Salt

Comparative studies on suitable mineral salts were carried out with results shown in Figure 1C,G. The results revealed that K_2HPO_4, $MgSO_4{\cdot}7H_2O$ and $CaCO_3$ had distinct positive effects both on cell growth and antifungal metabolites production of *S. alfalfae* XN-04. The results of concentration screening test showed that the optimum concentrations of K_2HPO_4, $MgSO_4{\cdot}7H_2O$ and $CaCO_3$ were 0.2 g/L, 0.4 g/L and 1.0 g/L, respectively (Figure 2D–F).

3.1.4. Trace Elements

The results showed that the maximum biomass and inhibition rate was achieved with $FeSO_4{\cdot}7H_2O$ (at 0.1 g/L), and was therefore, selected for further studies (Figures 1D,H and 2G).

Taking all this into account, seven factors (soluble starch, KNO_3, soybean cake powder, K_2HPO_4, $MgSO_4{\cdot}7H_2O$, $CaCO_3$ and $FeSO_4{\cdot}7H_2O$) were chosen for further optimization and their initial concentrations were determined.

3.2. Plackett–Burman Design (PBD)

To evaluate the most significant variables for biomass accumulation, PBD design was construed. A total of 12 different medium were prepared and the experimental runs were carried out based on the experimental matrix of PBD, and the observed responses are shown in Table 3. There was a variation from 2.33 to 4.75 g/L in mycelium dry weight (Table 3).

Table 3. The design and results of PBD.

Run	Code Number							Mycelial Dry Weight (g/L)
	X_1	X_2	X_3	X_4	X_5	X_6	X_7	
1	−1	−1	−1	1	−1	1	1	3.39
2	1	1	1	−1	−1	−1	1	2.75
3	−1	−1	1	−1	1	1	−1	2.52
4	1	−1	1	1	1	−1	−1	4.75
5	1	1	−1	1	1	1	−1	4.54
6	1	1	−1	−1	−1	1	−1	3.73
7	−1	1	1	1	−1	−1	−1	3.14
8	1	−1	−1	−1	1	−1	1	4.34
9	1	−1	1	1	−1	1	1	3.54
10	−1	1	−1	1	1	−1	1	3.28
11	−1	1	1	−1	1	1	1	2.33
12	−1	−1	−1	−1	−1	−1	−1	3.50

Regression analysis and ANOVA of the PBD were performed and represented in Table 4. The results showed that the regression was significant (p-value 0.0195), with the p-values for soluble starch, soybean cake powder and K_2HPO_4 were 0.0055, 0.0204 and 0.0261, respectively (Table 4). Values of prob > F less than 0.0500 indicated the model term was significant, and in this case X_1, X_2 and X_4 were significant model terms. Therefore, they were considered to be the most significant variables that affected mycelial dry weight. The other variables had small effects and low confidence levels ($p > 0.05$) which were considered mathematically insignificant (Table 4). The results of PBD also clearly showed that soluble starch, soybean cake powder and K_2HPO_4 had coefficient estimates of 0.46, −0.31 and 0.29, respectively, which meant that soluble starch and K_2HPO_4 exhibited a positive effect, while soybean cake powder acted negatively (Table 4). According to these results, X_1, X_2 and X_4 were selected for further optimization. The next step in the optimization should increase the concentration of soluble starch and K_2HPO_4 in the fermentation medium, and reduce the concentration of soybean cake powder.

Table 4. Analysis of variance of PBD.

Source	Factor	Freedom	Coefficient Estimate	Sum of Squares	Mean Square	F Value	Prob > F
Model		7	3.48	6.15	0.88	10.41	0.0195 *
X_1	Soluble starch	1	0.46	2.51	2.51	29.77	0.0055 **
X_2	KNO_3	1	−0.19	0.43	0.43	5.09	0.0871
X_3	Soybean cake powder	1	−0.31	1.17	1.17	13.89	0.0204 *
X_4	K_2HPO_4	1	0.29	1.00	1.00	11.89	0.0261 *
X_5	$MgSO_4{\cdot}7H_2O$	1	0.14	0.24	0.24	2.89	0.1645
X_6	$CaCO_3$	1	−0.14	0.24	0.24	2.89	0.1645
X_7	$FeSO_4{\cdot}7H_2O$	1	−0.21	0.54	0.54	6.42	0.0644
			$R^2 = 0.9479$				

* indicates a significant difference at $p < 0.05$ level; ** indicates a significant difference at $p < 0.01$ level.

3.3. The Steepest Ascent Experiment

In order to further approach the maximum response value region for each main factor for subsequent response surface analysis, the design of the steepest ascent test was performed. In the steepest ascent experiment, the change range of the concentrations of soluble starch, soybean cake powder and K_2HPO_4 were selected as 1.5 g/L, 1.5 g/L and 0.015 g/L (Table 5). All other components were fixed at those combinations of the maximum concentration of mycelial dry weight according to the results obtained from PDB (Run 4). The steepest ascent test design and its response values are shown in Table 5. The results revealed that the mycelial dry weight showed a trend of increasing and then decreasing, indicating that the test design was reliable. When the 2nd experimental was run, the mycelial dry weight reached the maximum (5.31 g/L), which was the region of the maximum response value of the three factors. Therefore, this point was chosen to set up basal concentrations for BBD.

Table 5. The design and results of the steepest ascent experiment.

Run	X_1 (g/L)	X_3 (g/L)	X_4 (g/L)	Mycelial Dry Weight (g/L)
1	25.00	25.00	0.250	4.42
2	26.50	23.50	0.265	5.31
3	28.00	22.00	0.280	5.06
4	29.50	20.50	0.295	4.99
5	31.00	19.00	0.310	4.27
6	32.50	17.50	0.325	4.07
7	34.00	16.00	0.340	3.73

3.4. Box–Behnken Design (BBD)

To further increase mycelial dry weight, the interactive effects of the most important variables, i.e., soluble starch (X_1), soybean cake powder (X_3) and K_2HPO_4 (X_4), were examined by RSM using BBD. The experimental design and the corresponding results are described in Table 6. The response (Y) fits with the ternary quadratic equation (final equation in terms of coded factors):

$$Y = +5.99 - 0.54 \times X_1 + 0.081 \times X_3 + 0.66 \times X_4 + 0.71 \times X_1X_3 + 0.93 \times X_1X_4 + 0.58 \times X_3X_4 - 9.92 \times X_{12} - 2.05 \times X_{32} - 1.05X_{42} \quad (2)$$

Table 6. The design and results of BBD.

Run	Code Number			Actual Value (g/L)	Predicted Value (g/L)
	X_1	X_3	X_4		
1	0	0	0	6.25	5.99
2	0	1	1	4.26	4.20
3	0	0	0	5.89	5.99
4	0	0	0	5.81	5.99
5	−1	−1	0	4.35	4.19
6	1	0	1	5.17	5.07
7	0	1	−1	1.82	1.73
8	−1	0	−1	5.17	5.07
9	0	−1	1	2.79	2.88
10	0	0	0	5.78	5.99
11	1	−1	0	1.68	1.69
12	1	0	−1	1.96	1.89
13	1	1	0	3.12	3.28
14	0	−1	−1	2.67	2.73
15	0	0	0	6.22	5.99
16	−1	0	1	4.22	4.29
17	−1	1	0	2.94	2.93

Table 6 showed the predicted responses of BBD on the basis of the above equation. This model equation's importance was statistically evaluated by the F test for the ANOVA, and the results were predicted in Table 7. The ANOVA regression model demonstrated a determination coefficient (R_2) of 0.9926, which meant 99.26% variability in the response can be explained by this model (Table 7). The value of the adjusted determination coefficient (R_{2adj}) was 0.9831 (Table 7). The higher value of R_{2adj} showed the strong significance of the model. R_2 and R_{2adj} were in close agreement. Moreover, a low value of coefficient of variation (5.11%) indicated the reliability and precision of these experiments executed (Table 7). An adequate precision value (26.776) measured the signal-to-noise ratio, and a ratio > 4.0 was desirable (Table 7). In this model, the very low *p*-value (<0.0001) for biomass is lower than 0.05, which revealed that the quadratic model we established was significant (Table 7). Based on the results of ANOVA table, it was found that the factors X_1, X_4, X_1X_3, X_1X_4, X_3X_4, X_{12}, X_{32} and X_{42} were all significant model terms (Table 7). The value of lack-of-fit was not significant ($p > 0.05$) for this model which suggested that the experimental data were in solid agreement with predicted responses (Table 7).

The 3D response surface and the 2D contour plots are generally the graphical representations of the regression equation. The elliptic order of contour in Figure 3A suggested that the interaction between soluble starch and soybean cake powder was significant. Similarly, there was a strong interaction between soluble starch and K_2HPO_4 (Figure 3B). Moreover, the circular order of contour of Figure 3C suggested that the interaction between soybean cake powder and K_2HPO_4 had a less significant effect. The 3D response surface aided in the visual determination of optimum levels of each variable as they interact. Three-dimensional response surface plots depicted mycelial dry weight with respect to soluble starch versus soybean cake powder (Figure 3D). Due to the interaction response of soluble starch with soybean cake powder, biomass increased with the increasing concentration of soluble starch and soybean cake powder up to 26.26 g/L and 23.54 g/L, respectively (Figure 3D). Further continuous increases in their concentrations resulted in a decrease in biomass. Figure 3E represented the interaction effect of soluble starch and K_2HPO_4 on mycelial dry weight. With an increase in soluble starch (25–26.26 g/L) and K_2HPO_4 (0.25–0.269 g/L) concentration, the biomass increased, and then dropped. Figure 3F revealed that the maximum mycelial dry weight was produced at a high level of soybean cake powder (26.26 g/L) and K_2HPO_4 (0.269 g/L) in the design range.

Table 7. Analysis of variance of BBD.

Source	Freedom	Sum of Squares	Mean Square	*F* Value	Prob > *F*
Model	9	41.12	4.57	104.15	<0.0001 ***
X_1	1	2.33	2.33	53.18	0.0002 ***
X_3	1	0.053	0.053	1.20	0.3088
X_4	1	3.45	3.45	78.54	<0.0001 ***
X_1X_3	1	2.03	2.03	46.29	0.0003 ***
X_1X_4	1	3.48	3.48	79.29	<0.0001 ***
X_3X_4	1	1.35	1.35	30.67	0.0009 ***
X_{12}	1	3.53	3.53	80.36	<0.0001 ***
X_{32}	1	17.74	17.74	404.35	<0.0001 ***
X_{42}	1	4.66	4.66	106.32	<0.0001 ***
Residual	7	0.31	0.044		
Lack of Fit	3	0.10	0.033	0.64	0.6258
Pure Error	4	0.21	0.052		
C. Total	16	41.43			
Standard deviation		0.21	R-squared		0.9926
Mean		4.10	Adjusted R-squared		0.9831
Coefficient of variation (C.V.%)		5.11	Predicted R-squared		0.9535
PRESS		1.92	Adequate precision		26.7760

*** indicates a significant difference at $p < 0.001$ level.

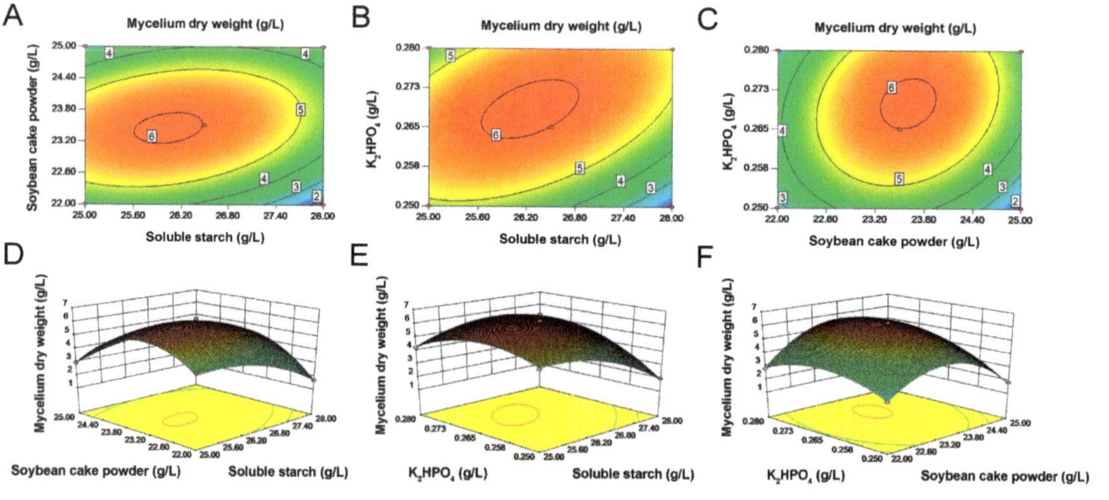

Figure 3. Contour plot described by the model on mycelial dry weight. (**A**) Effect of soluble starch and soybean cake powder. (**B**) Effect of soluble starch and K_2HPO_4. (**C**) Effect of soybean cake powder and K_2HPO_4. Response-surface curve of mycelial dry weight showing mutual interactions. (**D**) Soluble starch and soybean cake powder. (**E**) Soluble starch and K_2HPO_4. (**F**) Soybean cake powder and K_2HPO_4. Other variables, except for the two in each figure, were maintained at zero level in coded units. The numbers 2–6 in the figure indicate the mycelium dry weight at this location.

Figure 3. Contour plot described by the model on mycelial dry weight. (**A**) Effect of soluble starch and soybean cake powder. (**B**) Effect of soluble starch and K_2HPO_4. (**C**) Effect of soybean cake powder and K_2HPO_4. Response-surface curve of mycelial dry weight showing mutual interactions. (**D**) Soluble starch and soybean cake powder. (**E**) Soluble starch and K_2HPO_4. (**F**) Soybean cake powder and K_2HPO_4. Other variables, except for the two in each figure, were maintained at zero level in coded units. The numbers 2–6 in the figure indicate the mycelium dry weight at this location.

3.5. Validation of the Optimized Medium

Based on the results of optimized medium composition, the maximum mycelial dry weight of *S. alfalfae* XN-04 was predicted to be 6.12 g/L when the concentrations of X_1, X_3 and X_4 were 26.26 g/L, 23.54 g/L and 0.269 g/L, respectively. The result predicted by RSM

3.5. Validation of the Optimized Medium

Based on the results of optimized medium composition, the maximum mycelial dry weight of *S. alfalfae* XN-04 was predicted to be 6.12 g/L when the concentrations of X_1, X_3 and X_4 were 26.26 g/L, 23.54 g/L and 0.269 g/L, respectively. The result predicted by RSM was validated by carrying out an experiment using best-predicted solutions for growth of *S. alfalfae* XN-04. Under the optimized conditions, the average mycelial dry weight of *S. alfalfae* XN-04 reached 6.61 g/L, which was close to the RSM predicted value, suggesting that the experimental and the predicted values were in good agreement (Figure 4A). Compared with GS medium, the optimized medium enhanced mycelial dry weight by 7.47-fold (Figure 4A). Furthermore, under optimal conditions, the methanol extracts of cells show a stronger suppressive effect against *Fov* (95.65%) (Figure 4B). Therefore, we concluded that the model established in this paper was accurate and reliable for predicting the biomass of *S. alfalfae* XN-04.

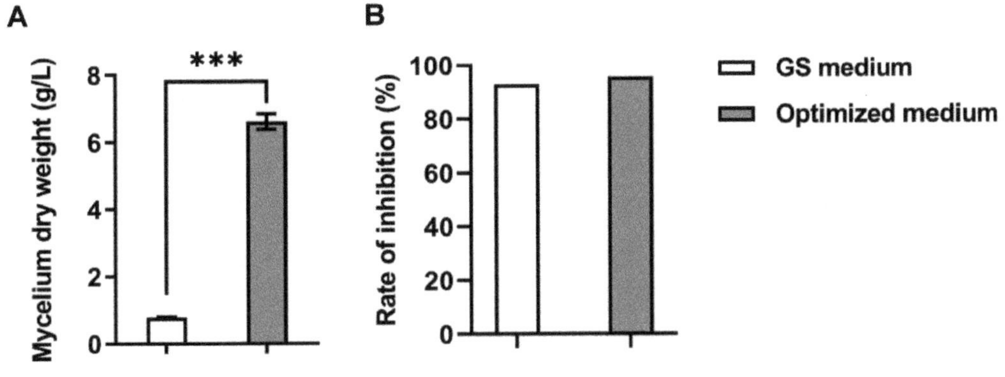

Figure 4. Validation of statistical optimization medium and the initial medium (GS medium). (**A**) Mycelial dry weight. (**B**) Rate of inhibition. Bars represent the SD of three replicates. *** indicates statistical significance based on the two-tailed test ($p < 0.0001$).

3.6. Purification and Structural Elucidation of the Antifungal Compound

A series of chromatographic procedures were performed to purify the main antifungal ingredient in the MeOH extract of *S. alfalfae* XN-04. A total of 10 fractions were obtained from the silica gel column chromatography (Figure S1), among which Fr. 5–9 exhibited antifungal activity (Table S1). Fr. 5 demonstrated strong antifungal activity, producing a radius of inhibition zone > 14 mm. Therefore, Fr. 5 was subjected to semi-preparative HPLC for further purification. The chromatogram was monitored at 363 nm, and a peak (compound **1**) was observed at the retention times of 10.08 (Figure 5A). Finally, compound **1** was obtained as a yellow amorphous solid. The results of LC-MS/MS showed that the most abundant high mass ion [M + H]+ was found at m/z 739.4619 (Figure S2). Compound **1** was identified as roflamycoin (molecular weight, 738; molecular formula, $C_{40}H_{66}O_{12}$) by comparing its NMR data (Table S1) with the previous study [32].

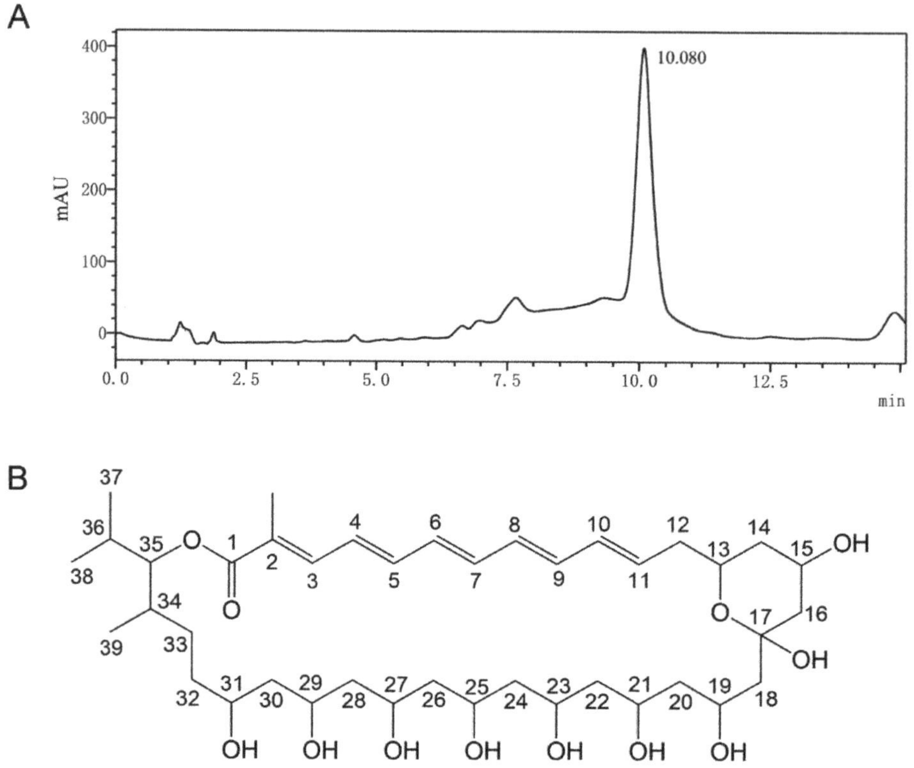

Figure 5. HPLC analysis and structure of roflamycoin produced by *S. alfalfae* XN-04. (**A**) HPLC analysis of Fr. 5 at 363 nm. (**B**) The structure of roflamycoin produced by *S. alfalfae* XN-04.

4. Discussion

Recent studies reported that actinomycetes, particularly *Streptomyces* spp., acted as effective biocontrol agents against multiple phytopathogens [33]. *Streptomyces* spp. can directly or indirectly benefit plants due to its ability to produce antibiotics, hydrolytic enzymes, or enzyme inhibitors [34]. A previous study indicated that *S. alfalfae* XN-04 can colonize the cotton plant root system, inhibiting the mycelial growth of *Fov* by production of hydrolytic enzymes and antifungal secondary metabolites [28]. However, the extremely low yield of antifungal metabolites and long fermentation period of *S. alfalfae* XN-04 have restricted its further research and market applications. In order to maximize the yield of antifungal metabolites and to provide data to support the large-scale and low-cost expansion of *S. alfalfae* XN-04, this study was conducted to optimize the fermentation medium components of *S. alfalfae* XN-04. Additionally, an antifungal compound was purified from the metabolites of *S. alfalfae* XN-04 and was identified as roflamycoin.

Carbon sources are essential components in constructing cellular materials and are also used as energy sources [35]. Nitrogen is an important element for nucleic acid and protein, which are the raw materials synthesized by microorganisms to create cellular metabolites [36]. In the process of microorganism fermentation, interaction between cell growth and secondary metabolites secretion is critically influenced by the growth-limiting nutrient at certain concentrations [37]. Therefore, selection as well as optimization of nutritional components is an indispensable step in large-scale production of probiotic biomass at a reasonable cost. Generally, secondary metabolites production in *Streptomyces* spp. is often stimulated by slowly assimilated complex carbohydrates (e.g., soluble starch, dextrins), but is suppressed by rapidly utilized carbon sources (e.g., glucose). Shakeel et al.

(2016) found that soluble starch was the best carbon for the cell growth of *Streptomyces platensis* 3–10 [37]. Jacob et al. (2014) reported that *Streptomyces nogalater* NIIST A30 was able to produce antibacterial metabolites at a higher level with starch as carbon source [38]. In this study, the results showed that the biomass and antifungal metabolites production of *S. alfalfae* XN-04 were significantly improved with the supplementation of soluble starch and soybean cake powder (Figure 1A,B,E,F). Moreover, the results also revealed that glucose had negative effect on the antifungal metabolites production of *S. alfalfae* XN-04 (Figure 1E). These findings appear to be consistent with the results in related previous studies [39]. A possible explanation of this phenomenon is that glucose may cause catabolite repression, in which the production of enzymes for secondary metabolites biosynthesis might be inhibited [40].

Minerals such as K^+, Ca^{2+}, Na^+ and Mg^{2+} have been known to play a crucial role in the metabolic process of all microorganisms, as it is essential for the formation of cell mass and acts as a cofactor for many biosynthetic enzymes to catalyze the necessary reactions [35]. Zhu et al. (2007) made an effort to ameliorate the production of avilamycin using *Streptomyces viridochromogenes* Tu57–1 and by adopting BBD in which they found a maximum production of 88.33 mg/L (2.80-fold increase) in the optimized medium with the ingredients such as $MgSO_4{\cdot}7H_2O$ 0.37 g/L, $CaCl_2{\cdot}2H_2O$ 0.39 g/L, soybean flour 21.97 g/L and soluble starch 37.22 g/L [41]. Peng et al. (2020) reported that the fungichromin production in *Streptomyces* sp. WP-1 was markedly increased in medium supplemented with magnesium phosphate or calcium phosphate [26]. In this study, the results showed that K_2HPO_4, $MgSO_4{\cdot}7H_2O$ and $CaCO_3$ had distinct positive effects both on microorganism growth and on antifungal metabolites production by *S. alfalfae* XN-04 (Figure 1C,G). Moreover, the results also revealed that the demand for Ca^{2+} of *S. alfalfae* XN-04 was at a high level (Figure 2F). This finding suggests that Ca^{2+} is a key factor in *S. alfalfae* XN-04 growth. This finding appears to be consistent with the results in related previous studies. A previous study reported that there was a high calcium content in spores of *Streptomyces* spp. and that for some species it was an essential triggering factor of spore germination [42]. Additionally, $CaCO_3$ used as medium substrate can balance the pH of the fermentation broth by reacting with the acid, which was produced by the process of microorganism fermentation to form neutral salts and CO_2, the latter escaping from the medium.

In order to clarify the interactions among factors selected by single-factor experiments, further work will be needed on mixing conditions. Generally, in non-growth associated fermentation, secondary metabolites production was substantially proportional to the quantity of biomass [18]. Therefore, it is essential to increase the total biomass of microorganisms, especially for *S. alfalfae* XN-04, since its antifungal metabolites were produced inside the cells. Taking in account the above theory, we focused on optimization of medium components, seeking a substantial increase in biomass without reducing the production of secondary metabolites. Therefore, mycelial dry weight was utilized as the response in subsequent experiment designs.

Statistical experimental design was an efficient method for optimizing the effects of each variable. The PBD enables screening for the most significant variables of microorganism fermentation, and that the steepest ascent experiment made the result approximate to the optimal region for easily carrying out the subsequent experiments. Finally, the RSM detected the optimum levels of significant factors. In 2004, Elibol tried to optimize a fermentation medium using a 24 full factorial CCD for the production of benzoisochromanequinone polyketide antibiotic actinorhodin with *Strptomyces coelicolor* A3 (2), where 200 mg/L of antibiotic yield was attained under the optimized medium composition, which was 32% higher than that of the unoptimized medium (148 mg/L) and was very close to the predicted yield (195 mg/L) [43]. The maximization of olivanic acid production using *Streptomyces olivaceus* MTCC 6820 in shake flask was executed in which 415 mg/L of yield was achieved in the medium optimized with CCD and it was eight times higher when compared with the normal unoptimized medium (50 mg/L) [44]. The results were in agreement with these results. In this study, the mycelial dry weight of *S. alfalfae* XN-04 reached 6.61 g/L (7.47-fold increase) in the medium optimized with RSM.

In conclusion, this study optimized the fermentation medium compositions for biomass yield and antifungal metabolites production of *S. alfalfae* XN-04. This study also reported an antifungal secondary metabolite, roflamycoin, which was purified from *S. alfalfae* XN-04 metabolites.

Supplementary Materials: The following supporting information can be downloaded at: https://www.mdpi.com/article/10.3390/microorganisms10091854/s1.

Author Contributions: Conceptualization, J.C. and Y.W.; methodology, J.C. and X.L.; software, J.C. and R.J.; validation, J.C. and L.H.; formal analysis, J.C. and Y.W.; investigation, J.C. and X.L.; resources, J.C.; data curation, J.C.; writing—original draft preparation, J.C. and Y.W.; writing—review and editing, J.C., X.L. and Y.W.; visualization, X.L.; supervision, J.C. and Y.W.; project administration, Y.W. funding acquisition, Y.W. All authors have read and agreed to the published version of the manuscript.

Funding: This research was funded by the Key Industry Chain Innovation Project of Shaanxi province (2020ZDLNY07-02, 2021ZDLNY03-05) and the National Key Research and Development Program of China Grant (2021YFD140100502).

Institutional Review Board Statement: Not applicable.

Informed Consent Statement: Not applicable.

Data Availability Statement: Not applicable.

Acknowledgments: We are grateful to John Richard Schrock (Emporia State University, USA) for proofreading this manuscript.

Conflicts of Interest: The authors declare no conflict of interest.

References

1. Raza, W.; Ling, N.; Zhang, R.F.; Huang, Q.W.; Xu, Y.C.; Shen, Q.R. Success evaluation of the biological control of Fusarium Wilts of cucumber, banana, and tomato since 2000 and future research strategies. *Crit. Rev. Biotechnol.* **2017**, *37*, 202–212. [CrossRef] [PubMed]
2. Palmieri, D.; Vitale, S.; Lima, G.; Di Pietro, A.; Turra, D. A bacterial endophyte exploits chemotropism of a fungal pathogen for plant colonization. *Nat. Commun.* **2020**, *11*, 5264. [CrossRef] [PubMed]
3. Hwang, K.S.; Kim, H.U.; Charusanti, P.; Palsson, B.O.; Lee, S.Y. Systems biology and biotechnology of *Streptomyces* species for the production of secondary metabolites. *Biotechnol. Adv.* **2014**, *32*, 255–268. [CrossRef] [PubMed]
4. Awla, H.K.; Kadir, J.; Othman, R.; Rashid, T.S.; Hamid, S.; Wong, M.Y. Plant growth-promoting abilities and biocontrol efficacy of *Streptomyces* sp. UPMRS4 against *Pyricularia Oryzae*. *Biol. Control* **2017**, *112*, 55–63. [CrossRef]
5. Cho, G.; Kim, J.; Park, C.G.; Nislow, C.; Weller, D.M.; Kwak, Y.S. Caryolan-1-ol, an antifungal volatile produced by *Streptomyces* spp., inhibits the endomembrane system of fungi. *Open Biol.* **2017**, *7*, 170075. [CrossRef]
6. Faheem, M.; Raza, W.; Zhong, W.; Nan, Z.; Shen, Q.R.; Xu, Y.C. Evaluation of the biocontrol potential of *Streptomyces goshikiensis* YCXU against *Fusarium oxysporum* f. sp. *niveum*. *Biol. Control* **2015**, *81*, 101–110. [CrossRef]
7. Grahovac, J.; Grahovac, M.; Dodic, J.; Bajic, B.; Balaz, J. Optimization of cultivation medium for enhanced production of antifungal metabolites by *Streptomyces hygroscopicus*. *Crop Prot.* **2014**, *65*, 143–152. [CrossRef]
8. Worsley, S.F.; Newitt, J.; Rassbach, J.; Batey, S.; Holmes, N.A.; Murrell, J.C.; Wilkinson, B.; Hutchings, M.I. *Streptomyces* endophytes promote host health and enhance growth across plant species. *Appl. Environ. Microbiol.* **2020**, *86*, e01053-20. [CrossRef]
9. Gong, Y.; Liu, J.Q.; Xu, M.J.; Zhang, C.M.; Gao, J.; Li, C.G.; Xing, K.; Qin, S. Antifungal volatile organic compounds from *Streptomyces setonii* WY228 control black spot disease of sweet potato. *Appl. Environ. Microbiol.* **2022**, *88*, e0231721. [CrossRef]
10. Ntemafack, A.; Ahmed, S.; Kumar, A.; Chouhan, R.; Kapoor, N.; Bharate, S.B.; Hassan, Q.P.; Gandhi, S.G. Plant growth promoting potential of butyl isobutyl phthalate and *Streptomyces* sp. from *Rumex dentatus* on rice. *Appl. Microbiol. Biotechnol.* **2022**, *106*, 2603–2617. [CrossRef]
11. Kasuga, K.; Sasaki, A.; Matsuo, T.; Yamamoto, C.; Minato, Y.; Kuwahara, N.; Fujii, C.; Kobayashi, M.; Agematu, H.; Tamura, T.; et al. Heterologous production of kasugamycin, an aminoglycoside antibiotic from *Streptomyces kasugaensis*, in *Streptomyces lividans* and *Rhodococcus erythropolis* L-88 by constitutive expression of the biosynthetic gene cluster. *Appl. Microbiol. Biotechnol.* **2017**, *101*, 4259–4268. [CrossRef] [PubMed]
12. Jiang, J.; Sun, Y.F.; Tang, X.; He, C.N.; Shao, Y.L.; Tang, Y.J.; Zhou, W.W. Alkaline pH shock enhanced production of validamycin a in fermentation of *Streptomyces hygroscopicus*. *Bioresour. Technol.* **2018**, *249*, 234–240. [CrossRef] [PubMed]

13. Yang, R.; Jiang, S.L.; Wen, X.D.; Song, X.C.; Wang, X.; Li, D.X.; Yin, Q.X.; Wu, X.; Wang, D.L.; Chen, Z. Antifungal activity and possible mode of action of ningnanmycin against tea gray blight disease pathogen *Pseudopestalotiopsis camelliae-sinensis*. *Phytopathology* **2021**, *111*, 1735–1742. [CrossRef] [PubMed]
14. Yang, M.L.; Zhang, W.; Lv, Z.Y.; Shi, L.M.; Zhang, K.C.; Ge, B.B. Evaluation of the inhibitory effects of wuyiencin, a secondary metabolite of *Streptomyces albulus* CK-15, against *Sclerotinia Sclerotiorum* in vitro. *Plant Dis.* **2022**, *106*, 156–164. [CrossRef]
15. Gomes, R.J.; Ida, E.I.; Spinosa, W.A. Nutritional supplementation with amino acids on bacterial cellulose production by *Komagataeibacter intermedius*: Effect analysis and application of response surface methodology. *Appl. Biochem. Biotechnol.* **2022**, *online ahead of print*. [CrossRef] [PubMed]
16. Vlajkov, V.; Anđelić, S.; Pajčin, I.; Grahovac, M.; Budakov, D.; Jokić, A.; Grahovac, J. Medium for the production of *Bacillus*-based biocontrol agent effective against aflatoxigenic *Aspergillus flavus*: Dual approach for modelling and optimization. *Microorganisms* **2022**, *10*, 1165. [CrossRef]
17. Iyyappan, J.; Bharathiraja, B.; Varjani, S.; PraveenKumar, R.; Kumar, S.M. Anaerobic biobutanol production from black strap molasses using *Clostridium acetobutylicum* MTCC11274: Media engineering and kinetic analysis. *Bioresour. Technol.* **2022**, *346*, 126405. [CrossRef]
18. Latha, S.; Sivaranjani, G.; Dhanasekaran, D. Response surface methodology: A non-conventional statistical tool to maximize the throughput of *Streptomyces* species biomass and their bioactive metabolites. *Crit. Rev. Microbiol.* **2017**, *43*, 567–582. [CrossRef]
19. Zalila-Kolsi, I.; Kessentini, S.; Tounsi, S.; Jamoussi, K. Optimization of *Bacillus amyloliquefaciens* BLB369 culture medium by response surface methodology for low cost production of antifungal activity. *Microorganisms* **2022**, *10*, 830. [CrossRef]
20. Santra, H.K.; Maity, S.; Banerjee, D. Production of bioactive compounds with broad spectrum bactericidal action, bio-film inhibition and antilarval potential by the secondary metabolites of the endophytic fungus *Cochliobolus* sp. APS1 isolated from the Indian medicinal herb *Andrographis paniculata*. *Molecules* **2022**, *27*, 1459. [CrossRef]
21. Venkateswarulu, T.C.; Prabhakar, K.V.; Kumar, R.B. Optimization of nutritional components of medium by response surface methodology for enhanced production of lactase. *3 Biotech.* **2017**, *7*, 186. [CrossRef] [PubMed]
22. Zhang, Y.; Huang, H.; Xu, S.; Wang, B.; Ju, J.; Tan, H.; Li, W. Activation and enhancement of Fredericamycin A production in deepsea-derived *Streptomyces somaliensis* SCSIO ZH66 by using ribosome engineering and response surface methodology. *Microb. Cell. Fact.* **2015**, *14*, 64. [CrossRef] [PubMed]
23. Amid, A.; Ismail, N.A.; Yusof, F.; Salleh, H.M. Expression, purification, and characterization of a recombinant stem bromelain from *Ananas comosus*. *Process. Biochem.* **2011**, *46*, 2232–2239. [CrossRef]
24. Almeida, D.G.; da Silva, R.D.F.S.; Luna, J.M.; Rufino, R.D.; Santos, V.A.; Sarubbo, L.A. Response surface methodology for optimizing the production of biosurfactant by *Candida tropicalis* on industrial waste substrates. *Front. Microbiol.* **2017**, *8*, 157. [CrossRef] [PubMed]
25. Singh, S.K.; Singh, S.K.; Tripathi, V.R.; Khare, S.K.; Garg, S.K. Comparative one-factor-at-a-time, response surface (statistical) and bench-scale bioreactor level optimization of thermoalkaline protease production from a *Psychrotrophic pseudomonasputida* SKG-1 isolate. *Microb. Cell. Fact.* **2011**, *10*, 114. [CrossRef] [PubMed]
26. Peng, C.; An, D.P.; Ding, W.X.; Zhu, Y.X.; Ye, L.; Li, J.Y. Fungichromin production by *Streptomyces* sp. WP-1, an endophyte from *Pinus dabeshanensis*, and its antifungal activity against *Fusarium oxysporum*. *Appl. Microbiol. Biotechnol.* **2020**, *104*, 10437–10449. [CrossRef] [PubMed]
27. Kim, Y.H.; Park, B.S.; Bhatia, S.K.; Seo, H.M.; Jeon, J.M.; Kim, H.J.; Yi, D.H.; Lee, J.H.; Choi, K.Y.; Park, H.Y.; et al. Production of rapamycin in *Streptomyces hygroscopicus* from glycerol-based media optimized by systemic methodology. *J. Microbiol. Biotechnol.* **2014**, *24*, 1319–1326. [CrossRef] [PubMed]
28. Chen, J.; Hu, L.F.; Chen, N.; Jia, R.M.; Ma, Q.; Wang, Y. The biocontrol and plant growth-promoting properties of *Streptomyces alfalfae* XN-04 revealed by functional and genomic analysis. *Front. Microbiol.* **2021**, *12*, 745766. [CrossRef]
29. Mei, X.Y.; Liu, Y.X.; Huang, H.C.; Du, F.; Huang, L.L.; Wu, J.Q.; Li, Y.W.; Zhu, S.S.; Yang, M. Benzothiazole inhibits the growth of *Phytophthora capsici* through inducing apoptosis and suppressing stress responses and metabolic detoxification. *Pestic. Biochem. Physiol.* **2019**, *154*, 7–16. [CrossRef]
30. Yan, L.; Zhang, Z.; Zhang, Y.; Yang, H.; Qiu, G.; Wang, D.; Lian, Y. Improvement of tacrolimus production in *Streptomyces tsukubaensis* by mutagenesis and optimization of fermentation medium using Plackett-Burman design combined with response surface methodology. *Biotechnol. Lett.* **2021**, *43*, 1765–1778. [CrossRef]
31. Kim, J.D.; Park, M.Y.; Jeon, B.J.; Kim, B.S. Disease control efficacy of 32,33-didehydroroflamycoin produced by *Streptomyces rectiviolaceus* strain DY46 against gray mold of tomato fruit. *Sci. Rep.* **2019**, *9*, 13533. [CrossRef] [PubMed]
32. Han, X.; Wang, J.; Liu, L.; Shen, F.; Meng, Q.; Li, X.; Li, Y.; Liu, D. Identification and predictions regarding the biosynthesis pathway of polyene macrolides produced by *Streptomyces roseoflavus* Men-myco-93-63. *Appl. Environ. Microbiol.* **2021**, *87*, e03157-20. [CrossRef] [PubMed]
33. Olanrewaju, O.S.; Babalola, O.O. *Streptomyces*: Implications and interactions in plant growth promotion. *Appl. Microbiol. Biotechnol.* **2019**, *103*, 1179–1188. [CrossRef] [PubMed]
34. Vurukonda, S.S.K.P.; Giovanardi, D.; Stefani, E. Plant growth promoting and biocontrol activity of *Streptomyces* spp. as endophytes. *Int. J. Mol. Sci.* **2018**, *19*, 952. [CrossRef]
35. Abhini, K.N.; Rajan, A.B.; Zuhara, F.; Sebastian, D. Response surface methodological optimization of l-asparaginase production from the medicinal plant endophyte *Acinetobacter baumannii* ZAS1. *J. Genet. Eng. Biotechnol.* **2022**, *20*, 22. [CrossRef]

36. Jose, P.A.; Jebakumar, S.R.D. Phylogenetic appraisal of antagonistic, slow growing actinomycetes isolated from hypersaline inland solar salterns at Sambhar salt Lake, India. *Front. Microbiol.* **2013**, *4*, 190. [CrossRef]
37. Shakeel, Q.; Lyu, A.; Zhang, J.; Wu, M.D.; Chen, S.W.; Chen, W.D.; Li, G.Q.; Yang, L. Optimization of the cultural medium and conditions for production of antifungal substances by *Streptomyces platensis* 3–10 and evaluation of its efficacy in suppression of clubroot disease (*Plasmodiophora brassicae*) of oilseed rape. *Biol. Control* **2016**, *101*, 59–68. [CrossRef]
38. Jacob, J.; Rajendran, R.U.; Priya, S.H.; Purushothaman, J.; Amma, D.K.B.N.S. Enhanced antibacterial metabolite production through the application of statistical methodologies by a *Streptomyces nogalater* NIIST A30 isolated from Western Ghats forest soil. *PLoS ONE* **2017**, *12*, e0175919. [CrossRef]
39. Messis, A.; Bettache, A.; Brahami, A.; Kecha, M.; Benallaoua, S. Optimization of antifungal production from a novel strain *Streptomyces* sp. TKJ2 using response surface methodology. *Med. Chem. Res.* **2014**, *23*, 310–316. [CrossRef]
40. Srivastava, A.; Singh, V.; Haque, S.; Pandey, S.; Mishra, M.; Jawed, A.; Shukla, P.K.; Singh, P.K.; Tripathi, C. Response surface methodology-genetic algorithm based medium optimization, purification, and characterization of cholesterol oxidase from *Streptomyces rimosus*. *Sci. Rep.* **2018**, *8*, 10913. [CrossRef]
41. Zhu, C.H.; Lu, F.P.; He, Y.N.; Zhang, J.K.; Du, X. Statistical optimization of medium components for avilamycin production by *Streptomyces viridochromogenes* Tu57-1 using response surface methodology. *J. Ind. Microbiol. Biotechnol.* **2007**, *34*, 271–278. [CrossRef] [PubMed]
42. González-Quiñónez, N.; Corte-Rodríguez, M.; Álvarez-Fernández-García, R.; Rioseras, B.; López-García, M.T.; Fernández-García, G.; Montes-Bayón, M.; Manteca, A.; Yagüe, P. Cytosolic copper is a major modulator of germination, development and secondary metabolism in *Streptomyces coelicolor*. *Sci. Rep.* **2019**, *9*, 4214. [CrossRef] [PubMed]
43. Elibol, M. Optimization of medium composition for actinorhodin production by *Streptomyces coelicolor* A3 (2) with response surface methodology. *Process. Biochem.* **2004**, *39*, 1057–1062. [CrossRef]
44. Singh, V.; Tripathi, C.K.M. Production and statistical optimization of a novel olivanic acid by *Streptomyces olivaceus* MTCC 6820. *Process. Biochem.* **2008**, *43*, 1313–1317. [CrossRef]

microorganisms

MDPI

Article

Evaluation of Endospore-Forming Bacteria for Suppression of Postharvest Decay of Apple Fruit

Anissa Poleatewich [1,*], Paul Backman [2] and Haley Nolen [3]

1 Department of Agriculture, Nutrition, and Food Systems, University of New Hampshire, Durham, NH 03824, USA
2 Department of Plant Pathology and Environmental Microbiology, Penn State University, University Park, PA 16802, USA
3 Department of Molecular, Cellular, and Biomedical Sciences, University of New Hampshire, Durham, NH 03824, USA
* Correspondence: anissa.poleatewich@unh.edu

Abstract: The use of microbial biocontrol agents for control of postharvest disease has been the subject of intensive research over the past three decades resulting in commercialization of several biocontrol products. The objective of this research was to test endospore-forming bacteria collected from apple leaves for suppression of bitter rot and blue mold on apple. Bacteria were collected from abandoned, low-input, organic, and conventionally managed orchards in Pennsylvania and were screened for their ability to produce endospores, hydrolyze chitin, reduce pathogen growth in vitro, and suppress postharvest disease in vivo. Several isolates reduced bitter rot lesion size on 'Rome Beauty' from 40–89% compared to untreated controls. *Bacillus megaterium* isolates, A3-6 and Ae-1, resulted in the greatest suppression of bitter rot lesion size. One isolate, A3-2, suppressed blue mold lesion size. Scanning electron microscopy of inoculated apple wounds suggests parasitism as a mode of action explains the suppression of bitter rot lesion size by isolate A3-6. Of the top seventeen isolates exhibiting biocontrol potential, 70% were collected from abandoned or unmanaged locations. This research demonstrates abandoned apple orchards can be a source of new biocontrol agents for control of postharvest diseases of apple.

Keywords: *Bacillus*; bitter rot; blue mold; biological control; apple

Citation: Poleatewich, A.; Backman, P.; Nolen, H. Evaluation of Endospore-Forming Bacteria for Suppression of Postharvest Decay of Apple Fruit. *Microorganisms* **2023**, *11*, 81. https://doi.org/10.3390/microorganisms11010081

Academic Editors: Tomislav Cernava, Beibei Ge and Kyungseok Park

Received: 28 November 2022
Revised: 16 December 2022
Accepted: 22 December 2022
Published: 28 December 2022

1. Introduction

Production of fresh fruits and vegetables presents a unique challenge as most produce is harvested over a relatively short period of time and then stored for weeks to months to avoid exceeding demand and to provide consumers with a product year-round. Postharvest decay by microorganisms represents a significant limitation in our ability to store fruit [1]. The economic losses incurred from storage diseases are considerable given the initial investment related to crop production and management of pathogens and pests in the field prior to harvest. To date, the use of synthetic fungicides (as pre- or postharvest treatments) remains an important strategy for managing postharvest decays [2,3]. Development of fungicide resistant strains has been documented for several postharvest pathogens [4,5]. Human health concerns, due to pesticide residues on surfaces directly consumed as food, have also limited the available options for control of postharvest pathogens [6,7]. Several countries have banned or significantly restricted postharvest applications of fungicides [3,8]. As a result, there is an ongoing effort to develop alternative control methods to reduce fungicide dependence, reduce environmental risks, and improve consumer confidence in food safety.

Considerable attention has focused on the use of biocontrol for management of postharvest diseases of fruits and vegetables, and there have been several review articles on the topic over the past 30 years [1–3,9–13]. The uniqueness of the postharvest environment offers an advantage as environmental variables are more stable compared to field conditions, and to an extent, parameters can be adjusted to favor survival of biocontrol agents (BCAs). The challenge is new products must not only be safe but effective and economical. Several microorganisms have been patented for postharvest biocontrol, many which are applied to harvested fruit as a dip or drench. Over the past two decades, postharvest biocontrol has evolved into a significant area of research. Pioneering work conducted by Wilson and Pusey [12,14] in the 1980s soon led to a wealth of research and development of commercial products. Although numerous products were pursued, most were met with limited success [2,9]. Currently, there are four commercial products available for postharvest use in the United States [15]: Bio-Save™ 10 and 11 containing two strains of *Pseudomonas syringae* (JET Harvest Solutions, Longwood, FL, USA), Lalfresh S containing *Clonostachys rosea* strain J1446 (*Gliocladium catenulatum*) (Lallemand Plant Care, Montreal, QC, Canada)., and the yeast-based product NEXY (Agrauxine by Lesaffre) containing *Candida oleophila* strain O. In their 2016 review, Droby and colleagues [9] highlighted several reasons contributing to poor market longevity of postharvest biocontrol products, including the need to consider the microhabitat in which BCAs are expected to reside when attempting to select and establish BCAs. Colonization of fruit surfaces and interaction with existing microflora are important considerations when screening BCAs [16]. Because of their popularity as biocontrol agents for soil-borne diseases [17], many researchers have attempted to employ soil/root-inhabiting *Bacillus* strains against foliar diseases without proper consideration of the necessary adaptations to survive in the phyllosphere or carposphere [18]. Therefore, selection of bacteria adapted to the target niche may increase the likelihood of long-term colonization and effective control of postharvest diseases. For example, BCAs for postharvest application should have the ability to survive at low temperatures typical of cold storage conditions [1]. Additionally, sampling from niches that are representative of the target application niche [19] may increase efficacy. For this reason, apple orchards may serve as a source for novel BCAs for postharvest biocontrol on apple. Furthermore, abandoned apple orchards—no longer exposed to chemical pesticides—may serve as a resource to identify novel BCAs or consortia of biocontrol microbes [20].

The objectives of this study were to (1) collect native bacteria from apple fruit and foliage from managed and abandoned apple trees, (2) screen these isolates for chitinase production, dual-culture antagonism, and suppression of fruit rots in vivo, and (3) further characterize the efficacy and modes of action of candidate isolates using scanning electron microscopy.

2. Materials and Methods

2.1. Isolation and Culture of Biocontol Bacteria

Healthy apple (*Malus domestica*) leaves and fruit were collected randomly from three orchards maintained under a standard fungicide, a 'reduced risk' fungicide program, no spray, or a certified organic program at the Penn State University Fruit Research and Extension Center (PSU FREC). Samples were also collected from feral trees in forested areas in Centre County, PA and from three abandoned orchards (two orchards unmanaged for one year and one orchard unmanaged for five years) in Adam's County, PA in September 2006 and June 2007. Within the conventionally managed orchards at the PSU FREC, samples were collected from cv. Golden Delicious, Rome Beauty, and Red Delicious trees grown on Malling 26 (M.26) rootstock. Samples were also collected from cv. GoldRush and Enterprise grown on M.26 rootstock in an organic demonstration orchard. Cultivars in the abandoned orchards and feral areas were unknown. Samples were stored in a cooler with ice during transport to the laboratory.

For isolation of bacteria, two leaves or two plugs of fruit tissue (collected with a 44.2 mm diameter cork borer) were placed in sterile 101 mm × 152 mm stomacher filter bags (Secure-T 80; Labplas, Sainte-Julie, QC, Canada) with 10 mL of 0.1 M potassium phosphate buffer and agitated at seven strokes per second for 30 s in a stomacher blender

(Bagmixer 100 MiniMix; Intersciences St. Nom la Bretèche France). The leaf tissue was further triturated using a pestle. 50 µL of the suspension was dilution plated on each of two media, tryptic soy agar (TSA) amended with Benomyl and apple juice yeast extract agar (AJYE), using a spiral plater (Autoplate 4000; Spiral Biotech Inc., Norwood, MA, USA). To select for endospore-forming bacteria, three 1 mL aliquots of the agitated solution were pipetted into 1.7 mL microtubes and heat treated in a 75 °C water bath for 15 min. The heated solutions were plated on tryptic soy agar (TSA) and apple juice-yeast extract agar (AJYA). After incubation, bacterial colonies were sub-cultured onto yeast extract dextrose calcium carbonate agar (YDC) to obtain pure cultures.

2.2. *Preliminary Screening for Antagonism and Disease Suppression*

2.2.1. Chitin Hydrolysis Assay

The bacteria isolated from apple leaves and fruit were first characterized for the ability to hydrolyze chitin using a conventional plate method. Colloidal chitin was prepared using a modified method of Kokalis-Burelle [21]. The percent chitin (w/v) of the suspension was determined using the oven dry weight of a 10 mL aliquot of the wet suspension. Aliquots of the settled chitin suspension were then blended for 30 s in a sterilized Waring laboratory blender (Model 7010S: Waring laboratory Science, Torrington, CT, USA), and the pooled solutions were diluted to a 1% (w/v) stock using 0.1 M potassium phosphate buffer. This stock suspension was stored at 4 °C and diluted to the desired concentration prior to use. Bacterial isolates were streaked onto 0.4% chitin nutrient agar (CNA). Isolates which produced a clearing zone were characterized as positive for chitin hydrolysis [22].

2.2.2. Dual-Culture Assay

An in vitro plate assay was used to screen isolates for the ability to reduce growth of two apple pathogens: *Venturia inaequalis* and *Colletotrichum acutatum*. A single-spore culture of *V. inaequalis* RSGD1.1.1 was obtained from Dr. M. Jimenez Gasco at Penn State University [23]. *C. acutatum* was isolated from infected fruit at the PSU FREC [24]. A mycelial plug of each pathogen was placed on potato dextrose agar (PDA) in a 100 mm Petri dish. A streak of the candidate bacterial isolate was placed 5 cm from the mycelial plug. Control plates consisted of the mycelial plug without bacteria. Three replicate plates were prepared for each pathogen–bacterial isolate combination. Plates were incubated at 20 °C, and radial growth was measured every two days for ten days. The final colony diameter perpendicular to the bacterium at the end of the experiment was also determined. Data were analyzed for statistical significance by PROC MIXED followed by a Dunnett's test using SAS (SAS Institute, version 9.2; Cary, NC, USA) to determine if the bacteria were able to significantly suppress fungal growth compared to the control.

2.2.3. In Vivo Antifungal Assay

Preliminary experiments were conducted to screen isolates for the ability to reduce severity of bitter rot disease (caused by *C. acutatum*) in vivo. The candidate bacterial isolates were grown in 100 mL of sterile tryptic soy broth in 250 mL Erlenmeyer flasks. Flasks were incubated for seven days at 28 °C and 120 rpm on a rotary incubator shaker (New Brunswick Scientific Model M1024-000, Edison, NJ, USA). Bacterial cells were harvested by centrifugation at 3800 rpm for 15 min at 4 °C using a Sorvall RT7 centrifuge (ThermoFisher Scientific, Waltham, MA, USA). The supernatant was discarded, and the bacterial pellets were re-suspended in 0.1 M potassium phosphate buffer and adjusted to a concentration of 1×10^7 CFU/mL. *C. acutatum* was prepared as described by Poleatewich et al. [25]. Apple cv. Rome Beauty fruit were harvested from trees planted in an experimental orchard at the PSU FREC and washed with a 5% commercial bleach solution and rinsed with tap water. All fruit were wounded at the equator to a depth of 4 mm using a sterilized six-penny nail. Nails were mounted through a rubber stopper to ensure uniformity of wound depth. Immediately after wounding, the wounds of 50 fruit per isolate were inoculated with 20 µL of the candidate biocontrol bacterial isolate or a sterile distilled water control. One hour after bacterial application, wounds of 25 fruit per treatment were inoculated with 20 μL of a 4.0 × 10^4 conidia/mL (800 spores/wound) spore suspension of *C. acutatum* as described by Poleatewich et al. [26]. The remaining 25 fruit per treatment (not inoculated with *C. acutatum*) served as controls to determine if the bacteria alone could cause symptoms on the fruit. Once the inoculum dried, the fruits were randomly placed on 20-count molded cardboard trays. The trays were stored in a dark walk-in growth room at 22 °C on plastic shelves. Fruits were evaluated for symptom severity seven days after pathogen challenge. To assess disease severity, the lesion diameter was measured laterally and horizontally across the wound site, and the total lesion area was estimated using the average of the two measurements. Data were analyzed for statistical significance by PROC MIXED followed by a Dunnett's test using SAS (SAS Institute, version 9.2; Cary, NC, USA) to determine if the bacteria were able to significantly suppress lesion size compared to the water control.

2.3. *Evaluation of Isolates for Disease Suppression on Fruit*

Based on the results from the preliminary in vitro and in vivo experiments, nine isolates were selected for further evaluation. Bacteria and pathogen cultures were grown as described above. Fruits were harvested from cv. Golden Delicious and Rome Beauty trees grafted on M.26 rootstock planted at the PSU FREC. Experiments were conducted on symptom-free Golden Delicious and Rome Beauty fruit in separate experiments. The experiment consisted of a 10 × 2 factorial with ten bacterial treatments (nine isolates + water control) and two pathogens. Each of the twenty treatments were applied to ten replicate fruits. For both cultivars, the fruits were prepared, wounded, and inoculated with the bacterial isolates as described above. One hour after bacterial application, wounds were inoculated with 20 μL of a 4.0 × 10^4 conidia/mL (800 spores/wound) spore suspension of *C. acutatum* or *P. expansum* as described by Poleatewich [26]. Once the inoculum dried, the fruit were placed in ten 20-count plastic trays and covered with a plastic lid (containing five holes for air circulation). The trays were stored in a dark walk-in growth room at 20 °C and ~24% relative humidity on plastic shelves. Fruits were evaluated for symptom severity seven days after pathogen challenge. To assess disease severity, the lesion diameter was measured laterally and horizontally across the wound site, and the total lesion area was estimated using the average of the two measurements. Data were analyzed for significance using the mixed procedure of SAS 9.2 with tray as the random variable followed by a Tukey test ($\alpha = 0.05$) to identify differences in mean lesion areas across treatments.

2.4. *Identificaion of Biocontrol Isolates*

Select bacterial isolates (chosen based on efficacy in preliminary screening assays) were streaked onto fresh YED agar one day before PCR amplification. This short growth period was used to ensure the bacterial cells were in the vegetative state. A small colony was removed from the surface of the YED plate with a 10 μL pipette tip and placed into a PCR tube containing 20 μL of master mix. The master mix contained 2 μL of 10× PCR buffer with 1.5 mM MgCl, 1.6 μL dNTP mic (200 μM each), 0.4 μL 530f primer (910 μM), 0.4 μL 1392 r primer (10 μM), and 0.2 μL Taq Polymerase (Gene Choice, San Diego, CA, USA) and consisted of a total volume of 20 μL. The universal primers used to amplify the small-subunit ribosomal RNA were 530f (5′-GTGCCAGCMGCCGCGG) and 1392 r (5′-ACGGGCGGTGTGTRC). PCR amplification was conducted using an Eppendorf Mastercycler Personal Thermal Cycler (Eppendorf AG, Hamburg, Germany) with the following cycle: 5 min at 95 °C followed by 35 cycles of 94 °C for 15 s, 72 °C for 15 s, and a final extension at 72 °C for 5 min. PCR products were cleaned using ExoSAP-IT (USB Corp., Santa Clara, CA, USA).

Sanger sequencing of PCR products was performed using an ABI Hitachi 3730 XL DNA Analyzer (Hitchi Ltd., Tokyo, Japan) at the Penn State Genomics Core Facility. Sequences were aligned using MAFFT alignment software version 7.305 b with default settings [27]. The resulting alignment was trimmed using trimAL version 1.4rev15 with a gap threshold of 0.5 [28]. Sequences were analyzed using the National Center for Biotechnology Information

(NBCI) Basic Local Alignment Search Tool (BLAST) search algorithm. The identities of the five apple isolates were determined by comparison with high-scored rRNA sequences in BLAST searches and sequences from *Bacillus*, *Brevibacillus*, and *Paenibacillus* type strains. The 16S rRNA sequence data from the five apple isolates and 37 strains were used to create phylogenetic trees. Phylogenetic trees were inferred and compared using IQTree (version 1.6.12, default settings) bootstrapping methods ultrafast bootstrap (UFBoot) and SH-aLRT test [29].

2.5. Colonization of Fruit Wounds by Biocontrol Isolates

A postharvest experiment was conducted to assess bacterial colonization of fruit wounds at room temperature and in cold storage. This experiment consisted of a 5 × 2 factorial with five bacterial isolates (A1-1, A3-6, Ae-1, FLS-1, and FLS-5) and two temperatures (20 °C and 2 °C) for a total of ten treatments. The experiment was conducted on cv. Golden Delicious fruit and repeated on cv. Rome Beauty fruit. Fruits were wounded to create two wounds per fruit, and 20 μL of 10^7 CFU/mL of bacteria was added to each wound on sixteen fruit per bacterial isolate (2.0 × 10^5 CFU/wound). Eight fruit per isolate treatment were placed in plastic trays with lids and stored at 20 °C, and eight fruit of each isolate treatment were stored in a walk-in cold room at 2 °C. Wound colonization of fruit stored at 20 °C was determined at one-, two-, four- and eight-days post inoculation (dpi) and one, seven, fourteen, and twenty-eight dpi for fruit stored at 2 °C. To assess colonization, wound tissues from two replicate fruit were excised using a #4 cork borer (44.2 mm^2) to a depth of approximately 4 mm. A single core was placed in a 101 mm × 152 mm stomacher filter bag with 5 mL of 0.1 M potassium phosphate buffer and triturated in a stomacher blender (Interscience, MiniMix, Saint Nom la Bretêche, France) for 30 s. 50 μL of the undiluted and 10-fold diluted triturate was plated in duplicate on yeast extract dextrose agar (YED). Plates were incubated at 20 °C for 24 h and enumerated. The minimum detectable population for this methodology was log 2.0 CFU/wound.

2.6. Cytological Investigation of Disease Suppression

Scanning electron microscopy (SEM) was used to observe interactions between the biocontrol bacteria and the postharvest pathogens, *C. acutatum* and *P. expansum*, in fruit wounds. This experiment consisted of a 5 × 2 × 2 factorial with five bacteria treatments (A3-6, Ae-1, FO-20, A3-2, and none), two pathogens (*C. acutatum* and *P. expansum*), and two sampling times (24 h and 4 dpi). The five bacterial isolates were chosen to represent positive and negative controls with respect to suppression of lesion size caused by *C. acutatum* and *P. expansum*. A sterilized six-penny nail was used to make four wounds on each Golden Delicious fruit. Wounds were then inoculated with 20 μL of the respective bacterial isolate (sterile water for no-bacteria controls) at a concentration of 10^7 CFU/mL. One hour after bacterial application, wounds were challenged with 20 μL of 3.0 × 10^4 conidia/mL suspension of either *C. acutatum* or *P. expansum*. A hemacytometer (American Optical, Buffalo, NY, USA) was used to adjust the concentration of the pathogen inoculum prepared as described earlier. Fruit was placed in 20-count molded plastic trays and placed in a dark walk-in growth room at 20 °C.

A 10 mL stock fixing solution was prepared with 5 mL of 0.2 M potassium phosphate buffer (pH 7.4), 1.6 mL of 16% paraformaldehyde, 0.6 mL of 25% glutaraldehyde, and sterile distilled water added to a final volume of 10 mL. The final concentration of the fixing solution was 25% paraformaldehyde and 1.5% gultaraldehyde. One ml of fixing solution was pipetted into each of ten vials. Wound tissue was excised using a sterilized #4 core borer (44.2 mm^2) to a depth of approximately 5 mm. Tissue samples were collected 24 h and 4 dpi and immediately immersed in the fixing solution for 24 h at 4 °C. Four tissue samples per treatment were placed in each vial (containing 1 mL of fixative). Following the 24-h fixation, samples were washed with 0.2 M potassium phosphate buffer (pH 7.4) in three five-minute wash cycles. After washing, samples were dehydrated in a gradient ethanol series at room temperature. Next, samples were critical-point dried using a Bal-Tec

CPD030 critical point dryer (Techno Trade, Manchester, NH, USA). The dried samples were immediately mounted on aluminum stubs (15 mm × 10 mm) using double-sided carbon tape and stored in a desiccator cabinet. The next-day samples were sputter coated with 10 nm of gold/palladium (2 min at 10 mAmps) using a Bal-Tec SCD-050 sputter coater. Following sample preparation, tissues were viewed using a JOEL JSM 5400 scanning electron microscope (JOEL; Peabody, MA, USA) at 10 kV, 0-degree tilt, and a 20 mm working distance.

3. Results

3.1. Isolation and Preliminary Screening of Bacterial Isolates from Apple

A total of 75 endospore-forming bacterial isolates were collected from feral trees and conventional and abandoned orchards in 2006 and 2007. Twenty-six isolates (35%) were positive for chitin hydrolysis as indicated by the production of a clearing zone surrounding the colony on 0.4% CNA. Eleven isolates (Ae-1, A1-1, A3-6, A3-2, A3-3, FLS-1, FLS-5, FO-1, WGD-5, WS-1, and WS-3) significantly ($p < 0.001$) reduced radial growth of *C. acutatum* hyphae on PDA (Table 1). Isolates A1-1, FLS-5, FO-1, and WS-1 resulted in the greatest suppression in radial growth of *C. acutatum* compared to no-bacteria control plates, ranging from 41% to 64% reduction in total colony size after the ten-day experiment. Isolates A1-1, A1-11, FO-1, and A3-F1 significantly ($p < 0.002$) reduced radial growth of *V. inaequalis* on PDA. Isolates A1-1 and FO-1 significantly reduced the growth of both pathogens assayed in vitro (Table 1). Six isolates (A1-1, A3-2, A3-3, A3-6, A3-F1, Ae-1, and FLS-1) reduced both growth in vitro and symptom severity in vivo of *C. acutatum* (Table 1).

Table 1. Results from preliminary in vitro and in vivo screening of bacterial isolates collected from apples in Pennsylvania. Shaded isolates suppressed growth both in vitro and in vivo.

Isolate ^{v}	Location ^{w}	Chitinase ^{x}	In Vitro Plate Assay *C. acutatum* ^{y}	In Vitro Plate Assay *V. inaequalis*	In Vivo Assay ^{z} *C. acutatum*
A1-1	abandoned	+	+	+	*
A1-11	abandoned	–	–	+	-
A2-4	abandoned	+	–	–	*
A3-2	abandoned	+	+	–	**
A3-3	abandoned	+	+	–	**
A3-4	abandoned	+	–	–	*
A3-6	abandoned	–	+	–	**
A3-F1	abandoned	+	–	+	**
Ae-1	abandoned	–	+	–	**
FC-2	FREC no spray	+	–	–	-
FLS-1	FREC Reduced risk	+	+	–	*
FLS-5	FREC Reduced risk	+	+	–	-
FO-1	FREC Organic	+	+	+	-
FO-20	FREC Organic	+	–	–	*
WGD-5	natural	+	+	–	-
WS-1	natural	–	+	–	-
WS-3	natural	+	+	–	-

^{v} Results are shown for isolates that suppressed growth in the in vitro or in vivo experiments. ^{w} Isolates were collected from abandoned orchards, feral trees in forested natural areas, and research orchards at the Penn State Fruit Research and Extension Center (PSU FREC). ^{x} Isolates marked as + produced a clearing zone on 0.4% chitin nutrient agar. Isolates marked as – did not produce a clearing zone. ^{y} Isolates marked as + significantly reduced pathogen lesion size ($\alpha = 0.05$) compared to the control. Isolates marked as – were not significantly different from the control. ^{z} Isolates were significantly different from the water control at $\alpha = 0.05$ (*) or $\alpha = 0.01$ (**).

3.2. Suppression of C. acutatum and P. expansum in Fruit Wounds

Two isolates tested (A3-6 and Ae-1) resulted in significant suppression of bitter rot lesion size on cv. Golden Delicious ($p < 0.001$) and Rome Beauty ($p < 0.01$) fruit, and one isolate (A3-2) suppressed blue mold lesion size on cv. Golden Delicious fruit ($p = 0.049$) (Figure 1). None of the isolates tested resulted in significant suppression of lesion size caused by both pathogens.

isolate (A3-2) suppressed blue mold lesion size on cv. Golden Delicious fruit ($p = 0.049$) (Figure 1). None of the isolates tested resulted in significant suppression of lesion size caused by both pathogens.

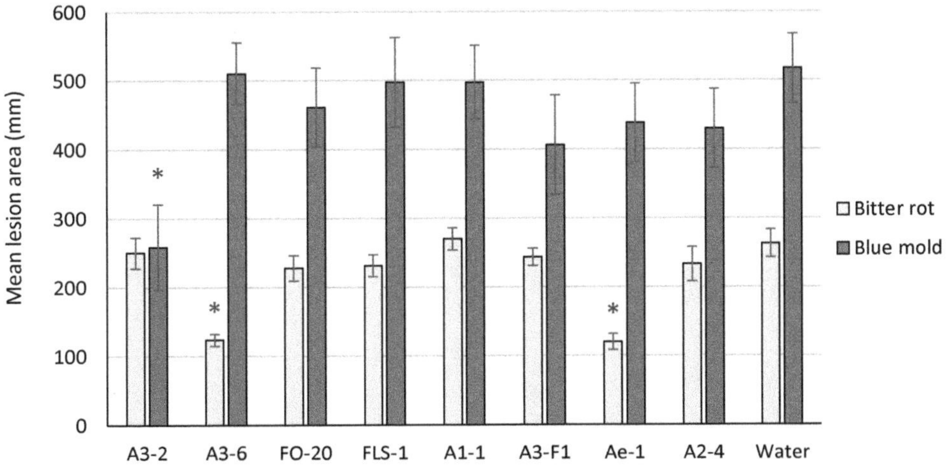

Figure 1. Mean lesion area on cv. Golden Delicious fruit six days post inoculation with *C. acutatum* (bitter rot) or *P. expansum* (blue mold). For each pathogen tested, bars with an asterisk (*) were significantly different from the water control at $\alpha = 0.05$. Error bars represent the standard error of the mean.

3.3. Isolate Identification

Using phylogenetic analyses, it was determined that isolates A1-1, A3-6, Ae-1, and FLS-5 belong to the genus *Bacillus*, and isolate FLS-1 belong to the genus *Brevibacillus* (Figure 2). Isolate A1-1 and FLS-5 were determined to be members of the *B. cereus* group. Isolate FLS-5 was most similar to *B. cereus*. Isolate A1-1 was most similar to *B. mycoides* and *B. weihenstephanensis*. Based on the colony morphology on nutrient media, isolate A1-1 was concluded to be *B. mycoides*. Isolates A3-6 and Ae-1 were most similar to *B. megaterium*. Sequence data from the five apple isolates were submitted to GenBank (accession OP396281-88).

3.4. Colonization of Fruit Wounds

All the bacterial isolates tested were able to colonize wounds on cv. Golden Delicious and Rome Beauty fruit at 20 °C and 2 °C (Figure 3). Approximately 2.0×10^5 CFU/mL were applied to each wound. Colonization levels by isolates FLS-5 and A1-1 were similar at the two temperature treatments. Isolate FLS-5 averaged between log 5.3 and log 6.8 CFU/wound at 20 °C and between log 5.3 and log 6.3 CFU/wound at 2 °C. Colonization by isolates A3-6 and Ae-1 was greater at 20 °C compared to 2 °C on both cultivars tested, ranging between log 3.6 and log 5.0 at 20 °C and between log 2.2 and log 3.8 CFU/wound at 2 °C.

3.5. Cytological Investigation of Disease Suppression

Suppression of bitter rot and blue mold symptoms by application of the bacterial isolates into wounds of cv. Golden Delicious fruit was investigated by SEM. Isolates A3-6 and Ae-1 were chosen for their ability to suppress *C. acutatum*, and isolate A3-2 was chosen for its ability to suppress *P. expansum*. Isolate FO-20 has not shown significant suppression of either pathogen in previous experiments and was thus chosen as a negative control. Observations of fruit inoculated in parallel with SEM observations confirmed isolates A3-6 and Ae-1 inoculated fruit developed lesions that were smaller than the control ($p < 0.05$), and FO-20 and A3-2 had no effect on lesion size caused by *C. acutatum* (Supplemental Figure S1A). Observations of wounded tissue colonized with isolate A3-6 showed attachment of bacterial cells to the hyphae of *C. acutatum* in samples taken at 4 dpi (Figure 4). Closer examination revealed sections of the hyphae where bacterial cells had attached were damaged and collapsed (Figure 4), while sections with no attached bacteria appeared normal. Wound tissue samples inoculated with isolate Ae-1 and challenged with *C. acutatum*, appeared to have higher numbers of bacterial cells covering the wound site at 4 dpi (Figure 4). Closer examination indicated Ae-1 did not attach or cause damage to hyphae of *C. acutatum* (Figure 4).

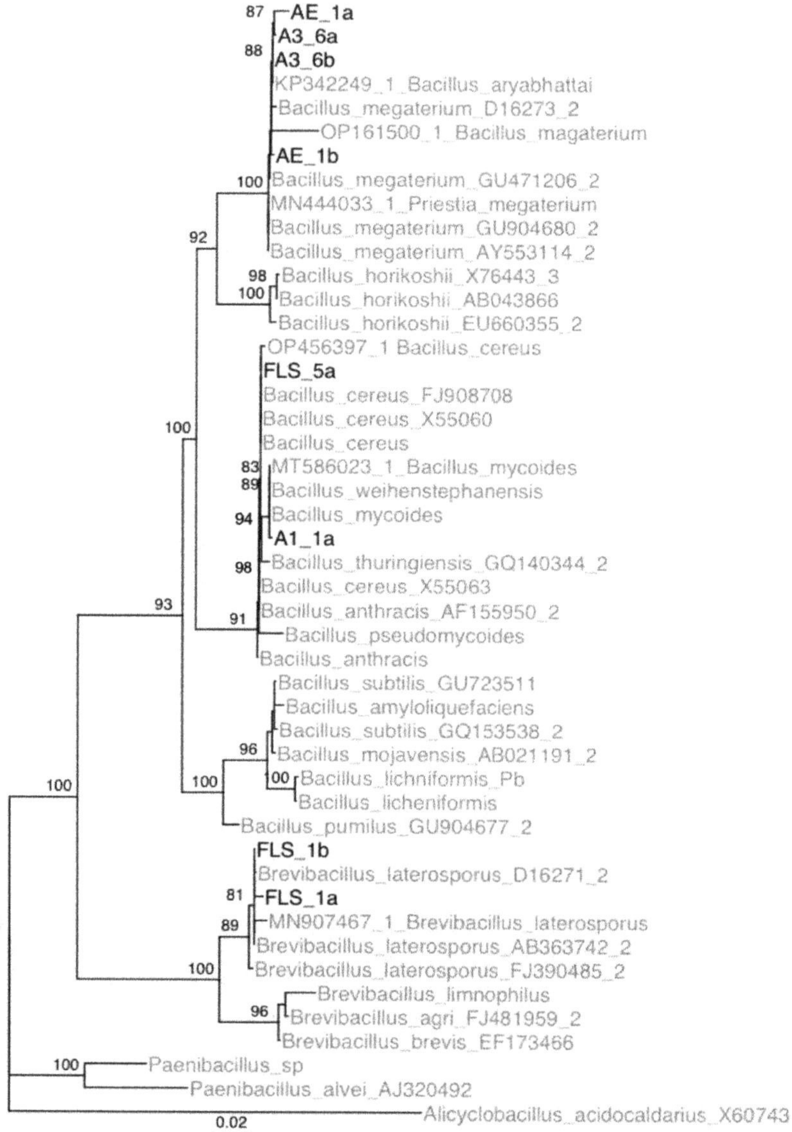

Figure 2. Inferred IQTree ultrafast bootstrap (UFBoot) of the isolates collected from Pennsylvania and type strains of *Bacillus* species based on 16S rDNA sequences. Numbers located at each node are bootstrap values. Note, bootstrap values below 80 are not shown on the tree. Isolates from this study are shown in black, and isolates obtained from GenBank are in gray.

ranging between log 3.6 and log 5.0 at 20 °C and between log 2.2 and log 3.8 CFU/wound at 2 °C.

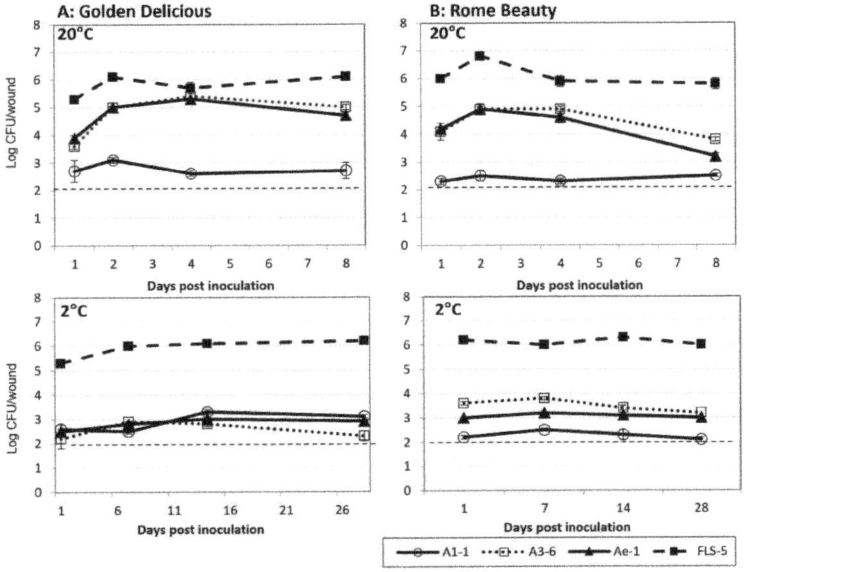

Figure 3. Colonization of fruit wounds by bacterial isolates A1-1, A3-6, Ae-1, and FLS-5 at 20 °C and 2 °C on (**A**) Golden Delicious and (**B**) Rome Beauty. Data are reported as log CFU/wound. Wounds were inoculated with 2 × 10⁴ CFU/wound. Error bars represent the standard error of the mean. The dashed line indicates the minimum detection level of the experiment.

none of the isolates attached or appeared to cause any physical damage to hyphae of *P. expansum* in apple wounds (Figure 5). Conidiophores bearing conidia were only observed in the control wounds.

3.5. Cytological Investigation of Disease Suppression

n of the bacterial
EM. Isolates A3-6
te A3-2 was cho-
n significant sup-
sen as a negative
ns confirmed iso-
than the control
cutatum (Supple-
ate A3-6 showed
n at 4 dpi (Figure
ells had attached
hed bacteria ap-
challenged with
g the wound site
cause damage to
A3-2 had signifi-
1B). Under SEM,

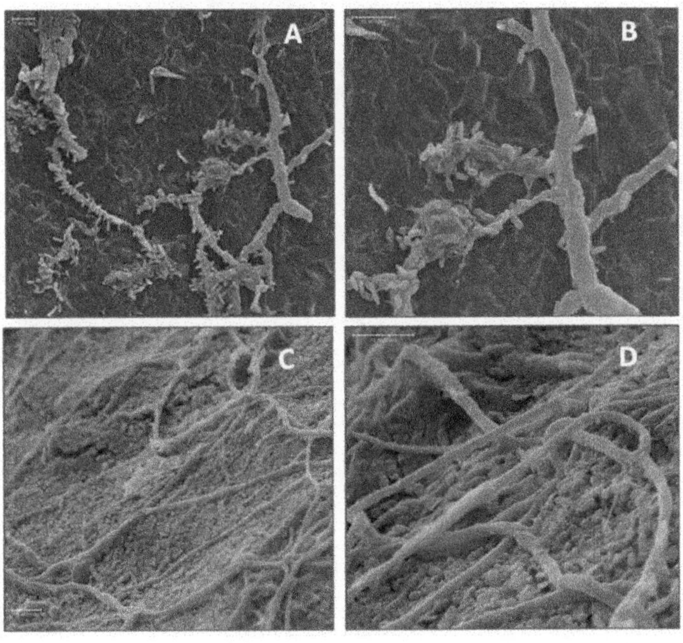

Figure 4. Scanning electron micrographs of biocontrol bacterial cells interacting with hyphae of *C. acutatum* in apple wounds four days after inoculation. (**A**) Attachment of biocontrol isolate A3-6 cells to hyphae of *C. acutatum* at 750×. (**B**) Attachment and degradation of *C. acutatum* hyphae by isolate A3-6 at 1500×. (**C**) Isolate Ae-1 in fruit wounds at 1000× and (**D**) 2000×.

For blue mold, inoculations confirmed fruit treated with isolate A3-2 had significantly smaller lesions compared to the control (supplemental Figure S1B). Under SEM, none of the isolates attached or appeared to cause any physical damage to hyphae of *P. expansum* in apple wounds (Figure 5). Conidiophores bearing conidia were only observed in water control wounds.

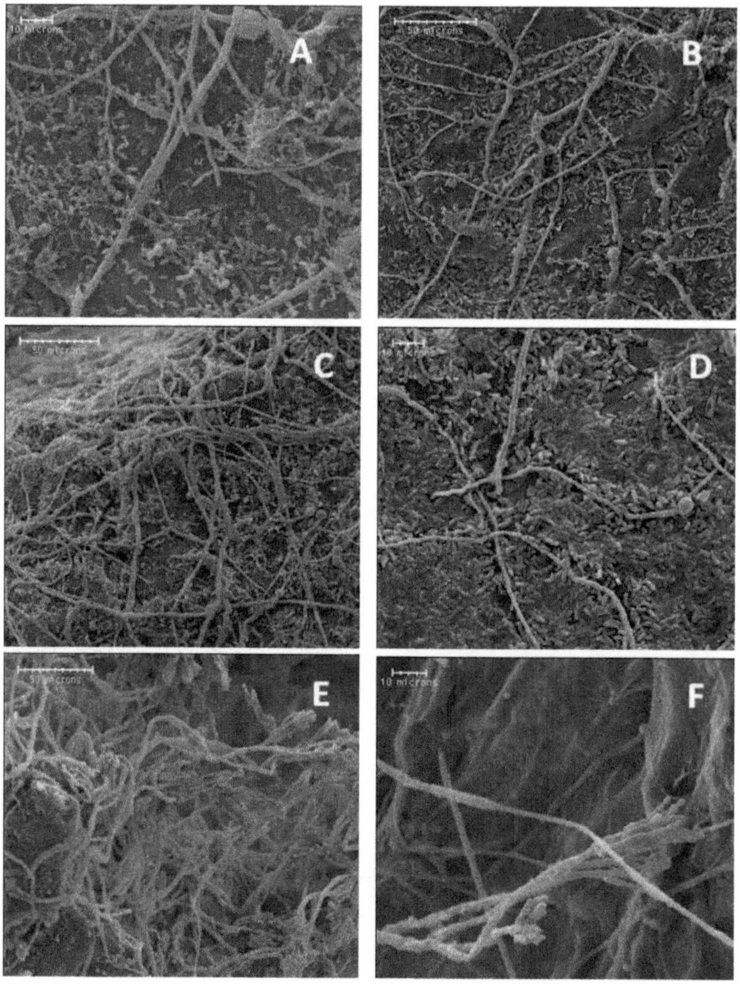

Figure 5. Scanning electron micrographs of biocontrol bacterial cells interacting with hyphae of *P. expansum* in apple wounds four days after inoculation. (**A**,**B**) Isolate A3-6 at 500× and 1000×, (**C**,**D**) isolate A3-2 at 500× and 1000×, and (**E**,**F**) no-bacteria control at 500× and 1000×.

4. Discussion

Several bacterial isolates in this study exhibited biocontrol potential in vitro and reduced severity of bitter rot and blue mold when applied to wounded apple fruit postharvest. Isolates A3-6 and Ae-1 resulted in the greatest suppression of bitter rot lesion area. Analyses indicate isolates Ae-1 and A3-6 were most similar to *B. megaterium*. The species *B. megaterium* has been shown to have biocontrol activity, including reduction in postharvest decay of peanut kernels caused by *Aspergillus flavus* [30] and *Septoria tritici* blotch of wheat [31]. Interestingly, *B. megaterium* isolates have also shown potential to inhibit growth of the human pathogen, *Listeria monocytogenes* [32]. In our previous research, isolate A3-6 also suppressed apple scab on leaves and fruit and bitter rot on apple fruit in storage [24]. While fruit treated with *B. megaterium* isolates A3-6 and Ae-1 had the smallest bitter rot lesions in this study, wound colonization levels were typically log 2 lower than *B. laterosporus* isolate FLS-5 at 2 °C and log 1 lower at 20 °C. These observations suggest the mechanism of bitter rot disease suppression by isolates A3-6 and Ae-1 may not involve competition for space or nutrients. This hypothesis was substantiated for isolate A3-6 in the SEM experiment in which hyphal tissues of *C. acutatum* had collapsed where the bacterial isolate had attached to the hyphae suggesting parasitism as a mode of action. In contrast, isolate Ae-1, which also suppressed bitter rot lesion size on fruit, did not appear to attach or damage hyphal tissues of *C. acutatum*. These observations suggest the mode of action for isolate Ae-1 is different from isolate A3-6 even though both isolates reduced fungal growth in the plate assays and did not produce a clearing zone on chitin-amended media, indicating the isolates do not produce chitinase. Bertagnolli et al. [33] identified several extracellular enzymes produced by *B. megaterium* strain B153-2-2, including protease and pectin lyase, capable of reducing growth of *Rhizoctonia solani* in vitro. In agreement with our study, Bertagnolli also indicated a lack of detectable chitinase activity by *B. megaterium*. Additional experimentation is needed to further elucidate the mechanism of disease suppression by *B. megaterium* isolate A3-6 and Ae-1, including characterization of secondary metabolites and their potential role in suppression of pathogen growth.

In this study, isolate *B. mycoides* A1-1 and *B. cereus* FLS-5 were positive for chitin hydrolysis and reduced growth of *C. acutatum* in vitro. In our previous research, A1-1 and FLS-5 significantly reduced apple scab severity on leaves and fruit and bitter rot in storage [24]. However, in this study isolate A1-1 did not consistently suppress bitter rot symptoms, and isolate FLS-5 was not effective in any in vivo experiments despite maintaining the highest population level in fruit wounds. Analyses indicate isolate A1-1 most closely resembles *B. mycoides* and *B. weihenstephanensis*. Mycoidal growth of isolate A1-1 on nutrient media allowed us to place this isolate in the sub-group *B. mycoides* as *B. weihenstepanesis* has a non-mycoidal colony morphology [34]. Phylogenetic analyses indicate isolate FLS-5 belongs to the *B. cereus* group. *B. cereus* is found in diverse habitats, is ubiquitous in soil [35], and can have biocontrol activity. However, some members of this group can cause two types of food poisoning, a diarrheal type and an emetic type, which are caused by different toxins [36]. Some *B. cereus* strains however, lack enterotoxin genes [37], and several have demonstrated biocontrol activity on several crops [21,38,39] and many are chitinolytic [38,40].

Isolate FLS-1 was positive for chitin hydrolysis, reduced growth of *C. acutatum* in vitro, and suppressed bitter rot symptoms in preliminary screenings but did not suppress symptoms in follow up experiments. Isolate FLS-1 was identified as *Brevibacillus laterosporus*, an aerobic spore-forming bacterium that has documented biocidal activity towards insects [41].

Several isolates in this study reduced pathogen growth in vitro but were not effective in vivo (Table 1). Similar observations have been reported in the literature as in vitro assays; while quick and inexpensive, are generally poor predictors of field performance [42] in part because of the artificial growth conditions compared to in vivo studies [19,43]. For example, Burr et al. [44] found no correlation between in vitro antibiosis and the ability of bacteria and yeasts to suppress apple scab on seedlings. Additionally, the selection of organisms solely based on in vitro tests (also known as dual-culture assays) may become problematic as potentially useful organisms, whose mechanism of disease suppression as anything other than antibiosis or direct parasitism will be overlooked. For example, the use of in vitro assays for screening may fail to identify inducers of induced systemic resistance (ISR) and systemic acquired resistance (SAR) [19,45].

In this study, we utilized chitin hydrolysis as a BCA-screening strategy. Many studies have focused on selecting and testing antagonistic bacteria, specifically those with chitinolytic properties for use as BCAs [21,40,46]. Assays to screen isolates for chitin hydrolysis have relied on conventional plate methods in which colloidal chitin is incorporated into an agar base [22]. Several species of the genus *Bacillus* have been shown to produce chitinase enzymes, including *B. circulans* [47], *B. licheniformis* [48], *B. macerans* [49], and *B. cereus* [38]. Because chitin is a major constituent of fungal cell walls, bacterial chitinase enzymes are thought to be useful for biocontrol of fungal plant pathogens [50], and several chitinolytic microorganisms have demonstrated biocontrol activity against fungal plant pathogens [40,51,52]. In this study, many of the isolates that were positive for chitinase production also demonstrated biocontrol potential (either via inhibition of pathogen growth or suppression of symptoms), although chitinase production was not necessary for biocontrol activity.

This research demonstrates apple orchards can be a source of new BCAs for control of postharvest diseases of apple. Furthermore, abandoned orchards that have returned to their "wild state" may serve as a resource to identify novel BCAs or consortia of biocontrol microbes. Of the top seventeen isolates exhibiting biocontrol potential in vitro or in vivo, twelve (70%) were collected from abandoned orchards or natural (unmanaged) locations. We also noticed, anecdotally, abandoned orchards were replete with healthy trees. These observations suggest abandoned orchards may contain unique microbial communities that are important for plant health, but further research is needed to understand these interactions. It is known that agricultural management and the persistent use of chemical pesticides and fertilizers in agroecosystems can alter plant-associated microbial communities [20,53,54]. There are limited examples however, of research comparing microbial community composition in abandoned and managed systems. Several studies have utilized abandoned agricultural land to study the effect of management practices on insect pests and their natural predators. For example, Altieri and Schmidt [55] found a negative correlation between plant-beneficial predator insect abundance and intensity of management practices. The rate of predation by beneficial species was highest in the abandoned orchard and the organic orchard planted with a cover crop. Furthermore, natural enemies tended to establish long term in organic orchards compared to conventional orchards. Taken together, this literature provides evidence that abandoned agroecosystems may harbor disease suppressive consortia of microbes. Further research is needed to test these hypotheses and identify implications for biocontrol in tree fruit production.

5. Conclusions

In conclusion, several isolates were identified in this research that suppress growth of *C. acutatum*, *P. expansum*, and *V. inaequalis* in vitro and reduced lesion size on apple fruit caused by *C. acutatum* and *P. expansum*. Future studies are needed to characterize these isolates, determine application rates, and determine compatibility with orchard IPM practices. This research also demonstrates abandoned apple orchards can be a source of new BCAs for suppression of postharvest diseases of apple.

Supplementary Materials: The following supporting information can be downloaded at: https://www.mdpi.com/article/10.3390/microorganisms11010081/s1, Figure S1: Mean lesion area on wounded Golden Delicious fruit.

Author Contributions: Conceptualization, A.P. and P.B.; methodology, A.P.; software, A.P. and H.N.; validation, A.P., P.B. and H.N.; formal analysis, A.P. and H.N.; investigation, A.P.; resources, P.B.; data curation, A.P. and H.N.; writing—original draft preparation, A.P.; writing—review and editing, A.P., P.B. and H.N.; visualization, A.P. and H.N.; supervision, P.B.; project administration, P.B.; funding acquisition, P.B. All authors have read and agreed to the published version of the manuscript.

Funding: This project was supported by the Penn State Department of Plant Pathology and Environmental Microbiology, the United States Agency for International Development (USAID) SANREM CRSP, and the generous support of the American people through Cooperative Agreement No. EPP-A-00-04-00013-00.

Data Availability Statement: Data available upon request. Sequence data of identified isolates were deposited to GenBank under accession IDs OP596281-88.

Acknowledgments: We would like to thank the faculty and staff at the PSU FREC, including Jim Travis, Noemi Halbrendt, and Brian Lehman for their assistance and advice. We also thank Liza DeGenring and Barry Rock for reading the drafts of this manuscript.

Conflicts of Interest: The authors declare no conflict of interest. The funders had no role in the design of the study; in the collection, analyses, or interpretation of data; in the writing of the manuscript; or in the decision to publish the results.

References

1. Sharma, R.R.; Singh, D.; Singh, R. Biological control of postharvest diseases of fruits and vegetables by microbial antagonists: A review. *Biol. Control* **2009**, *50*, 205–221. [CrossRef]
2. Droby, S.; Wisniewski, M.; Macarisin, D.; Wilson, C. Twenty years of postharvest biocontrol research: Is it time for a new paradigm? *Postharvest Biol. Technol.* **2009**, *52*, 137–145. [CrossRef]
3. Wisniewski, M.; Droby, S.; Norelli, J.; Liu, J.; Schena, L. Alternative management technologies for postharvest disease control: The journey from simplicity to complexity. *Postharvest Biol. Technol.* **2016**, *122*, 3–10. [CrossRef]
4. Li, H.X.; Xiao, C.L. Characterization of Fludioxonil-Resistant and Pyrimethanil-Resistant Phenotypes of *Penicillium expansum* from Apple. *Phytopathology* **2008**, *98*, 427–435. [CrossRef]
5. Zhao, H.; Kim, Y.; Huang, L.; Xiao, C. Resistance to thiabendazole and baseline sensitivity to fludioxonil and pyrimethanil in *Botrytis cinerea* populations from apple and pear in Washington State. *Postharvest Biol. Technol.* **2010**, *56*, 12–18. [CrossRef]
6. Kim, K.; Kabir, E.; Jahan, S.A. Exposure to pesticides and the associated human health effects. *Sci. Total Environ.* **2017**, *575*, 525–535. [CrossRef]
7. Ragsdale, N.N.; Sisler, H.D. Social and Political Implications of Managing Plant Diseases with Decreased Availability Of Fungicides In The United States. *Annu. Rev. Phytopathol.* **1994**, *32*, 545–557. [CrossRef]
8. Gullino, M.L.; Kuijpers, L.A.M. Social and Political Implications of Managing Plant Diseases with Restricted Fungicides In Europe. *Annu. Rev. Phytopathol.* **1994**, *32*, 559–581. [CrossRef]
9. Droby, S.; Wisniewski, M.; Teixidó, N.; Spadaro, D.; Jijakli, M.H. The science, development, and commercialization of postharvest biocontrol products. *Postharvest Biol. Technol.* **2016**, *122*, 22–29. [CrossRef]
10. Janisiewicz, W.J.; Korsten, L. Biological Control of Postharvest Diseases of Fruits. *Annu. Rev. Phytopathol.* **2002**, *40*, 411–441. [CrossRef]
11. Spadaro, D.; Gullino, M.L. State of the art and future prospects of the biological control of postharvest fruit diseases. *Int. J. Food Microbiol.* **2004**, *91*, 185–194. [CrossRef] [PubMed]
12. Wilson, C.L. Potential for Biological Control of Postharvest Plant Diseases. *Plant Dis.* **1985**, *69*, 375. [CrossRef]
13. Wisniewski, M.E.; Wilson, C.L. Biological Control of Postharvest Diseases of Fruits and Vegetables: Recent Advances. *HortScience* **1992**, *27*, 94–98. [CrossRef]
14. Pusey, P.L.; Wilson, C.L. Postharvest Biological Control of Stone Fruit Brown Rot by *Bacillus subtilis*. *Plant Dis.* **1984**, *68*, 753–756. [CrossRef]
15. National Pesticide Information Center. NPIC Product Research Online Database. Available online: http://npic.orst.edu/ (accessed on 10 November 2022).
16. Droby, S.; Wisniewski, M. The fruit microbiome: A new frontier for postharvest biocontrol and postharvest biology. *Postharvest Biol. Technol.* **2018**, *140*, 107–112. [CrossRef]
17. Jacobsen, B.J.; Zidack, N.K.; Larson, B.J. The Role of *Bacillus*-Based Biological Control Agents in Integrated Pest Management Systems: Plant Diseases. *Phytopathology* **2004**, *94*, 1272–1275. [CrossRef]
18. Legein, M.; Smets, W.; Vandenheuvel, D.; Eilers, T.; Muyshondt, B.; Prinsen, E.; Samson, R.; Lebeer, S. Modes of Action of Microbial Biocontrol in the Phyllosphere. *Front. Microbiol.* **2020**, *11*, 1619. [CrossRef]
19. Raymaekers, K.; Ponet, L.; Holtappels, D.; Berckmans, B.; Cammue, B.P. Screening for novel biocontrol agents applicable in plant disease management—A review. *Biol. Control* **2020**, *144*, 104240. [CrossRef]
20. Aranda, S.; Montes-Borrego, M.; Jiménez-Díaz, R.M.; Landa, B.B. Microbial Communities Associated with the Root System of Wild Olives (*Olea Europaea* L. Subsp. *Europaea* Var. *Sylvestris*) Are Good Reservoirs of Bacteria with Antagonistic Potential against *Verticillium dahliae*. *Plant Soil* **2011**, *343*, 329–345. [CrossRef]
21. Kokalis-Burelle, N.; Backman, P.A.; Rodríguez-Kábana, R.; Ploper, L.D. Potential for biological control of early leafspot of peanut using *Bacillus cereus* and chitin as foliar amendments. *Biol. Control* **1992**, *2*, 321–328. [CrossRef]
22. O'Brien, M.; Colwell, R.R. A rapid test for chitinase activity that uses 4-methylumbelliferyl-N-acetyl-beta-D-glucosaminide. *Appl. Environ. Microbiol.* **1987**, *53*, 1718–1720. [CrossRef] [PubMed]
23. Sitther, V.; Haro, P.A.G.; Molineros, J.E.; Garzon, C.D.; Jiménez-Gasco, M.M. Genetic diversity of apple- and crabapple-infecting isolates of *Venturia inaequalis* in Pennsylvania, the United States, determined by microsatellite markers. *For. Pathol.* **2018**, *48*, e12405. [CrossRef]
24. Poleatewich, A.M.; Ngugi, H.K.; Backman, P.A. Assessment of Application Timing of *Bacillus* spp. to Suppress Pre- and Postharvest Diseases of Apple. *Plant Dis.* **2012**, *96*, 211–220. [CrossRef]

25. Poleatewich, A.M.; Travis, J.W.; Backman, P.A. Evaluation of Endospore-Forming Bacteria for Control of Bitter Rot. *Plant Dis. Manag. Rep.* **2008**, *2*, PF056. [CrossRef]
26. Poleatewich, A.M.; Travis, J.W.; Backman, P.A. Evaluation of Preharvest and Postharvest Applications of Endospore-Forming Bacteria for Control of Bitter Rot, 2009. *Plant Dis. Manag. Rep.* **2010**, *4*, PF015. [CrossRef]
27. Katoh, K.; Standley, D.M. MAFFT Multiple Sequence Alignment Software Version 7: Improvements in Performance and Usability. *Mol. Biol. Evol.* **2013**, *30*, 772–780. [CrossRef]
28. Capella-Gutiérrez, S.; Silla-Martínez, J.M.; Gabaldón, T. trimAl: A tool for automated alignment trimming in large-scale phylogenetic analyses. *Bioinformatics* **2009**, *25*, 1972–1973. [CrossRef]
29. Hoang, D.T.; Chernomor, O.; Von Haeseler, A.; Minh, B.Q.; Vinh, L.S. UFBoot2: Improving the Ultrafast Bootstrap Approximation. *Mol. Biol. Evol.* **2018**, *35*, 518–522. [CrossRef]
30. Kong, Q.; Shan, S.; Liu, Q.; Wang, X.; Yu, F. Biocontrol of *Aspergillus flavus* on peanut kernels by use of a strain of marine *Bacillus megaterium*. *Int. J. Food Microbiol.* **2010**, *139*, 31–35. [CrossRef]
31. Kildea, S.; Ransbotyn, V.; Khan, M.R.; Fagan, B.; Leonard, G.; Mullins, E.; Doohan, F.M. *Bacillus megaterium* shows potential for the biocontrol of *septoria tritici* blotch of wheat. *Biol. Control* **2008**, *47*, 37–45. [CrossRef]
32. Sharma, R.; Gal, L.; Garmyn, D.; Bisaria, V.S.; Sharma, S.; Piveteau, P. Evidence of Biocontrol Activity of Bioinoculants Against a Human Pathogen, *Listeria monocytogenes*. *Front. Microbiol.* **2020**, *11*, 350. [CrossRef] [PubMed]
33. Bertagnolli, B.; Soglio, F.D.; Sinclair, J. Extracellular enzyme profiles of the fungal pathogen *Rhizoctonia solani* isolate 2B-12 and of two antagonists, *Bacillus megaterium* strain B153-2-2 and *Trichoderma harzianum* isolate Th008. I. Possible correlations with inhibition of growth and biocontrol. *Physiol. Mol. Plant Pathol.* **1996**, *48*, 145–160. [CrossRef]
34. Lechner, S.; Mayr, R.; Francis, K.P.; Pruss, B.M.; Kaplan, T.; Wiessner-Gunkel, E.; Stewart, G.S.A.B.; Scherer, S. *Bacillus weihenstephanensis* sp. nov. is a new psychrotolerant species of the *Bacillus cereus* group. *Int. J. Syst. Evol. Microbiol.* **1998**, *48*, 1373–1382. [CrossRef] [PubMed]
35. Jensen, G.B.; Hansen, B.M.; Eilenberg, J.; Mahillon, J. The hidden lifestyles of *Bacillus cereus* and relatives. *Environ. Microbiol.* **2003**, *5*, 631–640. [CrossRef]
36. Stenfors Arnesen, L.P.; Fagerlund, A.; Granum, P.E. From soil to gut: *Bacillus cereus* and its food poisoning toxins. *FEMS Microbiol. Rev.* **2008**, *32*, 579–606. [CrossRef]
37. Melnick, R.; Testen, A.; Poleatewich, A.; Backman, P.; Bailey, B. Detection and expression of enterotoxin genes in endophytic strains of *Bacillus cereus*. *Lett. Appl. Microbiol.* **2012**, *54*, 468–474. [CrossRef]
38. Pleban, S.; Chernin, L.; Chet, I. Chitinolytic activity of an endophytic strain of *Bacillus cereus*. *Lett. Appl. Microbiol.* **1997**, *25*, 284–288. [CrossRef]
39. Silva, H.S.A.; Romeiro, R.S.; Filho, R.C.; Pereira, J.L.A.; Mizubuti, E.S.G.; Mounteer, A. Induction of Systemic Resistance by *Bacillus cereus* Against Tomato Foliar Diseases Under Field Conditions. *J. Phytopathol.* **2004**, *152*, 371–375. [CrossRef]
40. Huang, C.-J.; Wang, T.-K.; Chung, S.-C.; Chen, C.-Y. Identification of an Antifungal Chitinase from a Potential Biocontrol Agent, *Bacillus cereus* 28-9. *BMB Rep.* **2005**, *38*, 82–88. [CrossRef]
41. de Oliveira, E.J.; Rabinovitch, L.; Monnerat, R.G.; Passos, L.K.J.; Zahner, V. Molecular Characterization of *Brevibacillus laterosporus* and Its Potential Use in Biological Control. *Appl. Environ. Microbiol.* **2004**, *70*, 6657–6664. [CrossRef]
42. Andrews, J.H. Biological Control in the Phyllosphere. *Annu. Rev. Phytopathol.* **1992**, *30*, 603–635. [CrossRef] [PubMed]
43. Parnell, J.J.; Berka, R.; Young, H.A.; Sturino, J.M.; Kang, Y.; Barnhart, D.M.; Dileo, M.V. From the Lab to the Farm: An Industrial Perspective of Plant Beneficial Microorganisms. *Front. Plant Sci.* **2016**, *7*, 1110. [CrossRef] [PubMed]
44. Burr, T.; Matteson, M.; Smith, C.; Corral-Garcia, M.; Huang, T.-C. Effectiveness of Bacteria and Yeasts from Apple Orchards as Biological Control Agents of Apple Scab. *Biol. Control* **1996**, *6*, 151–157. [CrossRef]
45. Bargabus, R.; Zidack, N.; Sherwood, J.; Jacobsen, B. Characterization of systemic resistance in sugar beet elicited by a non-pathogenic, phyllosphere-colonizing *Bacillus mycoides*, biological control agent. *Physiol. Mol. Plant Pathol.* **2002**, *61*, 289–298. [CrossRef]
46. Kavino, M.; Harish, S.; Kumar, N.; Saravanakumar, D.; Samiyappan, R. Effect of chitinolytic PGPR on growth, yield and physiological attributes of banana (Musa spp.) under field conditions. *Appl. Soil Ecol.* **2010**, *45*, 71–77. [CrossRef]
47. Watanabe, T.; Oyanagi, W.; Suzuki, K.; Tanaka, H. Chitinase system of *Bacillus circulans* WL-12 and importance of chitinase A1 in chitin degradation. *J. Bacteriol.* **1990**, *172*, 4017–4022. [CrossRef]
48. Takayanagi, T.; Ajisaka, K.; Takiguchi, Y.; Shimahara, K. Isolation and characterization of thermostable chitinases from *Bacillus licheniformis* X-7u. *Biochim. et Biophys. Acta (BBA)-Protein Struct. Mol. Enzym.* **1991**, *1078*, 404–410. [CrossRef]
49. Cody, R.M. Distribution of chitinase and chitobiase in bacillus. *Curr. Microbiol.* **1989**, *19*, 201–205. [CrossRef]
50. Senol, M.; Nadaroglu, H.; Dikbas, N.; Kotan, R. Purification of Chitinase enzymes from *Bacillus subtilis* bacteria TV-125, investigation of kinetic properties and antifungal activity against *Fusarium culmorum*. *Ann. Clin. Microbiol. Antimicrob.* **2014**, *13*, 35. [CrossRef]
51. Kobayashi, D.Y.; Reedy, R.M.; Bick, J.; Oudemans, P.V. Characterization of a Chitinase Gene from *Stenotrophomonas maltophilia* Strain 34S1 and Its Involvement in Biological Control. *Appl. Environ. Microbiol.* **2002**, *68*, 1047–1054. [CrossRef]
52. Kobayashi, D.Y.; Palumbo, J.D. Bacterial Endophytes and Their Effects on Plants and Uses in Agriculture. In *Microbial Endophytes*; CRC Press: Boca Raton, FL, USA, 2000; pp. 213–250. [CrossRef]

53. Glenn, D.; Bassett, C.; Dowd, S. Effect of pest management system on 'Empire' apple leaf phyllosphere populations. *Sci. Hortic.* **2015**, *183*, 58–65. [CrossRef]
54. Walter, M.; Frampton, C.M.; Boyd-Wilson, K.S.H.; Harris-Virgin, P.; Waipara, N.W. Agrichemical impact on growth and survival of non-target apple phyllosphere microorganisms. *Can. J. Microbiol.* **2007**, *53*, 45–55. [CrossRef] [PubMed]
55. Altieri, M. The dynamics of colonizing arthropod communities at the interface of abandoned, organic and commercial apple orchards and adjacent woodland habitats. *Agric. Ecosyst. Environ.* **1986**, *16*, 29–43. [CrossRef]

microorganisms

Article

Effects of Rhizosphere Microbial Communities on Cucumber *Fusarium* wilt Disease Suppression

Fan Yang [1,†], Huayan Jiang [1,2,†], Gaozheng Chang [1], Shen Liang [1], Kai Ma [1], Yuxin Cai [1], Baoming Tian [1,2,*] and Xuanjie Shi [1,*]

1 Institute of Horticulture, Henan Academy of Agricultural Sciences, Graduate T&R Base of Zhengzhou University, Zhengzhou 450002, China; xiaoyuefuxiang@163.com (F.Y.); jhy7327@163.com (H.J.)
2 School of Agricultural Sciences, Zhengzhou University, Zhengzhou 450001, China
* Correspondence: tianbm@zzu.edu.cn (B.T.); shixuanjie@hnagri.org.cn (X.S.)
† These authors have contributed equally to this work.

Abstract: Cucumber *Fusarium* wilt is a worldwide soil-borne disease that seriously restricts the yield and quality of cucumber. The rhizosphere soil microbiome, as the first line of defense against pathogens invading plant roots, plays a key role in rhizosphere immune formation and function. The purpose of this study was to reveal the key microecological factors and dominant microbial flora affecting cucumber resistance and susceptibility to *Fusarium* wilt by analyzing the physical and chemical properties and microbial flora of rhizosphere soil with different degrees of susceptibility and resistance to cucumber *Fusarium* wilt, thereby laying a foundation to establish cucumber resistance to the *Fusarium* wilt rhizosphere core microbiome. Firstly, Illumina Miseq sequencing technology was used to evaluate the physical and chemical properties and microbial groups of cucumber rhizosphere soil at different health levels, and the key environmental factors and microbial factors related to cucumber *Fusarium* wilt were screened out. Subsequently, PICRUSt2 and FUNGuild were used to predict the functions of rhizosphere bacteria and fungi. Combined with functional analysis, the possible interactions among soil physical and chemical properties, cucumber rhizosphere microorganisms, and *Fusarium* wilt were summarized. The results showed that the available potassium content in the rhizosphere soil of healthy cucumber decreased by 10.37% and 0.56%, respectively, compared with the rhizosphere soil of severely susceptible cucumber and mildly susceptible cucumber. Exchangeable calcium content increased by 25.55% and 5.39%; the α diversity Chao1 index of bacteria and fungi in the rhizosphere soil of healthy cucumber was significantly lower than that in the rhizosphere soil of seriously infected cucumber, and the MBC content of its physical and chemical properties was also significantly lower than that in the rhizosphere soil of seriously infected cucumber. There was no significant difference in the Shannon and Simpson diversity indexes between healthy cucumber rhizosphere soil and seriously infected cucumber rhizosphere soil. The results of the β diversity analysis showed that the bacterial and fungal community structure of healthy cucumber rhizosphere soil was significantly different from that of severely and mildly infected cucumber rhizosphere soil. At the genus level, through statistical analysis, LEfSe analysis, and RDA analysis, the key bacterial and fungal genera with potential biomarker values were screened out as *SHA_26*, *Subgroup_22*, *MND1*, *Aeromicrobium*, *TM7a*, *Pseudorhodoplanes*, *Kocuria*, *Chaetomium*, *Fusarium*, *Olpidium*, and *Scopulariopsis*, respectively. The bacteria *SHA_26*, *Subgroup_22*, and *MND1* related to cucumber *Fusarium* wilt inhibition belong to Chloroflexi, Acidobacteriota, and Proteobacteria, respectively. *Chaetomiacea* belongs to Sordariomycates. The results of functional prediction showed that changes to the KEGG (Kyoto Encyclopedia of Genes and Genomes) pathway in the bacterial microbiota were concentrated in tetracycline biosynthesis, selenocompound metabolism, lipopolysaccharide biosynthesis, etc., which were mainly involved in the metabolism of terpenoids and polyketides, energy metabolism, metabolism of other amino acids, glycan biosynthesis and metabolism, lipid metabolism, cell growth and death, transcription, metabolism of cofactors and vitamins, and biosynthesis of other secondary metabolites. The difference in fungi was mainly dung saprotroph–ectomycorrhizal–soil saprotroph–wood saprotroph. Through the correlation analysis and functional predictions of the key environmental factors, microbial flora, and cucumber health index in cucumber rhizosphere soil, we

Citation: Yang, F.; Jiang, H.; Chang, G.; Liang, S.; Ma, K.; Cai, Y.; Tian, B.; Shi, X. Effects of Rhizosphere Microbial Communities on Cucumber *Fusarium* wilt Disease Suppression. *Microorganisms* **2023**, *11*, 1576. https://doi.org/10.3390/microorganisms11061576

Academic Editor: Elisa Gamalero

Received: 28 March 2023
Revised: 1 June 2023
Accepted: 12 June 2023
Published: 14 June 2023

determined that the inhibition of cucumber *Fusarium* wilt was a synergistic effect of environmental factors and microbial flora, and a model diagram was drawn to briefly explain its mechanism. This work will provide a basis for the biological control of cucumber *Fusarium* wilt in the future.

Keywords: cucumber *Fusarium* wilt; *Fusarium oxysporum* f. sp. *cucumerinum*; high-throughput sequencing; rhizosphere microbial community

1. Introduction

Cucumber *Fusarium* wilt is a worldwide soil-borne disease that occurs in both facilities and open fields, seriously restricting cucumber production and quality. Cucumber *Fusarium* wilt is extremely destructive. It was reported that the disease may be aggravated by climate warming [1]. Cucumber *Fusarium* wilt is caused by *Fusarium oxysporum* f. sp. *cucumerinum* (FOC) [2]. This pathogen can produce chlamydospores, which are difficult to remove once they colonize the soil [3], enabling this pathogen to survive in the soil for many years. This disease has become an important factor restricting the development of modern agriculture. Chemical control is the most economical and effective control method at present, but excessive use of chemicals will seriously disturb rhizosphere microorganisms, make the ecological environment system more fragile, and lead to loss of agricultural biodiversity, which will yield environmental pollution, endanger human and animal health, and also cause "3R" problems such as pathogen resistance, pesticide residue, and pest resurgence. Therefore, the search for new efficient biological pesticides to replace chemical pesticides has become a current focus area of research in the field.

The prevention and control of cucumber *Fusarium* wilt from the perspective of soil microecological balance should be the future development direction. Existing studies have shown that as a core area for plant–soil–microorganism interactions, the rhizosphere is the only vector for soil-borne pathogens to invade plant roots, and rhizosphere microorganisms are considered to be the first line of defense against pathogen invasion [4–6]. Rhizosphere microorganisms are, in general, all microorganisms that are tightly attached to a few millimeters of soil on the surfaces of plant roots. These microorganisms are mainly composed of bacteria and represent one of the most active areas of microorganisms in the soil system [7]. To a certain extent, the relationship between rhizosphere microorganisms determines whether pathogenic bacteria can successfully invade plants [8]. KWAK et al. [9] found that compared to susceptible tomato varieties, resistant varieties could obviously recruit *Flavobacterium* to enrich the rhizosphere and used metagenomic sequencing and selective medium separation techniques to cultivate *Flavobacterium TRM1*. The strain was inoculated into the root of susceptible tomato, which significantly improved the resistance of the tomato to *Fusarium* wilt. By comparing the physical and chemical properties and microbial communities in the rhizosphere soil of healthy watermelon, watermelon infected with *Fusarium oxysporum*, and dead watermelon, Meng et al. [10] found that rhizosphere soil NH^{4+}, NO^{3-}, and hemicellulose can be used as indicators of plant health. At the same time, the relative abundance of potential antagonistic microorganisms in the rhizosphere soil of healthy plants was found to be the highest, while the relative abundance of potential plant pathogens was the lowest. This special microbial composition may be an important factor inhibiting the expression of *Fusarium* wilt disease and impacting plant health. Other studies have shown that the soil with inhibitory effects on strawberry fusarium wilt has beneficial microflora and active secondary metabolism [11]. Shen et al. [12] found that the *Pseudomonas* and *Tumebacillus* genera were enriched in healthy soil. In addition, SHI et al. [13] studied the relationship between the occurrence of potato scab and the composition and function of microbial communities in potato topsoil, speculating that *Stenotrophomonas, Agrobacterium, Sphingobium,* and *Streptomyces* may aggravate the incidence of scab, while *Geobacillus, Curtobacterium,* and *unclassified Geodermatophilaceae* may slow down the incidence of scab. Therefore, it is crucial to determine the resistance of rhizosphere microorganisms to pathogen invasion by analyzing the composition of rhizosphere microorganisms, the degree of correlation between microorganisms, and the types and functions of related antagonistic microorganisms. However, it remains unclear whether there are differences in the composition and diversity of rhizosphere bacteria between resistant and susceptible cucumber varieties and whether there is a correlation between bacterial differences and cucumber resistance to *Fusarium* wilt. With the continuous improvement and development of DNA sequencing technology, it is possible to conduct more comprehensive and accurate research on plant rhizosphere microorganisms. In addition, various algorithms based on 16S rRNA gene sequence assignment can be used to identify biological indicators in disease-suppressed soils.

It is unclear how specific microorganisms in the cucumber rhizosphere soil affect plant health and the development of soil-borne diseases. Therefore, the focus of this research was to detect the wilt-resistant soil of natural cucumbers and study the relationship between its disease inhibition characteristics and rhizosphere soil microorganisms. Our objectives were (1) to study the differences between the soil bacteria and pathogenic bacteria; (2) to identify potential key microbial groups, such as antagonistic microorganisms and pathogenic microorganisms; (3) to analyze the differences in microbial functions between disease-inhibiting soil and pathogenic soil; and (4) determine the microbial indicators related to plant health. This study examined the microbiota of a long-term monocrop cucumber field in Anyang, Henan province, China. Illumina Miseq sequencing technology was used to analyze the rhizosphere microbial community of healthy, mildly infected, and severely infected cucumbers. The key microbial groups involved in the inhibition of cucumber *Fusarium* wilt were identified, and the functional abundance of soil-inhibiting and pathogenic microorganisms was predicted based on gene sequence analysis. To determine the relationship between soil physicochemical properties and key microbial groups, the soil properties were analyzed. Finally, the relationship between major microbial groups and plant health was reviewed. The effect of soil microorganisms on the inhibition of cucumber *Fusarium* wilt pathogen was discussed to provide reference for the prevention and control of cucumber *Fusarium* wilt.

2. Materials and Methods

2.1. Basic Information on Sampling Sites

In May 2022, the incidence of cucumber *Fusarium* wilt was investigated in the cucumber planting area of Jingdian Town, Neihuang County, Anyang City, Henan Province, and areas matching the following conditions were selected as sampling sites: (1) the presence of a cucumber crop for more than 10 years; (2) a high incidence of more than 45% severe infection with cucumber wilt in soil; (3) 10% incidence of continuous crop cucumber infected with wilt in soil; (4) 5 % or lower, and the distance between the three sampling sites was close (less than 10 km in a straight line). According to the above conditions, Jingdian Town, Neihuang County, Anyang City, Henan Province (112° E, 34° N) was selected as the sampling site. The specific sampling site information is shown in Table 1.

Table 1. Basic information on sampling points.

Site	Test Abbreviation (Abbr)	Longitude (E)	Latitude (N)	Altitude (m)	Soil Type
Cucumber *Fusarium* wilt disease serious place	S	112.7611	34.6649	131	loam soil
Cucumber *Fusarium* wilt disease light place	R	112.7613	34.6653	131	loam soil
Places where cucumber wilt does not occur	HR	112.7615	34.6639	131	loam soil

2.2. Rhizosphere Soil Sampling

Samples were collected on 20 May 2022. To obtain cucumber rhizosphere soil, cucumber plants were removed and shaken vigorously. The soil that was tightly bound to the root systems was rinsed with sterile salt solution, and centrifugation of the suspension was performed for 10 min at 10,000× *g*, resulting in rhizosphere soil pellets [14]. Each cucumber plant's rhizosphere soil was mixed as one replicate, resulting in four replicates at each sampling site. The health index (HI) of cucumber was rated as "2" for healthy plants, "1" for lightly affected plants, and "0" for heavily affected plants. After soil samples were collected, the samples were stored in ice packs and transported within 12 h to the laboratory. A portion of the rhizosphere soil samples was stored at −80 °C until DNA extraction, and a portion was immediately used to perform physicochemical analysis.

2.3. Determination of Soil Chemical Properties and Extraction of Soil DNA

To assess soil conditions in the study area, three technical replicates were conducted to determine the physicochemical properties of HR, R, and S rhizosphere soils. Soil chemical properties, including soil pH, soil electrical conductivity (EC), available nitrogen content (AN), available phosphorus content (AP), available potassium content (AK), available iron content (AFe), exchangeable calcium content (ECa), available copper content (ACu), dehydrogenase activity (S-DHA), invertase activity (S-SR), urease activity (S-UR), and microbial biomass carbon (MBC) were determined according to the above method [15]. A DNeasy Power Soil Kit (QIAGEN GmbH, Hilden, Germany) was used for soil genomic DNA extraction. A NanoDrop spectrophotometer (Thermo Scientific, Wilmington, DE, USA) was used to determine the concentration and purity of the extracted DNA, and a 1% agarose gel was used to assess the DNA quality. For subsequent analysis, the isolated DNA was stored at −20 °C.

2.4. Tag Sequencing of Bacterial and Fungal Communities

Sequencing libraries for bacteria and fungi were constructed as described previously [16,17]. The study of bacterial and fungal communities was based on the sequencing of 16S rRNA genes and ribosomal DNA ITS region-end amplicons using the Illumina MiSeq platform of Personal Biotechnology Ltd. (Shanghai, China). The bacterial 16S rRNA gene fragment was amplified using bacterial primers 338F (5′-ACTCCTACGGGAGGCAGCAG-3′) and 806R (5′-GGACTACHVGGGTWTCTAAT-3′) for the V3-V4 hypervariable region [18]. The ITS region was targeted at primers 1737F (5′-GGAAGTAAAAGTCGTAACAAGG-3′) and 2043R (5′-GCTGCGTTCTTCATCGATGC-3′) [19]. PCR reaction conditions were as follows: pre-denaturation at 95 °C for 3 min, 95 °C for 30 s, 55 °C for 30 s, 72 °C 40 s, 35 cycles for fungi or 27 cycles for bacteria, AOA (ammonia-oxidizing archaea) and AOB (ammonia-oxidizing bacteria), and a final extension at 72 °C for 10 min and then at 10 °C until halted by user. In the PCR for the fungal sequences, 5 min were spent at 95 °C, followed by 35 cycles of 45 s at 94 °C, 50 s at 58 °C, 30 s at 68 °C, and a final extension of 10 min at 65 °C. An AxyPrep Gel Extraction Kit (Axygen, Union City, CA, USA) was used to recover PCR products from 2% agarose gels. DNA MiSeq was sequenced commercially using an Illumina MiSeq sequencer. The resulting DNA sequencing data were uploaded to the National Center for Biotechnology Information (Bethesda, MD, USA) Sequence Read Archive (SRA) database under BioProject accession numbers PRJNA921884 and PRJNA921900.

2.5. Bioinformatics and Statistical Analysis of Sequencing Data

Splitting of ~300 bp raw DNA sequences into each sample was based on unique barcodes and the trimming of adapter and primer sequences in QIIME (version 1.2.0) and USEARCH (version 9.1.13) [20]. After quality control, OTUs were clustered based on 97% pairwise identity, and chimeras were filtered using the UPARSE algorithm [21]. Further removal of mitochondrial and nonbacterial OTUs and OTUs with relative abundance below 0.01% allowed us to classify representative sequences for each OTU using the Ribosomal

Database Project classifier (RDP, version 11.5). The affiliations were classified using the RDP bacterial 16S database or the UNITE fungal ITS database [22].

The relative abundance of a given classification group for each sample was calculated by dividing the number of sequences belonging to the group by the total number of sequences. We first calculated the alpha diversity. Using alpha_rarefaction.py, the OTU table was diluted to read up to 80,985 readings. This value was used because it represents the lowest sequencing depth obtained from the sample. In order to calculate beta diversity, the unrefined "filtered_OTUtable" was first normalized with the R package metagenomeSeq (version 1.12). Currently, available sequencing technologies suffer from bias due to uneven sequencing depth, so we use cumulative and scaling (CSS) to avoid this problem [23]. Bray–Curtis and weighted and unweighted Unifrac phase difference matrices were calculated using standardized OTU tables, and principal coordinate analysis (PCoA) was performed using the Phyloseq package (v.1.10) to explore differences in microbial communities [24]. QIIME calculated the weighted UniFrac distance unweighted group averaging (UPGMA) clusters [25,26] to represent beta diversity. UPGMA clustering is a hierarchical clustering method using average links that can be employed to interpret a distance matrix. The relative abundance at the phylum level, family level, and genus level was calculated using OTU based on histograms generated for each taxonomic level using the Origin 9.9.0.225 software to compare the distribution of microorganisms in disease-suppressing and disease-causing soils. Using the Galaxy online interface, linear discriminant analysis (LDA), effect size (LEfSe), and genetic analysis were performed. Genes with an LD score of 3.5 were considered significant biomarkers for each treatment [27]. We also use redundancy analysis to test the relationships between microbial community structure and soil environmental variables with vegan package in the R program. We predicted the potential functions of bacterial communities using PICRUSt2 (systematic genetic Investigation of Communities by Reconstruction of observed States 2) in the KEGG pathway database [28]. According to the previous method [29], the FUNGuild (Fungi Functional Guild) V1.0 [30] online platform was used to classify the ecological functions of fungi. The correlation between soil physicochemical properties, key bacteria and fungi, and plant HI was analyzed using the Origin 9.9.0.225 software. The Krona 2.7.1 software was used to visualize the taxonomic composition of the samples.

2.6. Statistical Analyses

All statistical analyses were performed using IBM SPSS 20.0 (IBM Corporation, New York, NY, USA) and R software (Version 3.5.2). To analyze the effects of *Fusarium* wilt on soil properties and enzyme activity, we performed a one-way analysis of variance (ANOVA) and post hoc multiple comparison test (Tukey HSD). Kruskal–Wallis tests were used to determine the significance of microbial alpha diversity. As a statistical significance test, the Wilcoxon rank sum test was used for a comparison of the two groups. All statistical tests performed in this study were considered significant at $p < 0.05$, while $p < 0.01$ was considered extremely significant.

3. Results

3.1. Comparison of Physical and Chemical Properties of HR, R, and S

Twelve properties of rhizosphere soil were assessed in each group, including EC, AN, AP, AK, AFe, ECa, ACu, pH, S-DHA, S-SR, S-UR, and MBC, to determine the relationship between chemical properties, enzyme activities, and cucumber *Fusarium* wilt incidence. According to the results, HR, R, and S soil enzyme activities changed significantly in both physicochemical and enzyme activity properties (Table S1). The contents of AK, AN, AP, and AFe in S were significantly higher than those in HR and R, which increased by 9.39%, 8.89%, 12.00%, 20.71%, 27.12%, 37.48%, 19.54%, and 29.19%, respectively. However, the contents of ACu and ECa in S significantly decreased by 14.17%, 29.02%, 25.55%, and 21.31%, respectively. The results of soil enzyme activity showed that S-DHA, S-SR, and S-UR in S were significantly higher than those of the other two soils, which indirectly indicated that soil microbial biomass or energy exchange in S was higher than HR and R. Compared with HR and R, the MBC value of S increased by 22.45% and 38.78%, and the EC value increased by 39.91% and 41.28%, respectively, indicating that cucumber *Fusarium* wilt was related to the rhizosphere soil microbial biomass and soil EC value. There was no significant difference in the rhizosphere soil pH of cucumber under three different health conditions ($p > 0.05$), indicating that the occurrence of cucumber *Fusarium* wilt did not affect the rhizosphere soil pH. In conclusion, the results of soil physical and chemical properties showed that soil nutrient content, soil enzyme activity, and microbial community composition were all related to the occurrence of cucumber *Fusarium* wilt.

3.2. *Differences in the Composition of Microbial Communities in HR, R, and S*

An Illumina MiSeq sequencer was used to sequence 16S rRNA gene amplicons and ITS gene amplicons. We plotted the species accumulation curves of the bacterial and fungal community sequencing results to determine whether the sample size of this study was sufficient and estimated community richness. The results showed that when the number of samples exceeded 10, the number of identified new species basically reached the plateau, and the accumulation curve of bacteria tended to be flatter than that of fungi. In this study, 12 samples were sufficient to reflect the species composition of the community. The results also showed that the number of identified bacterial species was much higher than that of fungi, possibly because more bacteria than fungi were identified in the database. In order to study the common and unique species among the three samples, we plotted the Venn diagram. For the bacterial community, 8140 OTUs were generated (Figure 1B). Additionally, 1021 OTUs were generated for the fungal community (Figure 1G) and the α-diversity indices were calculated using QIIME2. The Chao1 index represented the richness of the microbial community, and the Shannon index and Simpson index represented the diversity of the microbial community. In the bacterial community, the Chao1 index of S (3299.79) was higher than that of HR and R (3188.32, 3126.10) (Table S2), but the difference was not significant (Figure 1C). In the comparison of the Shannon diversity index and Simpson diversity index, there was no significant difference in bacterial diversity between S and HR (Figure 1D,E), and the bacterial diversity of the two groups was significantly higher than that of the R group. In the fungal community, there were some differences in the Chao1 index (253.23, 265.34, 229.16), Shannon index (3.33, 3.18, 3.30), and Simpson index (0.79, 0.77, 0.82) of S, R, and HR, but there was no significant difference (Figure 1H–J).

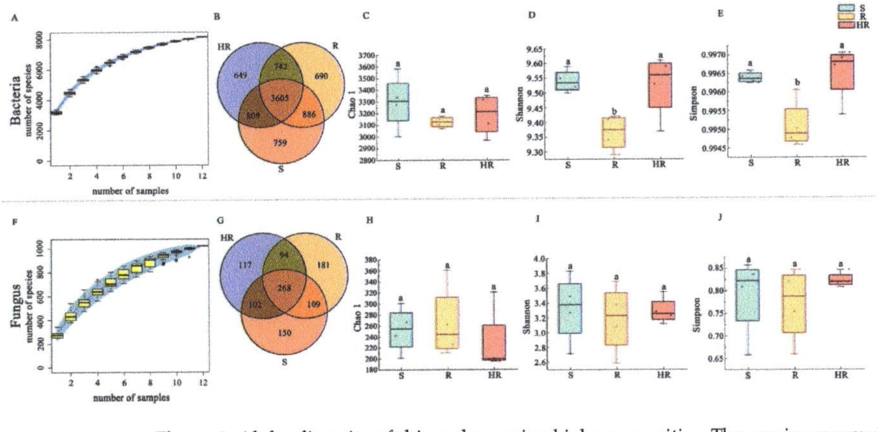

Figure 1. Alpha diversity of rhizosphere microbial communities. The species accumulation curves of bacteria (**A**) and fungi (**F**). Venn diagrams were created for bacteria (**B**) and fungi (**G**) identified in the three groups. Box plots show the variation of Chao1 index (**C**,**H**), Shannon index (**D**,**I**), and Simpson index (**E**,**J**) for bacteria and fungi from the three groups. Different letters within the graph indicate significant differences between the means ($p < 0.05$).

A PCoA analysis was performed using weighted unifrac and unweighted unifrac bacterial and fungal OTUs based on the combined sequences of 16S rRNA genes and ITS fragments in soil samples to visualize the similarities and dissimilarities between microbial communities (Figure 2A,B,D,E). UPGMA cluster analysis based on a sample distance matrix was used to show the similarities between samples (Figure 2C,F). Using weighted unifrac-based PCoA analysis (Figure 2A,D), only the microbial abundance in the sample was considered. The results showed that the bacterial community of HR could be distinguished from R and S, while R and S could not. The fungal abundance in the three sample groups was similar. Using PCoA analysis based on unweighted unifrac analysis (Figure 2B,E), only the microbial structure in the sample was considered, enabling the bacterial and fungal structures among HR, R, and S to be distinguished. This result indicates that there were differences in the bacterial and fungal community structures among HR, R, and S. UPGMA cluster analysis showed that the bacterial and fungal community structures were not similar between HR and R and S (Figure 2C,F). However, the fungal community structures of R and S were similar. The results of the α and β diversity analysis showed that the microbial community structure of HR was different from that of R and S.

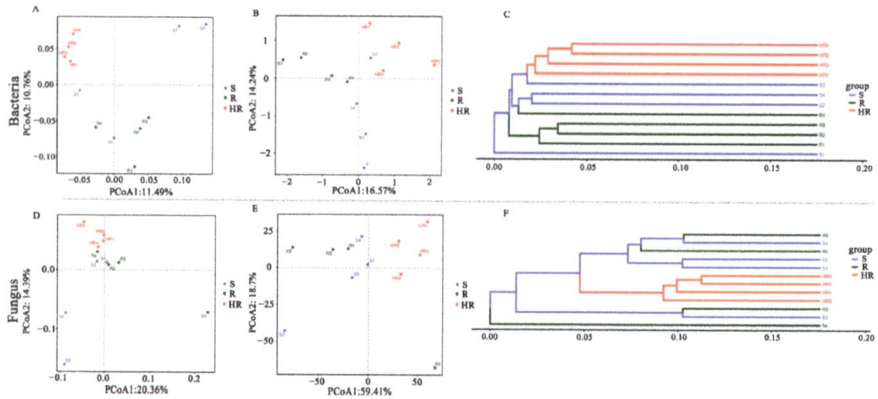

Figure 2. Beta diversity of the rhizosphere microbial community. Beta diversity of bacterial (**A**,**B**) and fungal (**D**,**E**) communities was rep coordinate analysis (PCoA) using weighted and unweighted unifrac. UPGMA clustering analysis based on Bray–Curtis for bacterial (**C**) a ties.

Figure 2. Beta diversity of the rhizosphere microbial community. Beta diversity of bacterial (**A**,**B**) and fungal (**D**,**E**) communities was represented using principal coordinate analysis (PCoA) using weighted and unweighted unifrac. UPGMA clustering analysis based on Bray–Curtis for bacterial (**C**) and fungal (**F**) communities.

3.3. Screening of Key Bacteria and Fungi in HR, R, and S

All bacterial OTUs were assigned to 50 bacterial phyla. Specifically, 13 bacterial phyla had an average relative abundance of more than 1%, and, together, these phyla accounted for more than 94% of the total sequences recovered. In order, the five most abundant bacterial phyla were Proteobacteria (24.96%), Actinobacteriota (18.60%), Chloroflexi (11.36%), Acidobacteriota (8.82%), and Bacteroidota (6.08%). The relative abundance of Actinobacteriota in HR (16.65%) was lower than that of R (19.43%) and S (19.73%), while the relative abundance of Acidobacteriota (10.64%) was higher than that of R (7.68%) and S (8.14%) (Figure 3A). At the order level, the relative abundance of Burkholderiales in HR (6.67%) was higher than that in R (4.84%) and S (5.72%) (Figure 3B). At the family level, the relative abundance of Nitrosomonadacea was higher in HR (5.17%) and R (7.77%) than in S (4.12%), and the relative abundance of Microbacterium was higher in S (3.81%) than in HR (2.47%) and R (3.75%) (Figure 3C). At the genus level, the relative abundance of *MND1* in HR (2.16%) was higher than that in R (1.34%) and S (1.55%) (Figure 3D). The LEfSe analysis method (Figure 4A) revealed 67 significantly different species among S, R, and HR. A total of 23 bacterial species were screened as dominant in S, with 11 bacterial genera in R and 33 bacterial genera in HR. At the genus level, we observed the relative abundance of the *Latescibacterota* of Latescibacterota (LDA = 3.33, $p = 0.024$), the *Aeromicrobium* of Actinobacteria (LDA = 3.09, p = 0.021), the *Dongia* of Proteobacteria (LDA = 2.85, p = 0.018), the *TM7a* of Patescibacteria (LDA = 2.44, p = 0.021), uncultured Proteobacteria (LDA = 2.43, p = 0.020), the *Stella* of Proteobacteria (LDA = 2.42, p = 0.015). the *Pseudorhodoplanes* of Proteobacteria (LDA = 3.34, p = 0.015) and the *Kocuria* of Actinobacteriota (LDA = 2.29, p = 0.024) in S were higher than those of HR and R. The relative abundance of the *Marine_Group_II* (LDA = 3.22, p = 0.007), the *Allorhizobium_Neorhizobium_Pararhizobium_Rhizobium* of Proteobacteria (LDA = 2.96, p = 0.023), the *Magnetospirillum* of Proteobacteria (LDA = 2.95, p = 0.005), and the *Patulibacter* of Actinobacteriota (LDA = 2.62, p = 0.021) in R were higher than those in the other two soils. In HR, the relative abundance of the *Subgroup_22* of Acidobacteriota (LDA = 3.57, p = 0.018), the *MND1* of Proteobacteria (LDA = 3.51, p = 0.015), the *NB1_j* of NB1_j (LDA = 3.47, p = 0.01), uncultured Acidobacteriota (LDA = 3.35, p = 0.01), the *Subgroup_17* of Acidobacteriota (LDA = 3.16, p = 0.024), the *Azoarcus* of Proteobacteria (LDA = 2.97, p = 0.012), the *Subgroup_11* of Acidobacteriota (LDA = 2.61, p = 0.023), uncultured Chloroflexi (LDA = 2.55, p = 0.023), and the *SHA_26* of Chloroflexi (LDA = 2.48, p = 0.023) was higher than that in the other two soils. Figures 4C,D and S1 show the taxonomic hierarchy of the sample taxa from the phylum to the genus level. The relative abundance of genera with significant differences between groups in the three groups was analyzed. The results showed that although the *Latescibacterota* of Latescibacterota was the dominant bacteria in S in the LEfSe analysis, it only showed the highest relative abundance in S3; these bacteria presented lower abundance in other sample groups of S, even lower than the relative abundance in HR (Figure 4C). Similarly, the relative abundance of the Stella of Proteobacteria was only highest in S2 (Figure 4C). These two prominent bacterial genera in S were not statistically significant. The relative abundance of the dominant bacteria *Allorhizobium_Neorhizobium_Pararhizobium_Rhizobium* of Proteobacteria in R was only the highest in R1 (Figure 4C), so it was not statistically significant. The *Subgroup_11* of Acidobacteriota in HR was also not statistically significant (Figure 4C), so it was not involved in the subsequent analysis.

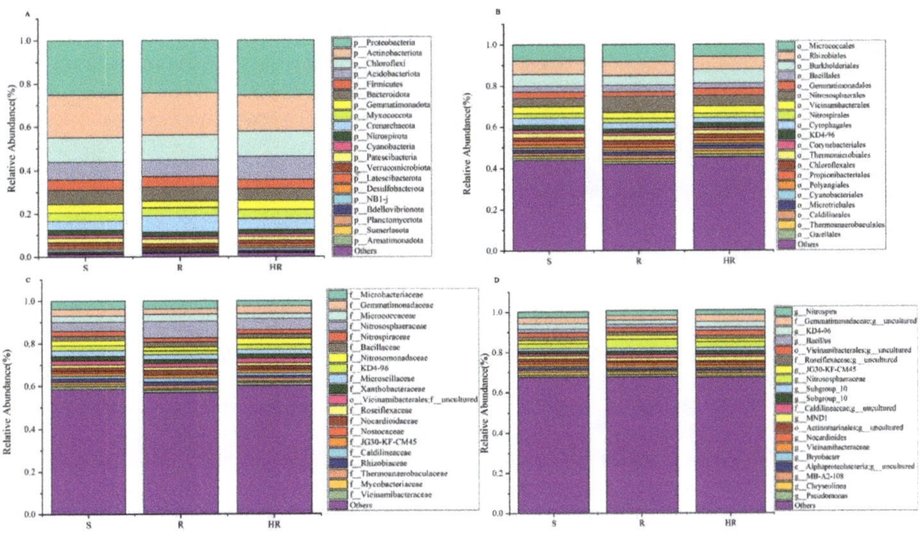

Figure 3. A histogram (relative abundance) of bacterial community composition at the phylum (**A**), order (**B**), family (**C**), and genus (**D**) levels (top 20).

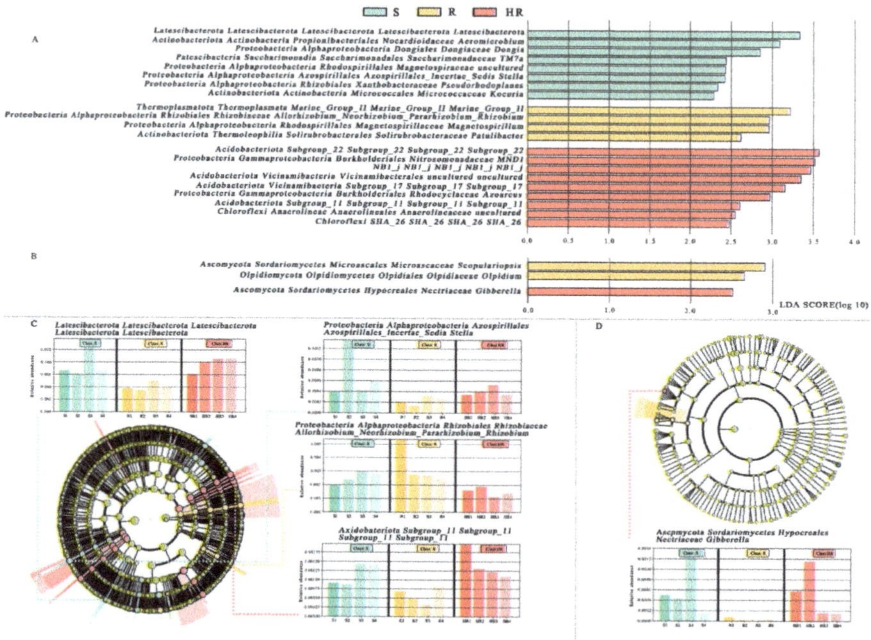

Figure 4. The linear discriminant analysis effect value (LEfSe) of bacterial (**A**) and fungal (**B**) communities was greater than 2.0 when the P value was less than 0.05. The histogram showed the LDA score of microbial groups. Groups that meet the LDA significance threshold >2.0 are displayed. The taxonomic cladogram showed the main bacterial (**C**) and fungal (**D**) taxa in the sample community from phylum to genus (from inside to outside). The size of the node corresponds to the average relative abundance of the classification unit; hollow nodes represent groups with no significant difference between groups, while nodes of other colors indicate that these groups show significant differences between groups, and the groups represented by colors are more abundant in the sample. The histogram showed the relative abundance of bacteria (**C**) and fungi (**D**) in the three sample groups.

Compared with rhizosphere bacteria, the fungal community diversity was lower. All fungal OTUs belonged to 10 fungal phyla. Among them, Ascomycota had the highest relative abundance, accounting for more than 84% of the total recovered sequences. At the phyla level, the relative abundance of Ascomycota in HR (84.46%) was higher than that in S (80.75%). The relative abundance of Mortierellomycota (8.58%) and Basidiomycota (7.23%) in S was higher than that in HR (6.20%, 5.16%) and R (6.98%, 2.80%) (Figure 5A). At the order level, the relative abundance of Sordariales was higher in HR (31.84%) than in R (25.75%) and S (25.85%), and the relative abundance of Mortierellales was higher in S (8.62%) than in HR (6.22%) and R (6.94%) (Figure 5B). At the family level, the relative abundance of Chaetomiaceae in HR (31.47%) was higher than that in R (25.29%) and S (25.18%); the relative abundance of Pyronemataceae in HR (20.90%) was higher than that in R (11.80%) and S (7.11%); and the relative abundance of Mortierellaceae in S (8.70%) was higher than that in HR (6.24%) and R (6.97%) (Figure 5C). At the genus level, the relative abundance of *Chaetomium* was higher in HR (31.12%) than in R (24.42%) and S (24.11%), while the relative abundance of *Fusarium* was highest in S (1.05%) and lowest in HR (0.54%). The relative abundance of *Scopulariopsis* was higher in R (1.21%) and S (1.10%) than in HR (0.23%). *Monosporascus* was only present in S (Figure 5D). The LEfSe analysis method (Figure 4B) showed that there were no dominant fungi in S. There were four dominant fungi in R and one dominant fungus in HR. At the genus level, the *Scopulariopsis* of Ascomycota (LDA = 2.96, p = 0.025) and the *Olpidium* of Olpidiomycota (LDA = 2.613, p = 0.023) were the dominant bacteria in R. The relative abundance of the *Gibberella* of Ascomycota (LDA = 2.42, $p = 0.024$) in HR was higher than that in the other two soils. However, its relative abundance was only higher in HR3, so it was not statistically significant (Figure 4D).

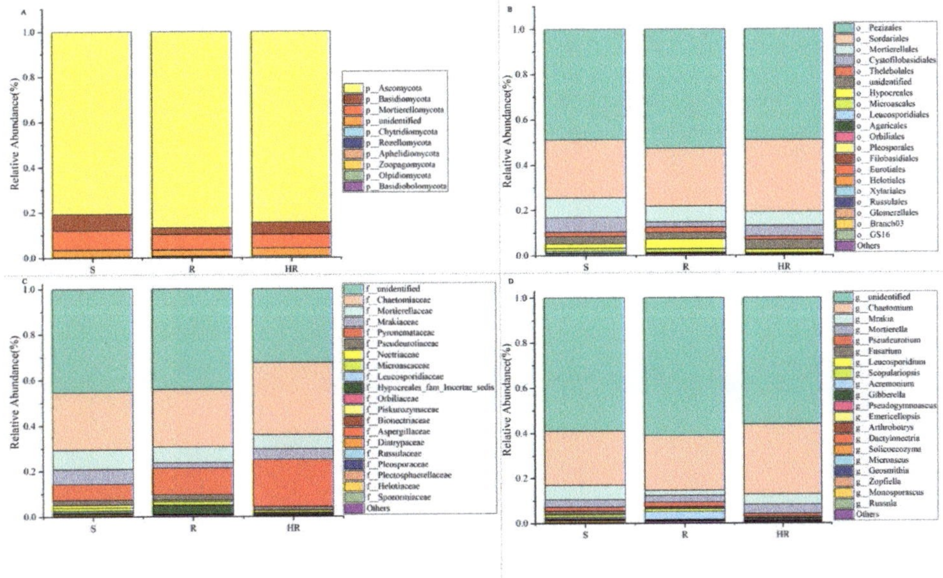

Figure 5. A histogram (relative abundance) of fungal community composition at the phylum (**A**), order (**B**), family (**C**), and genus (**D**) levels (top 20).

Figure 5. A histogram (relative abundance) of fungal community composition at the phylum (**A**), order (**B**), family (**C**), and genus (I

3.4. Correlation of Dominant Genera of Microbial Communities with Plant HI and Environmental Factors

In order to better understand the causes of microbial community changes, we performed a Spearman correlation analysis on the relationship between microbial community composition and soil physical and chemical properties (Figure 6). Spearman correlation analysis showed that soil ECa content was positively correlated with the relative abundance of plant HI and negatively correlated with AK (Figure 6A). Plant HI was significantly positively correlated with the relative abundance of *SHA_26*, and significantly negatively correlated with the relative abundance of *Aeromicrobium*, *TM7a*, *Pseudorhodoplanes*, and *Kocuria* (Figure 6B). Plant HI was positively correlated with the relative abundance of *Chaetomium* and negatively correlated with *Fusarium* (Figure 6C). RDA analysis between the soil physicochemical properties and key bacteria and fungi showed that there was a positive correlation between ECa, *Subgroup_22*, and *MND1* (Figure 6D,E). Aeromierobium had a strong positive correlation with AK and EC, respectively. ACu had a strong positive correlation with *Fusarium*, *Olpidium*, and *Scopulariopsis*. There was also a positive correlation between plant HI and the relative abundance of *Burkholderiales* and *MND1* (Figure 6B). Plant HI was significantly and positively correlated with the relative abundance of *Chaetomium* and negatively correlated with *Fusarium* (Figure 6C). Two diagrams were drawn according to the results of the correlation analysis (Figure 6D,E). The above results showed a direct correlation between soil factors and microbial changes, which also indicated that these key rhizosphere bacteria and fungi may have certain synergistic and antagonistic effects in cucumber rhizosphere soil.

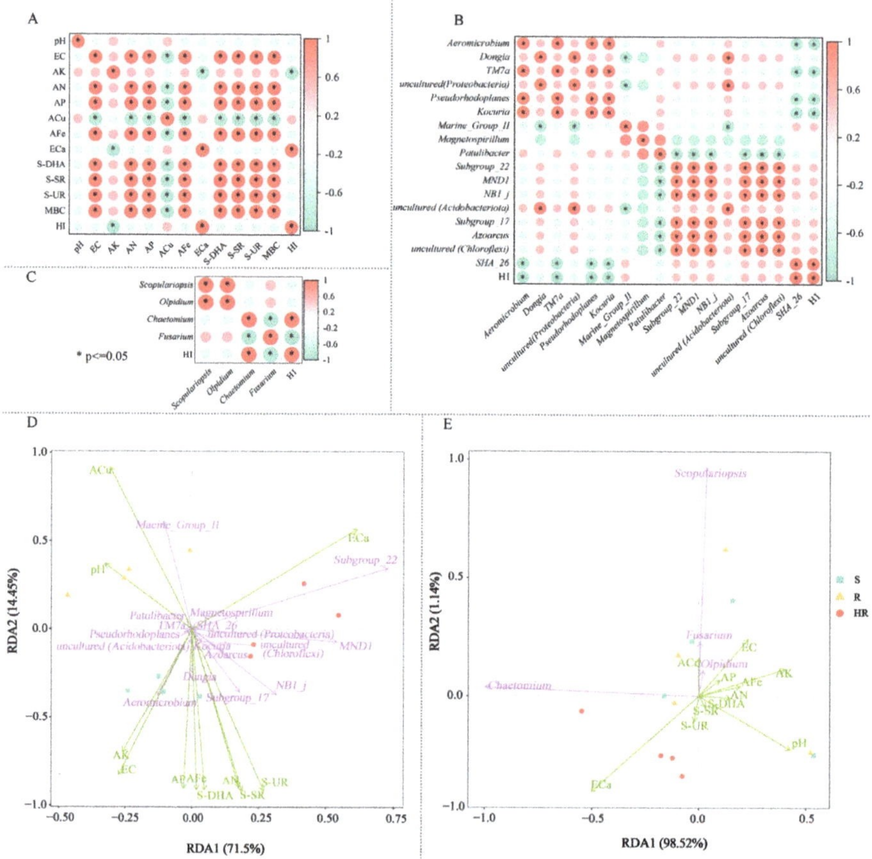

Figure 6. The correlation analysis heatmap of soil factors of HR, R, and S with screened bacteria (**A**) fungi (**B**), key bacteria, and fungi (**C**). Redundancy analysis (RDA) of the relationship between soil bacterial (**D**) and fungal (**E**) community structure and soil factors under different groups. In the correlation analysis, the legend shows the correlation coefficient value, red represents the positive correlation, blue represents the negative correlation; color depth indicates the strength of the correlation; * $p < 0.05$. In RDA analysis, the green arrows represent key bacteria and fungi at the genus level, whereas soil factors are shown with red arrows. Correlation between soil factors (ACu, AFe, ECa, EC, OM, MBC, S-DHA, S-UR, S-SR) and RDA axes are shown by both length and angle of arrows.

Figure 6. The correlation analysis heatmap of soil factors of HR, R, and S with screened bacteria (A) fungi (B), key bacteria, and fungi (
(RDA) of the relationship between soil bacterial (D) and fungal (E) community structure and soil factors under different groups. In the

3.5. Analysis of Microbial Community Function of HR, R, and S

To investigate the potential changes in soil microorganisms in the three sample groups, we compared the functional abundance of bacterial and fungal communities based on a sequence analysis of 16S rRNA genes' ITS markers. PICRUSt2 and FUNGulid were applied to predict microbial functions, thereby providing a basis for understanding the microbial community itself and its potential interactions with the host cucumber. By carefully studying the KEGG homologs in microbial metabolism (Figure 7A), the results showed that in the three-level prediction, the highest relative abundance of the KEGG pathway was in the biosynthesis of ansamycins, valine, leucine, and isoleucine biosynthesis, biosynthesis of vancomycin group antibiotics, C5-branched dibasic acid metabolism, and fatty acid biosynthesis. In the level 2 predictions, the overall relative abundance of carbohydrate metabolism, amino acid metabolism, and metabolism of cofactors and vitamins was higher. In the level

1 prediction, metabolism accounted for most of the whole. Then, we used the STAMP 2.1.3 software to analyze the differences in KEGG pathways among the three groups (Figure S2). The results showed that the difference between HR and R and S was greatest, while that between R and S was smaller. Therefore, we focus on analyzing the functional prediction results between HR and R and between HR and S. A total of 67 pathways between HR and R showed significant differences ($p < 0.05$), in which six pathways showed extremely significant differences ($p < 0.01$). Lipopolysaccharide biosynthesis, tetracycline biosynthesis, and photosynthesis were enriched in HR, while tyrosine metabolism, biosynthesis of siderophore group nonribosomal peptides, and linoleic acid metabolism were significantly present in R. A total of 42 pathways between HR and S showed significant differences ($p < 0.05$), in which 5 pathways showed extremely significant differences ($p < 0.01$). Tetracycline biosynthesis and selenocompound metabolism were extremely significantly enriched in HR, and Xylene degradation, primary bile acid biosynthesis, and bisphenol degradation were enriched in S.

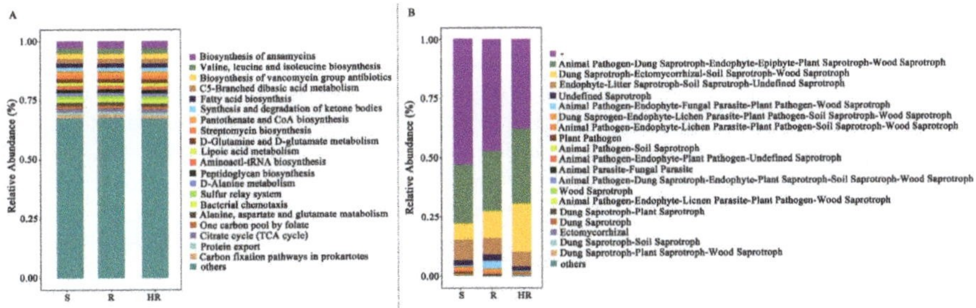

Figure 7. Picrust2 (**A**) and FUNGulid (**B**) annotated and predicted functions based on the bacterial and fungal community composition of

Figure 7. Picrust2 (**A**) and FUNGulid (**B**) annotated and predicted functions based on the bacterial and fungal community composition of HR, R, and S.

FUNGulid is a database of functional annotations of fungi that currently covers over 12,000 fungi. Fungulid was used to predict fungal function by annotating against the structural composition of fungal communities (Figure 7B). We used STAMP 2.1.3 software to analyze the functional prediction results between HR and R and HR and S (Figure S3). The results showed that the functional abundance of animal pathogen–plant pathogen–undefined saprotroph in R was significantly higher than that in HR ($p = 0.05$). The functional abundance of dung saprotroph–ectomycorrhizal–soil saprotroph–wood saprotroph was significantly higher than that of S ($p = 0.013$), which was more than three times that of HR.

3.6. Possible Mechanism of Rhizosphere Microorganisms Inhibiting Cucumber Fusarium Wilt

According to the correlation among soil physicochemical properties, microbial community, and plant HI in Figure 7, combined with RDA analysis between key bacteria and fungi and soil physical and chemical properties, we selected soil ECa content and AK content as key environmental factors. Bacteria *SHA_26*, *Subgroup_22*, *MND1*, *Aeromicrobium*, *TM7a*, *Pseudorhodoplanes* and *Kocuria*, fungi *Chaetomium*, *Fusarium*, *Olpidium*, and *Scopulariopsis* were the key microbial factors. The key factors related to cucumber *Fusarium* wilt inhibition were soil ECa content, *SHA_26*, *Subgroup_22*, *MND1*, and *Chaetomium*. To predict the functions of bacterial and fungal communities, we determined the relative functional abundance of tetracycline biosynthesis, selenocompound metabolism, lipopolysaccharide biosynthesis, cell cycle–Caulobacter, biosynthesis of vancomycin group antibiotics, fatty acid biosynthesis, polyketide sugar unit biosynthesis, pantothenate and CoA biosynthesis, indole alkaloid biosynthesis, RNA polymerase, carbon fixation pathways in prokaryotes, oxidative phosphorylation, and pantothenate and CoA biosynthesis in HR, which was significantly higher than that in R and S. The effects of rhizosphere bacteria and fungi on FOC may occur through the following four pathways: (1) competitive nutrition and spatial location with FOC; (2) toxin production of FOC; (3) changes to plant root exudates affecting the growth of FOC; and (4) microbe-associated molecular patterns (MAMPs) produced by the interaction between microorganisms and hosts activating detection of the plant immune system.

4. Discussion

The cucumber is one of the world's top ten vegetables and an important cash crop in China, cultivated throughout the country in all seasons of the year. At present, soil-borne diseases are the most common and serious threat to cucumber production facilities. Cucumber wilt is a destructive soil-borne disease caused by *Fusarium oxysporum* f. sp. *cucumerinum*, leading to significant damage to crop yield and quality and causing significant economic losses globally. Thus, studying rhizosphere soil physicochemical properties and the composition of cucumber microbial communities is important for the biological control of cucumber wilt.

4.1. Relationship between Rhizosphere Soil Physicochemical Properties and Plant Health

It is well known that abiotic soil factors contribute to the development of plant diseases. Soil physicochemical properties and enzyme activities have important effects on microbial activity. In this study, we determined whether there were significant differences in rhizosphere soil pH among three healthy cucumber plants and found no significant relationship between soil pH and the incidence of cucumber *Fusarium* wilt. Many studies have shown that the incidence of *Fusarium* wilt is low in high pH soils [31]. However, some studies have shown that the disease severity score decreases with a decrease in acidic pH [32,33], with no correlation between the severity of plant disease and pH [34]. It can be seen that the relationship between soil pH and plant disease occurrence may be affected by the plant and soil types studied. EC values that are too high can form reverse osmotic pressure, leaving the root system dry and brown and making the roots unable to absorb water and nutrition. According to Nam et al. [35], strawberry *Fusarium* wilt severity increases with an increase in EC values. According to some studies, salinity promotes the growth of *V. dahliae* in the root and shoot, which leads to an increase in pepper verticillium wilt severity [36–38]. It was suggested by Aviles et al. [39] that EC plays a key role in promoting the growth of Verticillium wilt in accordance with the findings of this study.

The results of the soil chemical property analysis showed that the contents of ACu and ECa in the rhizosphere soil of healthy and wilt-infected cucumber plants were significantly higher than those in severely wilted plants, whereas AK, AP, AN, and AFe contents were significantly lower than those in severely wilted plants. However, only ECa and AK were significantly correlated with the plant health index. This result indicates that the rhizosphere soil of severe *Fusarium* wilt plants lacks ECa nutrients and is rich in AK. In addition to improving plant biofilms, ECa can enhance plant resistance to environmental stress and improve the selective absorption capacity of plant cell biofilms. As a chemotactic agent of some important enzymes, ECa enhances the activity of α-amylase and phospholipase. Many plant diseases are closely related to calcium content. Jiang et al. [40] demonstrated that increasing the calcium concentration within a certain range can increase the peroxidase activity of tomato plants so as to better control tomato bacterial wilt. ECa may have the ability to control cucumber *Fusarium* wilt. Generally, the rhizosphere soils of susceptible plants contain lower levels of AK, AN, and AP [41–45]. Interestingly, in this study, the AK, AN, and AP contents of the rhizosphere soil of severely diseased plants were higher than those of the other two soils. There may also be a correlation between growth restriction and nutrient uptake by plants [46,47]. Further research is necessary to examine the relationship between these three factors and *Fusarium* wilt. The presence of AFe in the environment is conducive to the formation of chlorophyll and the metabolic process of nitrogen. Soluble organic matter such as root exudates, soil organic matter, and microbial activity metabolites complexed or chelated with AFe in the soil solution help to increase the iron concentration in the soil solution and prompt iron to move to the root system. According to this study, the amount of iron available in the rhizosphere soil of severely diseased plants was significantly higher than that in the other two soils, likely due to a high level of microbial activity. An imbalance of trace elements will yield a change in the microbial population structure of the soil and an increase in pathogenic microbes. In a study by He et al. [48], researchers examined the changes in trace elements in continuous cropping cucumber cultivars and found that effective iron ions increased as continuous cropping years increased and that *Fusarium* number increased linearly as continuous cropping stubbles increased.

Since the secretion of microbial activity in the soil is one of the main sources of soil enzymes, the number of microorganisms has a significant correlation with enzyme activity, but the results of different scholars' studies are inconsistent, which may be related to factors such as soil type, plant species, maintenance management, and soil microbial community. In this experiment, S-DHA, S-SR, and S-UR were found to be significantly higher in S compared to the other two soils, and Pearson correlation analysis showed that all three were positively correlated with the microbial population. The highest MBC in S indicated that this soil contained the highest total soil microbial population. This result suggests that the highest soil enzyme activity in the rhizosphere soil of severely susceptible cucumber plants may be related to metabolism caused by a high number of microbes.

4.2. Key Microbiota Associated with Plant Health

Soil microorganisms play an important role in the occurrence and prevention of plant diseases [49,50]. It was reported that when attacked by pathogens, plant roots resist pathogens by altering their root exudates [51] to attract beneficial microorganisms and enhance their microbial activity [11]. In terms of microbial diversity, abundance, and activity, Janvier et al. [52] observed that these factors may not be consistent indicators of soil health.

It is well known that plant pathogens can cause diseases by directly infecting plants. However, it was demonstrated that soil microbial communities play an important role in disease development as they interact directly with plant pathogens and regulate disease development in various ways [53,54]. It was observed that most of the antagonistic bacteria used to control pathogens belong to Proteobacteria [55], while Acidobacteria are usually associated with disease inhibition [56,57]. As shown in this study, Proteobacteria and Acidobacteriota were more abundant in healthy plants' rhizosphere soil compared to the two other groups. In addition, LEfSe analysis revealed that most of the dominant bacteria in the rhizosphere soil of healthy plants belonged to Acidobacteria and Proteobacteria (Figure 8A). The relative abundance of Chloroflexi *SHA_26* was significantly positively correlated with plant HI. At present, there is no research on *SHA_26*, so the function of the *SHA_26* genus is not understood. Chloroflexi is a type of bacteria that produces energy through photosynthesis. At the order level, the relative abundance of Burkholderiales in healthy plants was found to be higher (Figure 8A). Burkholderiales are members of the Amoebae subfamily that produce secondary metabolites and volatile organic compounds [58,59]. Burkholderiales have the ability to inhibit pathogens. In addition, they are an important plant growth-promoting variety of bacteria in the rhizosphere environment. It was reported that the rhizosphere biological control strain most prevalent in the wheat take-all disease inhibition period is Burkholderiales [60]. Previous studies showed that [52,61,62] Burkholderiales inhibit multiple pathogens in the rhizosphere soil of healthy plants. Burkholderiales may control *Fusarium* wilt in the rhizosphere soil of healthy plants. At the family level, Nitrosomonadaceae are the dominant bacteria in the rhizosphere soil of healthy plants. Nitrosomonadaceae are used for the bioremediation of toxic chemicals in the soil. The addition of Nitrosomonadaceae reduces the loss of nitrogen and the time needed to stabilize the nitrogen profile. Li et al. [62] reported that Nitrosomonadaceae were significantly and positively correlated with Panax ginseng yield. Additionally, the relative abundance of the *MND1* genus in the rhizosphere soil of healthy plants was higher than that in the other two groups. It is well known that the presence of many members of the *MND1* genus can improve soil quality. As emphasized by Sun et al. [63], *MND1* plays an important role in the mineralization of organic phosphorus and the dissolution of inorganic phosphorus. *MND1* belongs to the Nitrosomonadaceae family and is an ammonia-oxidizing bacterium involved in nitrogen fixation, ammonia oxidation, and other nitrogen cycle processes [64]. However, in this experiment, the contents of AN and AP in healthy soil were lower than those in R and S, so *MND1*'s involvement in inhibiting cucumber *Fusarium* wilt may be related to the other functions of this genus. Wang et al. [65] showed that *MND1* may be valuable for promoting the degradation of BP-3 in plant growth.

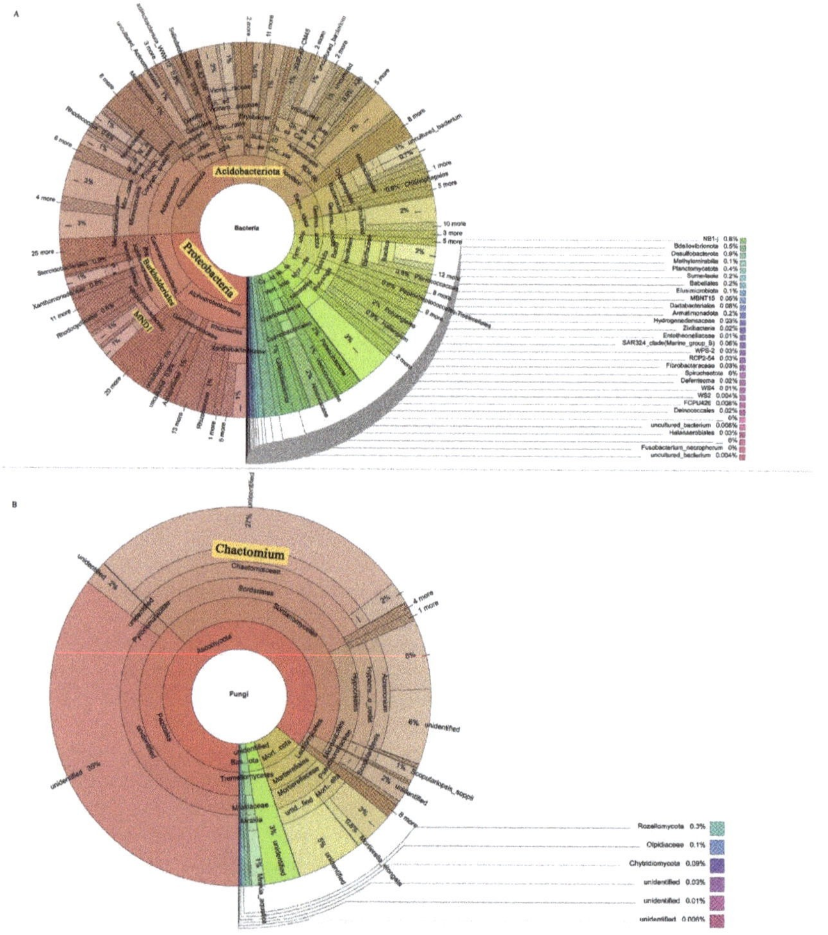

Figure 8. Species composition of bacteria (**A**) and fungi (**B**). The prominent part of the yellow background indicates the key microbial groups.

The relative abundance of *Chaetomium* accounted for a large proportion of all three groups, and we found a positive correlation between their relative abundance and HI values. *Chaetomium* belongs to the Sordariales of Sordariomycetes (Figure 8B). Members of the genus *Chaetomium* are widely distributed in the soil and can produce a large amount of cellulase and physiologically active secondary metabolites, which can effectively degrade cellulose, lignin, and other difficult-to-degrade macromolecular organic matter, as well as antagonize some microorganisms in the soil, thereby inhibiting the growth of pathogenic bacteria [66,67]. Li et al. [68] reported that some compounds isolated from *Chaetomium* showed significant inhibitory effects on the pathogenic fungus of Panax notoginseng root rot. The abundance of pathogens is a direct factor affecting plant health [69]. In this study, we found that the relative abundance of *Fusarium* was negatively correlated with HI values. In addition to *Fusarium*, several other fungi associated with plant pathogenicity, such as *Monosporascus*, *Leucosporidium*, and *Dactylonectri*, had the highest relative abundance in the rhizosphere soil of severely affected plants. *Monosporascus cannonballus* is a common pathogen causing the root rot of melon, which we found to be present only in the rhizosphere soil of severely damaged plants. A functional prediction by FUNGulid found that the rhizosphere soils of severely affected plants had the highest functional abundances of dung saprotroph, ectomycorrhizal saprotroph, soil saprotroph, and wood saprotroph. This result may be directly related to these plant pathogens.

Plants with a healthy rhizosphere, lightly afflicted plants, and severely afflicted plants differ significantly in their rhizosphere soil microbial community composition. In maintaining plant health, the rhizosphere microbiota plays a key role [61]. Some evidence indicates that microbial communities differ between suppressed soils and favorable soils, with suppressed soils showing a greater abundance of taxa associated with disease suppression [14]. The microbiome of the rhizosphere of healthy plants may, therefore, be able to control soil pathogens such as wilt. In this study, Acidobacteriota *Subgroup_22*, Chloroflexi *SHA_26*, and *MND1* were significantly enriched in the rhizosphere soil of healthy plants. Many studies showed that Acidobacteriota are associated with a variety of diseases. At present, there is no experimental evidence demonstrating that Chloroflexi *SHA_26* and *MND1* have a preventive effect on cucumber *Fusarium* wilt, but they are known to be beneficial to plant growth. *Chaetomium*, which has an inhibitory effect on pathogenic bacteria, is also abundant in the rhizospheres of healthy plants. However, it is not sufficient to interpret these differences as a result of soil–plant interactions. Studies are needed to determine how these microbial and chemical properties directly affect plant health. Finally, we combined correlation analysis to draw a possible mechanism map, revealing that environmental factors, rhizosphere bacteria, and fungi inhibit the occurrence of *Fusarium* wilt through four possible pathways (Figure 9). Whether the specific microbial communities and soil factors in the rhizosphere soil of healthy cucumbers have the ability to inhibit disease remains to be determined in future studies.

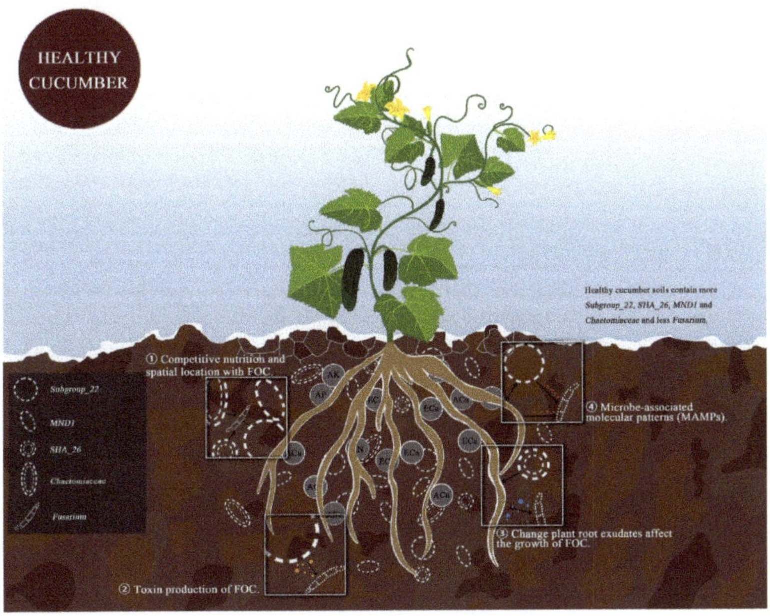

Figure 9. Rhizosphere soil bacteria and fungi combined with soil factors through four possible ways synergistic inhibition of cucumber Fusarium wilt possible mechanism diagram.

5. Conclusions

Despite using the same growth conditions for cucumber, there were significant differences observed in soil physical and chemical properties and microbial communities under the three different health conditions. ECa levels in the rhizosphere soil of healthy and susceptible plants may be able to control soil-borne diseases. In the rhizosphere soil of healthy cucumber plants, Acidobacteriota *Subgroup_22*, Chloroflexi *SHA_26*, *MND1*, and *Chaetomium* as pathogen antagonists presented the highest relative abundance, while the relative abundance of potential plant pathogens was the lowest. This result may be due to the fact that this special microbial composition inhibits the expression of *Fusarium* wilt and the development of healthy plants.

In summary, in the rhizosphere soil, we detected ECa, bacteria Acidobacteriota *Subgroup_22*, Chloroflexi *SHA_26*, *MND1*, and fungi *Chaetomiacea* were associated with *Fusarium* wilt inhibition. Ultimately, we conclude that the physicochemical properties and microbial community composition of cucumber rhizosphere soil in the three different health states can be used to judge plant health.

Supplementary Materials: The following supporting information can be downloaded at: https://www.mdpi.com/article/10.3390/microorganisms11061576/s1, Table S1: Soil physical and chemical properties of HR, R and S; Table S2: Bacterial alpha diversity index table; Table S3: Fungal alpha diversity index table; Figure S1: The taxo-nomic cladogram showed the main bacterial (A) and fungal (B) taxa in the sample community from phylum to genus (from inside to outside). The size of the node corresponds to the average relative abundance of the classification unit; hollow nodes represent groups with no significant difference between groups, while nodes of other colors indicate that these groups show significant differences between groups, and the groups represented by colors are more abundant in the sample. The histogram showed the relative abundance of bacteria (A) and fungi (B) in the three sample groups; Figure S2: Prediction and comparison of rhizosphere bacterial community functions. The changes of bacterial functional group composition were inferred by PICRUSt2. STAMP software was used to analyze the difference of KEGG function between HR and R (A) and HR and S (B), and Welch's two-sided t-test and Bonferroni multiple test correction method was used. The abscissa of the left column graph represents the average value of a certain function percentage, the ordinate represents the function name, and different colors represent different groups. The figure on the right represents the proportion of species abundance differences within the set confidence interval; Figure S3. Prediction and comparison of rhizosphere fungal community functions. The changes of bacterial functional group composition were inferred by FUNGuild. STAMP software was used to analyze the difference of KEGG function between HR and R (A) and HR and S (B), and Welch's two-sided t-test and Bonferroni multiple test correction method was used. The abscissa of the left column graph represents the average value of a certain function percentage, the ordinate represents the function name, and different colors represent different groups. The figure on the right represents the proportion of species abundance differences within the set confidence interval.

Author Contributions: F.Y. grew plants and collected the root materials. H.J. and F.Y. contributed to collection and analysis of sequence data. All authors (F.Y., H.J., G.C., S.L., K.M., Y.C., B.T. and X.S.) contributed to the experimental design and to the preparation of the manuscript. All authors have read and agreed to the published version of the manuscript.

Funding: This research was funded by the Scientific and Technological Breakthrough Foundation of Henan Province Project (212102110426, 222102110004), Major Science and Technology Special Projects in Henan Province (201111110500), the Technology System of Watermelon Industry in Henan Province (HARS-22-10-G1), and the Technology System of Bulk Vegetable Industry in Henan Province (HARS-22-07-G2).

Institutional Review Board Statement: Not applicable.

Informed Consent Statement: Not applicable.

Data Availability Statement: All raw sequence data have been made available in the NCBI Sequence Read Archive (SRA) database under the bioproject accession number PRJNA921884 and PRJNA921900.

Conflicts of Interest: The authors declare no conflict of interest.

References

1. Delgado-Baquerizo, M.; Guerra, C.A.; Cano-Diaz, C.; Egidi, E.; Wang, J.T.; Eisenhauer, N.; Singh, B.K.; Maestre, F.T. The proportion of soil-borne pathogens increases with warming at the global scale. *Nat. Clim. Chang.* **2020**, *10*, 550–554. [CrossRef]
2. Cohen, R.; Orgil, G.; Burger, Y.; Saar, U.; Elkabetz, M.; Tadmor, Y.; Edelstein, M.; Belausov, E.; Maymon, M.; Freeman, S.; et al. Differences in the responses of melon accessions to fusarium root and stem rot and their colonization by *Fusarium oxysporum* f. sp *radicis-cucumerinum*. *Plant Pathol.* **2015**, *64*, 655–663. [CrossRef]
3. Sun, L.; Song, S.; Fu, L.; Deng, X.H.; Wang, D.S.; Liang, X.L.; Li, R.; Shen, Q.R. Exploring a soil fumigation strategy based on ammonium bicarbonate to control *Fusarium* wilts of cucurbits. *Crop. Prot.* **2015**, *70*, 53–60. [CrossRef]
4. Mendes, R.; Garbeva, P.; Raaijmakers, J.M. The rhizosphere microbiome: Significance of plant beneficial, plant pathogenic, and human pathogenic microorganisms. *FEMS Microbiol. Rev.* **2013**, *37*, 634–663. [CrossRef]
5. Chapelle, E.; Mendes, R.; Bakker, P.A.H.; Raaijmakers, J.M. Fungal invasion of the rhizosphere microbiome. *ISME J.* **2016**, *10*, 265–268. [CrossRef]
6. Perez-Jaramillo, J.E.; Mendes, R.; Raaijmakers, J.M. Impact of plant domestication on rhizosphere microbiome assembly and functions. *Plant Mol. Biol.* **2016**, *90*, 635–644. [CrossRef] [PubMed]
7. Philippot, L.; Raaijmakers, J.M.; Lemanceau, P.; van der Putten, W.H. Going back to the roots: The microbial ecology of the rhizosphere. *Nat. Rev. Microbiol.* **2013**, *11*, 789–799. [CrossRef]
8. Yang, Z.; Dai, C.C.; Wang, X.X.; Li, X.G. Advance in Research on Rhizosphere Microbial Mechanisms of Crop Soil-borne Fungal Diseases. *Acta Pedol. Sin.* **2019**, *56*, 12–22.
9. Kwak, M.J.; Kong, H.G.; Choi, K.; Kwon, S.K.; Song, J.Y.; Lee, J.; Lee, P.A.; Choi, S.Y.; Seo, M.; Lee, H.J.; et al. Rhizoshere microbiome structure alters to enable wilt resistance in tomato. *Nat. Biotechnol.* **2018**, *36*, 1100–1109. [CrossRef]
10. Meng, T.; Wang, Q.; Abbasi, P.; Ma, Y. Deciphering differences in the chemical and microbial characteristics of healthy and *Fusarium* wilt-infected watermelon rhizosphere soils. *Appl. Microbiol. Biotechnol.* **2019**, *103*, 1497–1509. [CrossRef]
11. Cha, J.Y.; Han, S.; Hong, H.J.; Cho, H.; Kim, D.; Kwon, Y.; Kwon, S.K.; Crusemann, M.; Lee, Y.B.; Kim, J.F.; et al. Microbial and biochemical basis of a Fusarium wilt-suppressive soil. *ISME J.* **2016**, *10*, 119–129. [CrossRef] [PubMed]
12. Shen, Z.; Ruan, Y.; Xue, C.; Zhong, S.; Li, R.; Shen, Q. Soils naturally suppressive to banana *Fusarium* wilt disease harbor unique bacterial communities. *Plant Soil* **2015**, *393*, 21–33. [CrossRef]
13. Shi, W.; Li, M.; Wei, G.; Tian, R.; Li, C.; Wang, B.; Lin, R.; Shi, C.; Chi, X.; Zhou, B.; et al. The occurrence of potato common scab correlates with the community composition and function of the geocaulosphere soil microbiome. *Microbiome* **2019**, *7*, 18. [CrossRef] [PubMed]
14. Fu, L.; Ruan, Y.Z.; Tao, C.Y.; Li, R.; Shen, Q.R. Continous application of bioorganic fertilizer induced resilient culturable bacteria community associated with banana Fusarium wilt suppression. *Sci. Rep.* **2016**, *6*, 11. [CrossRef] [PubMed]
15. Shen, Z.; Zhong, S.; Wang, Y.; Wang, B.; Mei, X.; Li, R.; Ruan, Y.; Shen, Q. Induced soil microbial suppression of banana fusarium wilt disease using compost and biofertilizers to improve yield and quality. *Eur. J. Soil Biol.* **2013**, *57*, 27731. [CrossRef]
16. Kozich, J.J.; Westcott, S.L.; Baxter, N.T.; Highlander, S.K.; Schloss, P.D. Development of a Dual-Index Sequencing Strategy and Curation Pipeline for Analyzing Amplicon Sequence Data on the MiSeq Illumina Sequencing Platform. *Appl. Environ. Microbiol.* **2013**, *79*, 5112–5120. [CrossRef]
17. Caporaso, J.G.; Lauber, C.L.; Walters, W.A.; Berg-Lyons, D.; Lozupone, C.A.; Turnbaugh, P.J.; Fierer, N.; Knight, R. Global patterns of 16S rRNA diversity at a depth of millions of sequences per sample. *Proc. Natl. Acad. Sci. USA* **2011**, *108*, 4516–4522. [CrossRef]
18. Munyaka, P.M.; Eissa, N.; Bernstein, C.N.; Khafipour, E.; Ghia, J.E. Antepartum Antibiotic Treatment Increases Offspring Susceptibility to Experimental Colitis: A Role of the Gut Microbiota. *PLoS ONE* **2015**, *10*, e0142536. [CrossRef]
19. Jiao, S.; Chen, W.; Wang, J.; Du, N.; Li, Q.; Wei, G. Soil microbiomes with distinct assemblies through vertical soil profiles drive the cycling of multiple nutrients in reforested ecosystems. *Microbiome* **2018**, *6*, 13. [CrossRef] [PubMed]
20. Caporaso, J.G.; Kuczynski, J.; Stombaugh, J.; Bittinger, K.; Bushman, F.D.; Costello, E.K.; Fierer, N.; Gonzalez Peña, A.; Goodrich, J.K.; Gordon, J.I.; et al. QIIME allows analysis of high-throughput community sequencing data. *Nat. Methods* **2010**, *7*, 335–336. [CrossRef] [PubMed]
21. Edgar, R.C. UPARSE: Highly accurate OTU sequences from microbial amplicon reads. *Nat. Methods* **2013**, *10*, 996–998. [CrossRef]
22. Wang, Q.; Garrity, G.M.; Tiedje, J.M.; Cole, J.R. Naive Bayesian classifier for rapid assignment of rRNA sequences into the new bacterial taxonomy. *Appl. Environ. Microbiol.* **2007**, *73*, 5261–5267. [CrossRef]
23. Paulson, J.N.; Stine, O.C.; Bravo, H.C.; Pop, M. Differential abundance analysis for microbial marker-gene surveys. *Nat. Methods* **2013**, *10*, 1200–1202. [CrossRef] [PubMed]
24. McMurdie, P.J.; Holmes, S. phyloseq: An R Package for Reproducible Interactive Analysis and Graphics of Microbiome Census Data. *PLoS ONE* **2013**, *8*, 11. [CrossRef] [PubMed]
25. Lozupone, C.; Lladser, M.E.; Knights, D.; Stombaugh, J.; Knight, R. UniFrac: An effective distance metric for microbial community comparison. *ISME J.* **2011**, *5*, 169–172. [CrossRef]
26. Alabouvette, C. Fusarium-wilt suppressive soils from the Chateaurenard region—Review of a 10-year study. *Agronomie* **1986**, *6*, 273–284. [CrossRef]

27. Ma, M.; Jiang, X.; Wang, Q.; Ongena, M.; Wei, D.; Ding, J.; Guan, D.; Cao, F.; Zhao, B.; Li, J. Responses of fungal community composition to long-term chemical and organic fertilization strategies in Chinese Mollisols. *MicrobiologyOpen* **2018**, *7*, 12. [CrossRef] [PubMed]
28. Douglas, G.M.; Maffei, V.J.; Zaneveld, J.; Yurgel, S.N.; Brown, J.R.; Taylor, C.M.; Huttenhower, C.; Langille, M.G.I. PICRUSt2: An improved and customizable approach for metagenome inference. *bioRxiv* **2020**, 672295.
29. Louca, S.; Parfrey, L.W.; Doebeli, M. Decoupling function and taxonomy in the global ocean microbiome. *Science* **2016**, *353*, 1272–1277. [CrossRef]
30. Sun, R.; Dsouza, M.; Gilbert, J.A.; Guo, X.; Wang, D.; Guo, Z.; Ni, Y.; Chu, H. Fungal community composition in soils subjected to long-term chemical fertilization is most influenced by the type of organic matter. *Environ. Microbiol.* **2016**, *18*, 5137–5150. [CrossRef]
31. Senechkin, I.V.; van Overbeek, L.S.; van Bruggen, A.H.C. Greater Fusarium wilt suppression after complex than after simple organic amendments as affected by soil pH, total carbon and ammonia-oxidizing bacteria. *Appl. Soil Ecol.* **2014**, *73*, 148–155. [CrossRef]
32. Duffy, B.K.; Ownley, B.H.; Weller, D.M. Soil chemical and physical properties associated with suppression of take-all of wheat by Trichoderma koningii. *Phytopathology* **1997**, *87*, 1118–1124. [CrossRef] [PubMed]
33. Lacey, M.J.; Wilson, C.R. Relationship of common scab incidence of potatoes grown in Tasmanian ferrosol soils with pH, exchangeable cations and other chemical properties of those soils. *J. Phytopathol.-Phytopathol. Z.* **2001**, *149*, 679–683. [CrossRef]
34. Hiddink, G.A.; van Bruggen, A.H.C.; Termorshuizen, A.J.; Raaijmakers, J.M.; Semenov, A.V. Effect of organic management of soils on suppressiveness to *Gaeumannomyces graminis* var. *tritici* and its antagonist, *Pseudomonas fluorescens*. *Eur. J. Plant Pathol.* **2005**, *113*, 417–435. [CrossRef]
35. Nam, M.H.; Lee, H.C.; Kim, T.I.; Lee, E.M.; Yoon, H.S. Effect of Nutrition Solution pH and Electrical Conductivity on Fusarium Wilt on Strawberry Plants in Hydroponic Culture. *Res. Plant Dis.* **2018**, *24*, 26–32. [CrossRef]
36. Levin, A.G.; Lavee, S.; Tsror, L. The influence of salinity on Verticillium dahliae in stem cuttings of five olive cultivars. *J. Phytopathol.* **2007**, *155*, 587–592. [CrossRef]
37. Pascual, I.; Azcona, I.; Morales, F.; Aguirreolea, J.; Sanchez-Diaz, M. Growth, yield and physiology of Verticillium-inoculated pepper plants treated with ATAD and composted sewage sludge. *Plant Soil* **2009**, *319*, 291–306. [CrossRef]
38. Saadatmand, A.R.; Banihashemi, Z.; Sepaskhah, A.R.; Maftoun, M. Soil salinity and water stress and their effect on susceptibility to Verticillium wilt disease, ion composition and growth of pistachio. *J. Phytopathol.* **2008**, *156*, 287–292. [CrossRef]
39. Aviles, M.; Borrero, C. Identifying Characteristics of Verticillium Wilt Suppressiveness in Olive Mill Composts. *Plant Dis.* **2017**, *101*, 1568–1577. [CrossRef]
40. Jiang, J.F.; Li, J.G.; Dong, Y.H. Effect of calcium nutrition on resistance of tomato against bacterial wilt induced by Ralstonia solanacearum. *Eur. J. Plant Pathol.* **2013**, *136*, 547–555. [CrossRef]
41. Dominguez-Hernandez, J.D.; Negrin-Medina, M.A.; Rodriguez-Hernandez, C.M. Potassium Selectivity in Transported Volcanic Soils (Sorribas) under Banana Cultivation in Relation to Banana-Wilt Expression Caused by *Fusarium oxysporum* f. sp Cubense. *Commun. Soil Sci. Plant Anal.* **2010**, *41*, 1674–1692. [CrossRef]
42. Duan, C.; Xue, Q.; Hu, S.; Zhao, J.; Wei, Y.; Wang, L.; Shen, G.; Chen, Q. Microbial ecology of Fusarium wilt infected and healthy cucumber plant in root zone of continuous cropping soil. *J. Northwest A F Univ. Nat. Sci. Ed.* **2010**, *38*, 143–150.
43. Deng, X.; Li, Q.F.; Wu, C.Y.; Liu, J.K. Influence of the Number of Pathogen Causing Banana Fusarium Wilt and Soil Factors on the Infection Degree of Banana Plants. In Proceedings of the 3rd International Conference on Chemical Engineering and Advanced Materials (CEAM 2013), Guangzhou, China, 6–7 July 2013; Trans Tech Publications Ltd.: Guangzhou, China, 2013; pp. 1989–1993.
44. Hong, S.; Jv, H.L.; Yuan, X.F.; Geng, J.J.; Wang, B.B.; Zhao, Y.; Wang, Q.; Li, R.; Jia, Z.J.; Ruan, Y.Z. Soil Organic Nitrogen Indirectly Enhances Pepper-Residue-Mediated Soil Disease Suppression through Manipulation of Soil Microbiome. *Agronomy* **2022**, *12*, 2077. [CrossRef]
45. Zhang, X.Z.; Sa, R.N.; Gao, J.Y.; Wang, C.L.; Liu, D.W.; Zhang, Y.J. Preventive Effect of Vermicompost against Cucumber Fusarium Wilt and Improvement of Cucumber Growth and Soil Properties. *Int. J. Agric. Biol.* **2020**, *23*, 515–521.
46. Liu, X.; Zhang, Y.; Ren, X.J.; Chen, B.H.; Shen, C.W.; Wang, F. Long-term greenhouse vegetable cultivation alters the community structures of soil ammonia oxidizers. *J. Soils Sediments* **2019**, *19*, 883–902. [CrossRef]
47. Nayyar, A.; Hamel, C.; Lafond, G.; Gossen, B.D.; Hanson, K.; Germida, J. Soil microbial quality associated with yield reduction in continuous-pea. *Appl. Soil Ecol.* **2009**, *43*, 115–121. [CrossRef]
48. He, L.; Chen, Y.; Chen, J.; Tong, X.; Han, C. Changes of Microelements, Pathogenic Bacteria and Yield of Straw Nutrition Medium in Continuous Cropping of Cucumber. *J. Shenyang Agric. Univ.* **2010**, *41*, 13–17.
49. Liu, K.; McInroy, J.A.; Hu, C.H.; Kloepper, J.W. Mixtures of Plant-Growth-Promoting Rhizobacteria Enhance Biological Control of Multiple Plant Diseases and Plant-Growth Promotion in the Presence of Pathogens. *Plant Dis.* **2018**, *102*, 67–72. [CrossRef] [PubMed]
50. Ciancio, A.; Pieterse, C.M.J.; Mercado-Blanco, J. Editorial: Harnessing Useful Rhizosphere Microorganisms for Pathogen and Pest Biocontrol. *Front. Microbiol.* **2016**, *7*, 5. [CrossRef]
51. Doornbos, R.F.; van Loon, L.C.; Bakker, P. Impact of root exudates and plant defense signaling on bacterial communities in the rhizosphere. A review. *Agron. Sustain. Dev.* **2012**, *32*, 227–243. [CrossRef]

52. Janvier, C.; Villeneuve, F.; Alabouvette, C.; Edel-Hermann, V.; Mateille, T.; Steinberg, C. Soil health through soil disease suppression: Which strategy from descriptors to indicators? *Soil Biol. Biochem.* **2007**, *39*, 1–23. [CrossRef]
53. Sarma, B.K.; Yadav, S.K.; Singh, S.; Singh, H.B. Microbial consortium-mediated plant defense against phytopathogens: Readdressing for enhancing efficacy. *Soil Biol. Biochem.* **2015**, *87*, 25–33. [CrossRef]
54. Panke-Buisse, K.; Poole, A.C.; Goodrich, J.K.; Ley, R.E.; Kao-Kniffin, J. Selection on soil microbiomes reveals reproducible impacts on plant function. *ISME J.* **2015**, *9*, 980–989. [CrossRef] [PubMed]
55. Raaijmakers, J.M.; Paulitz, T.C.; Steinberg, C.; Alabouvette, C.; Moenne-Loccoz, Y. The rhizosphere: A playground and battlefield for soilborne pathogens and beneficial microorganisms. *Plant Soil* **2009**, *321*, 341–361. [CrossRef]
56. Sanguin, H.; Sarniguet, A.; Gazengel, K.; Moenne-Loccoz, Y.; Grundmann, G.L. Rhizosphere bacterial communities associated with disease suppressiveness stages of take-all decline in wheat monoculture. *New Phytol.* **2009**, *184*, 694–707. [CrossRef]
57. Trivedi, P.; Delgado-Baquerizo, M.; Trivedi, C.; Hamonts, K.; Anderson, I.C.; Singh, B.K. Keystone microbial taxa regulate the invasion of a fungal pathogen in agro-ecosystems. *Soil Biol. Biochem.* **2017**, *111*, 10–14. [CrossRef]
58. Depoorter, E.; Bull, M.J.; Peeters, C.; Coenye, T.; Vandamme, P.; Mahenthiralingam, E. Burkholderia: An update on taxonomy and biotechnological potential as antibiotic producers. *Appl. Microbiol. Biotechnol.* **2016**, *100*, 5215–5229. [CrossRef] [PubMed]
59. Groenhagen, U.; Baumgartner, R.; Bailly, A.; Gardiner, A.; Eberl, L.; Schulz, S.; Weisskopf, L. Production of bioactive volatiles by different *Burkholderia ambifaria* strains. *J. Chem. Ecol.* **2013**, *39*, 892–906. [CrossRef] [PubMed]
60. Li, M.Z.; Chen, Z.J.; Qian, J.; Wei, F.G.; Zhang, G.Z.; Wang, Y.; Wei, G.F.; Hu, Z.G.; Dong, L.L.; Chen, S.L. Composition and function of rhizosphere microbiome of *Panax notoginseng* with discrepant yields. *Chin. Med.* **2020**, *15*, 12. [CrossRef] [PubMed]
61. Cain, C.C.; Henry, A.T.; Waldo, R.H.; Casida, L.J.; Falkinham, J.O. Identification and characteristics of a novel *Burkholderia* strain with broad-spectrum antimicrobial activity. *Appl. Environ. Microbiol.* **2000**, *66*, 4139–4141. [CrossRef] [PubMed]
62. Sun, J.H.; Yang, L.; Wei, J.; Quan, J.N.; Yang, X.T. The responses of soil bacterial communities and enzyme activities to the edaphic properties of coal mining areas in Central China. *PLoS ONE* **2020**, *15*, 16. [CrossRef] [PubMed]
63. Bollmann, A.; Sedlacek, C.J.; Norton, J.; Laanbroek, H.J.; Suwa, Y.; Stein, L.Y.; Klotz, M.G.; Arp, D.; Sayavedra-Soto, L.; Lu, M.; et al. Complete genome sequence of *Nitrosomonas* sp. Is79, an ammonia oxidizing bacterium adapted to low ammonium concentrations. *Stand. Genom. Sci.* **2013**, *7*, 469–482.
64. Wang, W.J.; Liu, A.R.; Fu, W.T.; Peng, D.L.; Wang, G.; Ji, J.; Jin, C.; Guan, C.F. Tobacco-associated with *Methylophilus* sp. FP-6 enhances phytoremediation of benzophenone-3 through regulating soil microbial community, increasing photosynthetic capacity and maintaining redox homeostasis of plant. *J. Hazard. Mater.* **2022**, *431*, 15. [CrossRef]
65. Meiler, D.; Taylor, A. The effect of cochliodinol, a metabolite of *Chaetomium cochliodes*, on the respiration of microsopores of *Fusarium oxysporum*. *Can. J. Microbiol.* **1971**, *17*, 83–86. [CrossRef] [PubMed]
66. Park, J.H.; Choi, G.J.; Jang, K.S.; Lim, H.K.; Kim, H.T.; Cho, K.Y.; Kim, J.C. Antifungal activity against plant pathogenic fungi of chaetoviridins isolated from *Chaetomium globosum*. *FEMS Microbiol. Lett.* **2005**, *252*, 309–313. [CrossRef]
67. Li, W.; Yang, X.Q.; Yang, Y.B.; Duang, R.T.; Chen, G.Y.; Li, X.Z.; Li, Q.L.; Qin, S.H.; Li, S.Q.; Zhao, L.X.; et al. Anti-phytopathogen, multi-target acetylcholinesterase inhibitory and antioxidant activities of metabolites from endophytic *Chaetomium globosum*. *Nat. Prod. Res.* **2016**, *30*, 2616–2619. [CrossRef] [PubMed]
68. Berendsen, R.L.; Pieterse, C.M.; Bakker, P.A. The rhizosphere microbiome and plant health. *Trends Plant Sci.* **2012**, *17*, 478–486. [CrossRef] [PubMed]
69. Mendes, R.; Kruijt, M.; de Bruijn, I.; Dekkers, E.; van der Voort, M.; Schneider, J.H.M.; Piceno, Y.M.; DeSantis, T.Z.; Andersen, G.L.; Bakker, P.; et al. Deciphering the Rhizosphere Microbiome for Disease-Suppressive Bacteria. *Science* **2011**, *332*, 1097–1100. [CrossRef]

microorganisms

MDPI

Article

Synergistic Effects of *Clonostachys rosea* Isolates and Succinate Dehydrogenase Inhibitors Fungicides against Gray Mold on Tomato

Jiehui Song [1,2,†], Tengyu Lei [1,3,†], Xiaojuan Hao [3], Huizhu Yuan [1], Wei Sun [1] and Shuning Chen [1,*]

1 Key Laboratory of Pesticides Evaluation, Ministry of Agriculture, Institute of Plant Protection, Chinese Academy of Agricultural Sciences, Beijing 100193, China
2 Jiangsu Key Laboratory of Crop Genetics and Physiology & Co-Innovation Center for Modern Production Technology of Grain Crops, Agricultural College, Yangzhou University, Yangzhou 225009, China
3 College of Agriculture, Shanxi Agricultural University, Jinzhong 030801, China
* Correspondence: chenshuning@caas.cn
† These authors contributed equally to this work.

Abstract: Gray mold caused by *Botrytis cinerea* is a devastating disease in tomatoes. Site-specific fungicide application is still key to disease management; however, chemical control has many drawbacks. Here, the combined application of a biological agent, *Clonostachys rosea*, with newly developed succinate dehydrogenase inhibitors (SDHI) fungicides showed stronger synergistic effects than the application of SDHI fungicides alone on tomato gray mold control. *C. rosea* 67-1 has been reported as an efficient biological control agent (BCA) for *B. cinerea*. Little information is currently available about the combination of *C. rosea* and fungicides in the control of gray mold. By testing the sensitivity to fungicides with different action mechanisms, *C. rosea* isolates showed high tolerance to SDHI fungicides (1000 μg mL^{-1}) on PDA, and the conidial germination rate was almost not affected under 120 μg mL^{-1} of fluxapyroxad and fluopyram. In greenhouse experiments, the control effect of the combination of *C. rosea* and fluxapyroxad or fluopyram against tomato gray mold was significantly increased than the application of BCA or SDHI fungicides alone, and the combination allows a two-fold reduction of both the fungicide and BCA dose. Further, the biomass of *B. cinerea* and *C. rosea* on tomato plants was determined by qPCR. For *B. cinerea*, the trend of detection level for different treatments was consistent with that of the pot experiments, and the lowest biomass of *B. cinerea* was found when treated with *C. rosea* combined with fluxapyroxad and fluopyram, respectively. For *C. rosea*, qPCR assay confirmed its colonization on tomato plants when mixed with fluopyram and fluxapyroxad. These results indicated that combining *C. rosea* 67-1 with the SDHI fungicides could synergistically increase control efficacy against tomato gray mold.

Keywords: tomato gray mold; *Clonostachys rosea*; succinate dehydrogenase inhibitors (SDHI); synergistic effect

Citation: Song, J.; Lei, T.; Hao, X.; Yuan, H.; Sun, W.; Chen, S. Synergistic Effects of *Clonostachys rosea* Isolates and Succinate Dehydrogenase Inhibitors Fungicides against Gray Mold on Tomato. *Microorganisms* **2023**, *11*, 20. https://doi.org/10.3390/microorganisms11010020

Academic Editors: Tomislav Cernava, Beibei Ge and Kyungseok Park

Received: 17 November 2022
Revised: 17 December 2022
Accepted: 17 December 2022
Published: 21 December 2022

1. Introduction

Gray mold caused by *Botrytis cinerea* can be a devastating disease in tomatoes worldwide. It is also common with numerous other fruit, vegetables, and ornamental crops [1], which makes it difficult to control. Although cultural methods such as appropriate plant spacing, rational fertilization, and breeding disease-resistant varieties can reduce disease incidence, site-specific fungicide application is still crucial to disease management [2]. However, the polycyclic nature of the disease, abundant sporulation, high genetic variability, and short generation time of the pathogen contribute to a high risk for the development of resistance to site-specific fungicides used for control [3]. Several of the most serious issues of fungicide resistance have been reported in *B. cinerea*, including resistance to methyl benzimidazole carbamates, dicarboximides, succinate dehydrogenase inhibitors (SDHI), anilinopyrimidines, quinone outside inhibitors, phenylpyrroles, and sterol biosynthesis inhibitor class III fungicide hydroxyanilide, etc [4–6]. Besides the resistance risk, chemical control has certain drawbacks, such as phytotoxicity to tomato plants, toxicity to non-target organisms, and stringent requirements for correctly timing the fungicide application, which all hinder its usage and ability to effectively control disease [7,8].

The application of biological control agents (BCAs) to manage tomato gray mold is a promising alternative to synthetic fungicides [9–11]. Among them, *C. rosea* has been shown to be effective in controlling gray mold in several crops, both in field and greenhouse cultivations [12–15]. It protects plants against *B. cinerea* by inhibiting spore production and suppressing gray mold development [16]. The defense mechanisms of tomato plants against gray mold, including changes in the signaling molecule and defense enzyme activity, could also be induced when treated with *C. rosea* [16–19]. In recent years, *C. rosea* has been commercially available as a biofungicide, which can effectively control many plant diseases, including gray mold, and has been applied to millions of hectares in China. Although BCA may decrease the frequency and total amount of fungicide spraying, reducing residues and resistance risk, their effectiveness is usually inferior to that of chemical fungicides because of the complexity of the field environment [11,20]. Given the limitations of biological and chemical control strategies, combining *C. rosea* with newly developed fungicides may develop a practical method to control *B. cinerea* in tomato fields. Many studies reported the combination of BCAs and fungicides in controlling plant disease. However, few studies showed the combination of *C. rosea* and fungicides to control plant disease.

The SDHIs are the group that rapidly incorporates new broad-spectrum compounds in the market [21]. They have the specific function of preventing mitochondrial respiration by inhibiting the activity of mitochondrial respiration complex II, which consists of a flavoprotein (SdhA), ferritin (SdhB), and two membrane anchoring proteins (SdhC and SdhD) [22]. Regardless of the SDHI high fungicides efficacy, these fungicides are classified as the medium-to-high risk of resistance [23,24]. Resistance to carboxin, boscalid, penthiopyrad, and fluopyram was reported shortly after their registration [21,23,25–29]. Thus, resistance management practices, such as rotation with different FRAC code fungicides and reducing the rates of fungicide application, must be implemented for the sustained efficacy of SDHI fungicides against the gray mold of tomatoes.

C. rosea has been reported as an efficient biological control agent for *B. cinerea*. Its combination with fungicides may prolong the fungicides' life and provide a viable strategy for disease control. Thus, this study aimed to (1) evaluate the compatibility of the SDHI fungicides and the antagonistic isolate *C. rosea* and (2) determine the synergistic effect of combined application of SDHI fungicides fluxapyroxad as well as fluopyram with *C. rosea* 67-1 for control of tomato gray mold in the greenhouse.

2. Materials and Methods

2.1. Fungal Isolates and Pesticide

The *C. rosea* isolates (JLB-7-1, 67-1, SYP-4-2, SHW-1-1, YJS-3-2, GS6-1, NHH-48-2, BD-2-1) were provided by the Manhong Sun' lab. Among them, isolate 67-1 has been reported in previous studies [30,31]. *B. cinerea* isolates YN80 and YN81 were collected from a tomato from Yunnan province. Isolates were recovered from stock cultures stored with silica blue gel beans at −20 °C on dried filter paper discs (Fisher Scientific, Pittsburgh, PA, USA). All isolates used for inoculations were maintained on PDA (Potato Dextrose Agar: potato 200 g L^{-1}, glucose 15 g L^{-1}, agar 15 g L^{-1}, add deionized water to 1 L. Fresh potato was boiled in deionized water for 20 min, then filtered and the potato juice were taken to make medium) medium at 25 °C in darkness unless otherwise specified. The isolates from storage were grown for five days on PDA before being used for experiments.

Technical-grade carbendazim (98% a.i;Jiangsu Longdeng Chemical Co., Ltd., Suzhou, China), 96% tebuconazole (Guangxi Nanning Guangphthalein Agricultural Chemical Co., Ltd., Nanning, China), 98% pyraclostrobin (Shaanxi Diedu Medichem Co. Ltd., Xian, China), 97% Boscalid (Beijing Bailingwei Technology Co., Ltd., Beijing, China), 98% fluxapyroxad (BASF Corp., Research Triangle Park, NC, USA), 98% fluopimomide (Shandong Zhongnong United Biotechnology Co., Ltd., Jinan, China), and 98% fluopyram (ACMEC, Shanghai, China) were used in this study. Stock solutions were made by dissolving each fungicide in DMSO at the concentration of 10^5 μg a.i. mL^{-1}. The stock solutions were stored at 4 °C in darkness. Salicylhydroxamic acid (SHAM, 99% a.i.; Syngenta Biotechnology Co. Ltd., Shanghai, China) was added to pyraclostrobin-amended PDA at 100 μg mL^{-1} to suppress the alternative oxidase pathway [32]. Corresponding control dishes contained SHAM.

2.2. Fungicides Sensitivity Assessments of C. rosea and B. cinerea In Vitro

Sensitivity to carbendazim, tebuconazole, boscalid, and pyraclostrobin was assessed on fungicide-amended PDA at 0, 0.1, 0.3, 1, 3, 10, and 30 μg a.i. mL^{-1}. Furthermore, sensitivity to boscalid, fluxapyroxad, fluopimomide, and fluopyram was assessed on fungicide-amended PDA at 0, 0.1, 0.3, 1, 3, 10, 30, 100, 300, 1000, and 3000 μg a.i. mL^{-1}. To inoculate test plates, mycelial plugs were removed with a 5-mm cork borer from the margins of 5-day-old colonies and placed upside down on the centers of 9-cm plastic Petri dishes containing fungicide-amended or unamended media. Each isolate was tested in triplicate, and plates were incubated until the diameter reached 60 mm (around five days for *B. cinerea* and nine days for C. rosea). Fungicide sensitivity, as measured by the 50% effective concentration (EC_{50}) value, was calculated as described by Wong and Wilcox (2002) [33]. Briefly, the percent relative growth (RG) was calculated as (radial growth at fungicide concentration/radial growth on the non-amended control plate) × 100. The EC_{50} value was estimated by linear regression of the probit-transformed relative inhibition (RI) value (RI = 1 − RG) on log10 transformed-fungicide concentration. The EC_{50} value for each isolate was calculated as the mean of the three replicates.

2.3. Effect of SDHI Fungicides to C. rosea Conidia Germination

To determine the inhibition effect of SDHI fungicides boscalid, fluxapyroxad, fluopimomide, and fluopyram on *C. rosea* and *B. cinerea*, a spore germination rate test was conducted as described. To stimulate sporulation, *C. rosea* isolate 67-1 was inoculated in Czapek Dox Liquid Medium (Sigma-Aldrich, St.Louis, MO, USA) [34]. *B. cinerea* isolate YN80 was inoculated in a PDA medium. Conidia were harvested by flooding 1–2-week-old *C. rosea* and *B. cinerea* cultures with a sterile scraper and suspending them in sterile distilled water. The conidial concentration of *C. rosea* and B. cinerea was then quantified microscopically using a hemocytometer and diluted to a concentration of 1.0×10^6 conidia mL^{-1}. An aliquot of 200 μL of conidia suspension was plated on the YBA medium (10 g L^{-1} bacto-peptone (Sinopharm, Beijing, China), and 20 g L^{-1} sodium acetate (Sinopharm, Beijing, China), 10 g L^{-1} yeast extract (Sinopharm, Beijing, China), and 15 g L^{-1} agar (Sinopharm, Beijing, China)), then mixed with fungicide using a sterile glass spreader at the final concentrations of 0, 7.5, 15, 30, 60, and 120 μg mL^{-1}. After 18–24 h incubation at 25 °C in the dark, the number germinated per 100 conidia was counted, and the germination rate of conidia was calculated. The experiment was performed twice.

2.4. Greenhouse Experiments

Tomato (*Solanum lycopersicum* Dunal L.) seedlings (Jinpengwuxian, Xi'an Jinpeng Seedling Co., Ltd., Xi'an, China) were planted in 1 kg of autoclaved potting medium (field soil/peat/sand, 1:1:1 wt/vol/wt; one seedling per pot) and maintained under a 16-h photoperiod at 90% relative humidity and 25 °C room temperature. Forty-day-old tomato seedlings were used for the inoculation test. Nine treatments were applied to the seedlings to measure the synergistic effects of C. *rosea* and SDHI fungicides: 1. YN80 treatment, inoculated with mycelial plugs of *B. cinerea* isolate YN80 and sprayed with distilled water; 2. 67-1 treatment, sprayed with 6 mL of 10^7 conidia mL^{-1} conidia suspension of C. *rosea* isolate 67-1; 3. fluxapyroxad treatment, sprayed with 6 mL of 30 μg mL^{-1} a.i. fluxapyroxad; 4. fluopyram treatment, sprayed with 6 mL of 30 μg mL^{-1} a.i. fluopyram; 5. 67-1 combined with fluxapyroxad treatment, sprayed with the mixture of conidia suspension and fluxapyroxad (5×10^6 conidia mL^{-1} conidia suspension: 15 µg mL^{-1} a.i fluxapyroxad, 1:1); 6. 67-1 combined with fluopyram treatment, sprayed with the mixture of conidia suspension and fluopyram (5×10^6 conidia mL^{-1} conidial suspension:15 µg mL^{-1} a.i fluopyram, 1:1); 7. 67-1 rotate with fluxapyroxad treatment, sprayed the 5×10^6 conidia mL^{-1} conidia suspension first, and fluxapyroxad (15 µg mL^{-1} a.i) 24 h later; 8. 67-1 rotate with fluopyram treatment, sprayed the 5×10^6 conidia mL^{-1} conidia suspension first, and fluopyram (15 µg mL^{-1} a.i) 24 h later; 9. blank control, only sprayed with 8 mL of distilled water. Moreover, 0.1% Tween 80 was included in all spray treatments as a surfactant.

After 24 h, all of the above tomato seedlings treatments were inoculated with 5-mm-agar plugs of *B. cinerea* isolate YN80 on the leaves referred to Myresiotis et al. [32], except the blank control treatment. Each plant was inoculated with ten agar plugs, one plug for each leaf. Six pots were prepared for each treatment. After inoculation, tomato plants were immediately returned to the chamber to maintain a high relative humidity and an appropriate temperature. Seven days after inoculation, lesion diameters were measured at two perpendicular directions using a caliper, and the control efficacy of each treatment was calculated. The experiments were performed three times.

2.5. qPCR for Specific Quantification of C. rosea *and* B. cinerea

To measure the concentration of the DNA, standard plasmids were constructed. The DNA sequence for *B. cinerea* was amplified using the primers P1 (5′-GCTGTAATTTCAATGT GCAGAATCC-3′) and P2 (5′-GGAGCAACAATTAATCGCATTTC-3′) targeting the Bcos5 gene as reported by Duan et al. [35]. As for *C. rosea*, primers targeting β-tubulin-encoding genes were retrieved from Genbank (Accession number AF435066). Primers CLO-QF/CLO-QR (CAACAACAACGAGTGGGGAG/ATAAAAGACGGAGCGAAGAC) were designed and used in this study. PCR reactions were performed as follows: 95 °C for 5 min, and then 35 cycles of denaturation at 95 °C for 30 s, annealing at 60 °C for 30 s, extension at 72 °C for 30 s, with a final extension at 72 °C for 10 min. Then, purified PCR products were inserted into the cloning vector pClone007 Vector Kit (Tsingke Biotechnology, Beijing, China), and transformed into an *E. coli* DH5α competent cell. The transformed competent cells were coated in the LB medium (Luria-Bertani: tryptone (Sinopharm, Beijing) 10 g L^{-1}, yeast extract (Sinopharm, Beijing) 5 g L^{-1}, NaCl (Sinopharm, Beijing) 10 g L^{-1}, agar (Sinopharm, Beijing) 15 g L^{-1}) containing 200 µg mL^{-1} of ampicillin, and incubated at 37 °C to obtain the target cell after 12–16 h. The plasmid DNA was extracted from the target cell using a plasmid mini kit (Tsingke Biotechnology, Beijing, China). The plasmid DNA was used for preparing 10-fold dilution series of eight concentration points starting with about 10 ng/µL, as a "fungal DNA series". The initial stock solution contained around 3×10^8 target copies/µL, which was calculated by converting the stock concentration and the mass of the fragment into copy numbers. The concentration of plasmid DNA was quantified by spectrophotometry. The standard curve was prepared in fungal DNA series and amplified to obtain standard curves. Each standard curve was measured in three technical replicates. Standard curves were generated by plotting the logarithmic values of target copies versus the corresponding cycle threshold (Ct) values and fitted into a linear regression model. It was always checked that the R^2 of standard curves ranged from 0.99 to 1. Only Ct values inferior to 40 for *B. cinerea* and 35 for *C. rosea* were considered to avoid false positives, and each standard was measured in three technical replicates.

Following the method in Section 2.4, fifteen leaves (five for each plant) were collected from treatment "*B. cinerea* treatment", "*C. rosea* treatment", "fluxapyroxad", "fluopyram", "67-1 combined with fluxapyroxad treatment", "67-1 combined with fluopyram treatment" and then ground into a fine powder under liquid nitrogen. For each sample, 150 ± 2 mg was used for DNA extraction to detect fungal content by qPCR. The genomic DNA was subsequently extracted using the Plant Genomic DNA Kit (TIANMO BIOTECH, Beijing, China) according to the manufacturer's instructions.

As for the qPCR detection of *B. cinerea* and *C. rosea*, primers P1/ P2 and CLO-QF/CLO-QR for the construction of standard plasmid were used. All qPCR reactions were performed on QuantStudio™ 6 Flex Real-Time PCR System (Applied Biosystems, Waltham, MA, USA) in transparent Multiwell 96-well plates and sealed with adhesive foil. Twenty microliter reaction volume contained 10 μL TSINGKE TSE201 2×TSINGKE® Master qPCR Mix (SYBR Green I) (Tsingke Biotechnology Co., Ltd., China), 0.8 μL of each primer, 0.4 μL 50×ROX Reference Dye II (Tsingke Biotechnology Co., Ltd., China), 7 μL of DNAse-free water, and 1 μL of DNA sample (unless otherwise stated). The detection wavelength was 520 nm ±10 nm. The following thermal program was applied: an initial denaturation step of 94 °C for 5 min, followed by 40 amplification cycles of 15 s denaturing step (94 °C) and 60 s annealing-extension step (60 °C). All of the experiments were repeated independently twice. Three replications per sample were included in all of the experiments.

2.6. Statistical Analysis

Control efficacy = [(lesion diameter of the control − lesion diameter of the treatment)/lesion diameter of the control] × 100%. Results were represented as the mean values ± standard deviation. One-way analysis of variance (ANOVA) with a least significant difference (LSD) test in SPSS software (version 21.0; IBM SPSS Inc. Chicago, IL, USA) was used to evaluate the significant differences between treatments.

3. Results

3.1. In Vitro Mycelial Growth Inhibition of *C. rosea* and *B. cinerea* by Differernt Fungicides

To test the compatibility of *C. rosea* and fungicides, carbendazim, tebuconazole, pyraclostrobin, and boscalid were selected as representative fungicides for Methyl Benzimidazole Carbamates (MBCs), sterol demethylation inhibitors (DMIs), quinone outside inhibitors (QoIs), and SDHIs fungicides, respectively. The sensitivity of *C. rosea* isolates to those fungicides was tested (Figure 1). Overall, *C. rosea* isolates displayed the strongest tolerance to SDHI fungicide boscalid. Boscalid at 10 μg mL^{-1} or 30 μg mL^{-1} showed no suppressive activity against mycelium growth of *C. rosea* on PDA medium. In contrast, *C. rosea* isolates were quite sensitive to cabendazim and pyraclostrobin, with EC_{50} values of 0.34 μg mL^{-1}-1.66 μg mL^{-1} and 0.52 μg mL^{-1}-11.17 mL^{-1}, respectively. Tebuconazole also had an inhibitory effect on *C. rosea* mycelia for most of the isolates tested (except for isolate NHH−48-2), with EC_{50} values of 0.02 μg mL^{-1}-21.11 μg mL^{-1}. *C. rosea* isolate NHH-48-2 was tolerant to tebuconazole, with EC_{50} values of 102.86 μg mL^{-1} (Table 1).

Table 1. Fungicide sensitivities of the *Clonostachys rosea* and *Botrytis cinerea* isolates to carbendazim, tebuconazole, pyraclostrobin, boscalid, fluxapyroxad, fluopimomide, and fluopyram.

Species	Isolate	EC_{50} (μg mL^{-1}) [z] ± SE						
		Carbendazim	Tebuconazole	Pyraclostrobin	Boscalid	Fluxapyroxad	Fluopimomide	Fluopyram
Clonostachys rosea	JLB-7-1	1.66 ± 0.27	0.02 ± 0.01	11.17 ± 2.08	>1000	>1000	>1000	>1000
	67-1	1.04 ± 0.66	10.24 ± 1.71	0.52 ± 0.29	>1000	>1000	>1000	>1000
	SYP-4-2	0.34 ± 0.23	9.39 ± 1.63	0.59 ± 0.32	>1000	>1000	>1000	>1000
	SHW-1-1	0.73 ± 0.48	21.11 ± 11.20	4.46 ± 3.42	>1000	>1000	>1000	>1000
	YJS-3-2	0.50 ± 0.34	20.39 ± 14.54	0.74 ± 0.35	>1000	>1000	>1000	>1000
	GS6-1	0.89 ± 0.57	7.46 ± 5.95	0.52 ± 0.31	>1000	>1000	>1000	>1000
	NHH-48-2	0.71 ± 0.49	102.86 ± 53.70	3.08 ± 1.31	>1000	>1000	>1000	>1000
	BD-2-1	0.74 ± 0.52	16.87 ± 11.57	0.66 ± 0.47	>1000	>1000	>1000	>1000
Botrytis cinerea	YN80	0.01 ± 0.002	0.27 ± 0.16	31.95 ± 10.97	15.46 ± 4.50	1.75 ± 1.41	12.96 ± 5.85	1.12 ± 0.67
	YN81	0.03 ± 0.006	0.47 ± 0.17	22.69 ± 6.16	5.95 ± 3.98	0.40 ± 0.33	33.41 ± 7.34	1.92 ± 1.15

[z] EC_{50} = Effective concentration that inhibits 50% of fungal growth. SE, standard error.

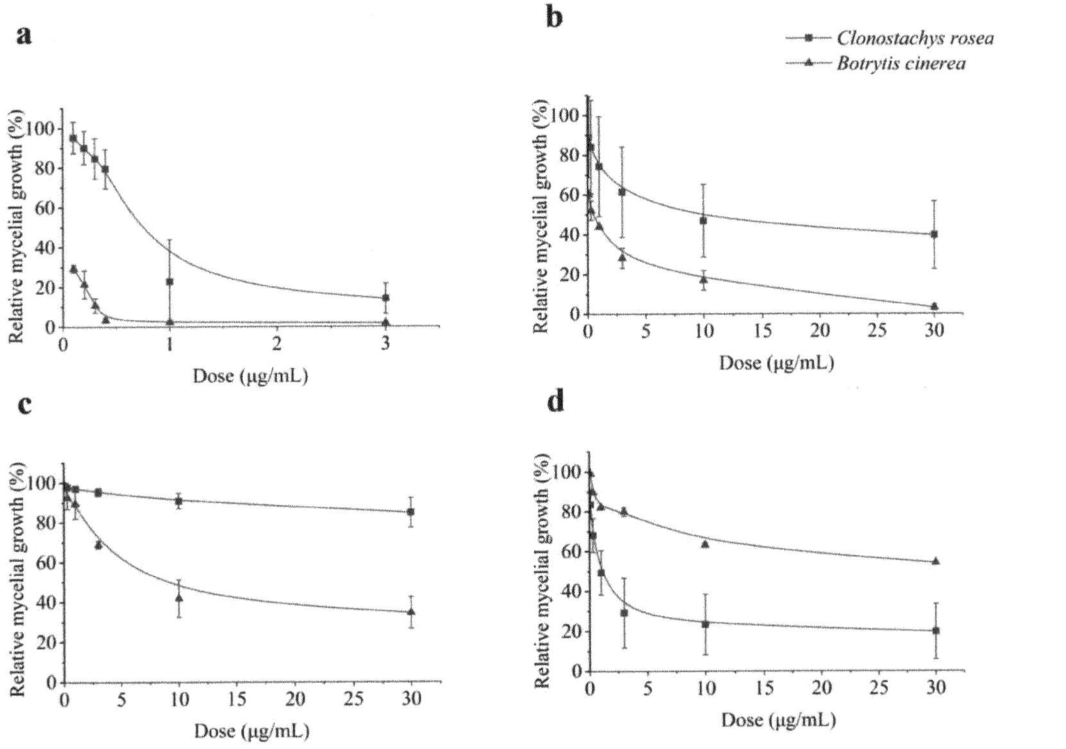

Figure 1. Contrasting in vitro relative growth of *Clonostachys rosea* and *Botrytis cinerea* on different concentrations of (**a**) carbendazim, (**b**) tebuconazole, (**c**) boscalid, and (**d**) pyraclostrobin in PDA medium. Mean and standard deviation from the average of eight *C. rosea* isolates (isolates JLB-7-67-1, SYP-4-2, SHW-1-1, YJS-3-2, GS6-1, NHH-48-2, and BD-2-1) and two *B. cinerea* (YN80 and YN81) isolates were collected.

To further explore the compatibility of SDHI fungicides with *C. rosea* isolates, more fungicides from the same categories were tested for their effects on *C. rosea* isolates. A more comprehensive range of concentration was tested for SDHI fungicides boscalid, fluxapyroxad, fluopimomide, and fluopyram from 0.1 μg mL^{-1} to 3000 μg mL^{-1}. All tested *C. rosea* isolates displayed strong tolerance to all SDHI fungicides tested. When treated with 100 μg mL^{-1} of SDHIs, the growth of mycelium was only suppressed by 9.11% to 28.20% (Figure 2). Even when treated with 3000 μg mL^{-1} of SDHIs, the mycelium could grow by 53.73% to 77.96% compared to the unamended control. In contrast, the *B. cinerea* isolates YN80 and YN81 were sensitive to all the SDHI fungicides tested, with EC_{50} less than 15.46 μg mL^{-1} (Table 1). Based on the EC_{50} value, fluxapyroxad and fluopyram were most effective against the *B. cinerea* isolates used in this study. Thus, those two fungicides were selected for the following experiments.

Figure 1. Contrasting in vitro relative growth of *Clonostachys rosea* and *Botrytis cinerea* on differe concentrations of (**a**) carbendazim, (**b**) tebuconazole, (**c**) boscalid, and (**d**) pyraclostrobin in PD medium. Mean and standard deviation from the average of eight *C. rosea* isolates (isolates JLB-7- 67-1, SYP-4-2, SHW-1-1, YJS-3-2, GS6-1, NHH-48-2, and BD-2-1) and two *B. cinerea* (YN80 and YN8 isolates were collected.

To further explore the compatibility of SDHI fungicides with *C. rosea* isolates, mo fungicides from the same categories were tested for their effects on *C. rosea* isolates. more comprehensive range of concentration was tested for SDHI fungicides boscali fluxapyroxad, fluopimomide, and fluopyram from 0.1 μg mL^{-1} to 3000 μg mL^{-1}. All teste *C. rosea* isolates displayed strong tolerance to all SDHI fungicides tested. When treate with 100 μg mL^{-1} of SDHIs, the growth of mycelium was only suppressed by 9.11% 28.20% (Figure 2). Even when treated with 3000 μg mL^{-1} of SDHIs, the mycelium cou grow by 53.73% to 77.96% compared to the unamended control. In contrast, the *B. ciner* isolates YN80 and YN81 were sensitive to all the SDHI fungicides tested, with EC_{50} le than 15.46 μg mL^{-1} (Table 1). Based on the EC_{50} value, fluxapyroxad and fluopyram we most effective against the *B. cinerea* isolates used in this study. Thus, those two fungicid were selected for the following experiments.

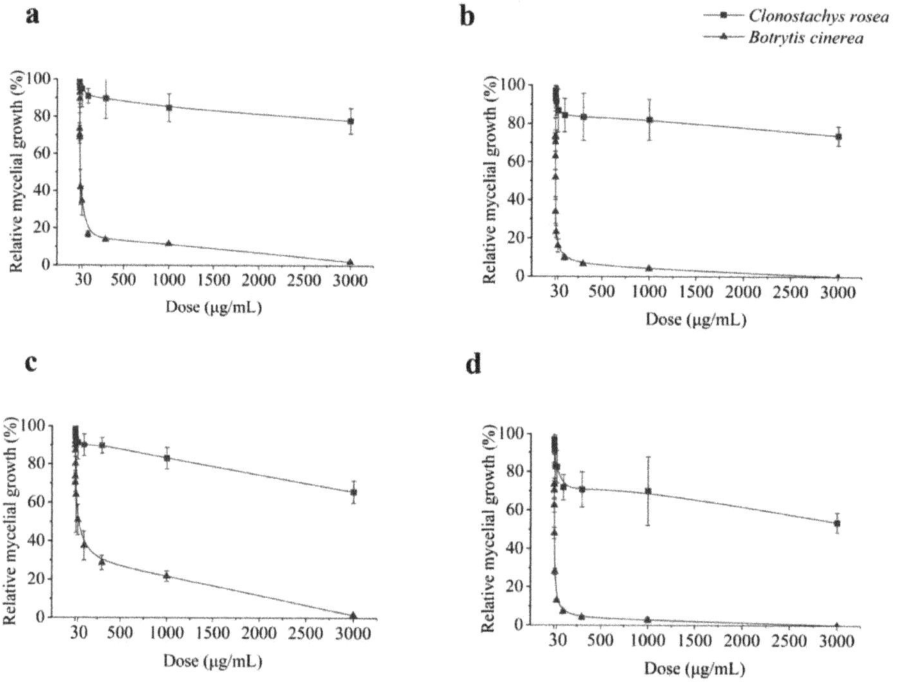

Figure 2. Contrasting in vitro relative growth of *Clonostachys rosea* and *Botrytis cinerea* on different concentrations of (**a**) boscalid, (**b**) fluxapyroxad, (**c**) fluopimomide, and (**d**) fluopyram. Mean and standard deviation from the average of eight *C. rosea* isolates (isolates JLB-7-1, 67-1, SYP-4-2, SHW-1-1, YJS-3-2, GS6-1, NHH-48-2, and BD-2-1) and two *B. cinerea* isolates (YN80 and YN81) were collected.

3.2. Inhibition Effect of Fungicides on the Germination Rate of C. rosea Conidium

The germination inhibition assays of SDHI fungicides were also conducted in our study. The SDHI fungicides had strong inhibitory activity on the spore germination of *B. cinerea*. The germination rate of YN80 was less than 10% when treated with 15 μg mL^{-1} of fluxapyroxad and fluopyram (Table 2). In contrast, the inhibitory activity of fluxapyroxad and fluopyram against *C. rosea* was very weak. A strong residual growth (with a germination rate above 95%) was observed for *C. rosea* isolate 67-1 when treated with 120 μg mL^{-1} of fluxapyroxad and fluopyram. Thus, good compatibility was observed for SDHI fungicides and *C. rosea* in vitro.

Table 2. In vitro germination rate of conidia of *Clonostachys rosea* and *Botrytis cinerea* under different fungicide concentrations.

Species [y]	Fungicide	Gemination Rate of Conidium at Different Fungicide Concentrations (%) [z]					
		Concentrations of Fungicides (μg mL^{-1})					
		0	7.5	15	30	60	120
Clonostachys rosea	Fluxapyroxad	99.99 ± 0.01	99.99 ± 0.01	99.99 ± 0.01	99.99 ± 0.01	99.99 ± 0.01	99.99 ± 0.01
	Fluopyram	99.99 ± 0.01	99.99 ± 0.01	99.99 ± 0.01	99.34 ± 0.47	97.70 ± 0.07	97.34 ± 0.41
Botrytis cinerea	Fluxapyroxad	95.71 ± 0.71	16.95 ± 1.86	8.34 ± 3.36	5.67 ± 3.26	3.61 ± 1..02	1.80 ± 0.08
	Fluopyram	95.71 ± 0.71	21.05 ± 0.76	8.62 ± 3.71	10.81 ± 0.50	4.62 ± 0.37	2.00 ± 0.08

[y] Isolate 67-1 represented the *Clonostachys rosea*, isolate YN80 represented the *Botrytis cinerea*. [z] Mean ± standard deviation; at least 200 conidia were examined microscopically to determine germination in each of three replicate plates 24 h at 22 °C.

3.3. Synergistic Effects of C. rosea Isolate 67-1 and SDHI Fungicides against Tomato Gray Mold in the Greenhouse

The data regarding the combined effects of *C. rosea* isolate 67-1 and SDHI fungicides against tomato gray mold in the greenhouse are presented in Figure 3 and Table 3. The average disease diameter in the control group was 2.67 cm in the greenhouse, indicating that *B. cinerea* was successfully inoculated and well developed (Figure 3). Overall, the combined application of *C. rosea* and SDHI fungicides, either in a mixture or in a rotation, significantly reduced the disease incidence and severity of tomato gray mold. The highest control efficacy of 77.07% was obtained with pretreatment of isolate 67-1 at 5×10^6 conidia mL^{-1} and then fluopyram at 15 μg mL^{-1}. The control efficacy of the combined application of isolate 67-1 with fluxapyroxad and fluopyram reached 70.91% and 71.94%, respectively. Sole treatment of fluxapyroxad and fluopyram at 30 μg mL^{-1} produced a significantly lower control efficacy of 52.28% and 58.31%, respectively, while *C. rosea* treatment 10^7 conidia mL^{-1} yielded a control efficacy of 46.42% (Table 3).

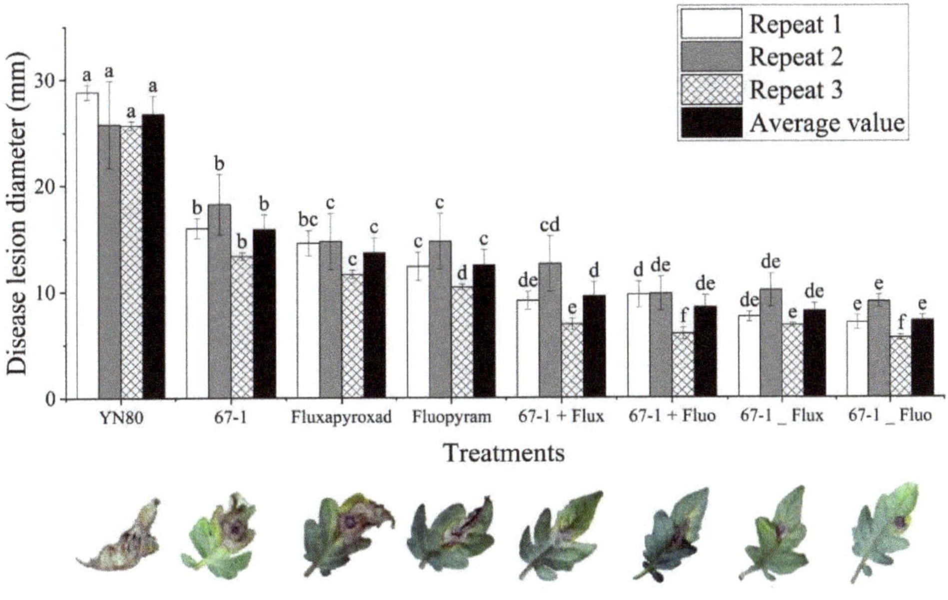

Figure 3. The disease lesion diameter of tomato gray mold when treated by *Clonostachys rosea* 67-1, fluxapyroxad, fluopimomide alone, in combination, or in rotation. Data are presented as the mean ± standard deviation (SD). The different lowercase letters indicate significant differences between different treatments in each repeat at the 5% level of probability. "YN80", only inoculated with mycelial plugs of *B. cinerea* isolate YN80; "67-1", sprayed with conidia suspension of *C. rosea* isolate 67-1; "Fluxapyroxad", sprayed with fluxapyroxad; "Fluopyram", sprayed with fluopyram; "67-1+Flux", sprayed with the mixture of 67-1 conidia suspension and fluxapyroxad; "67-1+Fluo" sprayed with the mixture of 67-1 conidia suspension and fluopyram; "67-1_Flux", sprayed the 67-1 conidia suspension first and fluxapyroxad 24 h later; "67-1_Fluo" sprayed the 67-1 conidia suspension first and fluxapyroxad 24 h later.

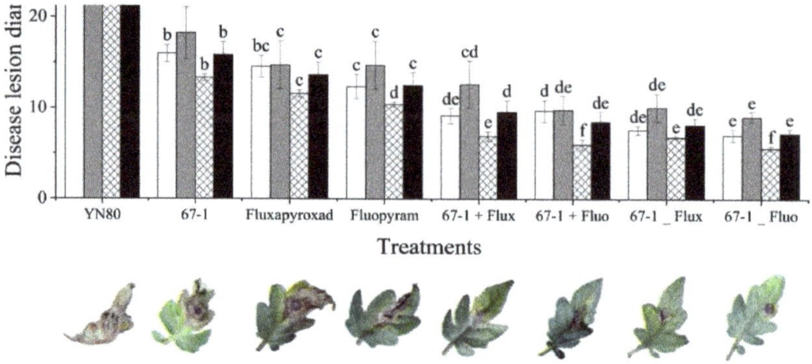

67-1+Fluo	71.94% ± 6.34% ab
67-1_Flux	73.65% ± 1.24% ab
67-1_Fluo	77.07% ± 2.26% a

Figure 3. The disease lesion diameter of tomato gray mold when treated by *Clonostachys rosea* 67-1, fluxapyroxad, fluopimomide alone, in combination, or in rotation. Data are presented as the mean ± standard deviation (SD). The different lowercase letters indicate significant differences between different treatments in each repeat at the 5% level of probability. "YN80", only inoculated with mycelial plugs of *B. cinerea* isolate YN80; "67-1", sprayed with conidia suspension of *C. rosea* isolate 67-1; "Fluxapyroxad", sprayed with fluxapyroxad; "Fluopyram", sprayed with fluopyram; "67-1+Flux", sprayed with the mixture of 67-1 conidia suspension and fluxapyroxad; "67-1+Fluo" sprayed with the mixture of 67-1 conidia suspension and fluopyram; "67-1_Flux", sprayed the 67-1 conidia suspension first and fluxapyroxad 24 h later; "67-1_Fluo" sprayed the 67-1 conidia suspension first and fluxapyroxad 24 h later.

[y] Treatment "67-1", sprayed with conidia suspension of *Clonostachys rosea* isolate 67-1; "Fluxapyroxad", sprayed with fluxapyroxad; "Fluopyram", sprayed with fluopyram; "67-1+Flux", sprayed with the mixture of 67-1 conidia suspension and fluxapyroxad; "67-1+Fluo" sprayed with the mixture of 67-1 conidia suspension and fluopyram; "67-1_Flux" [illegible]; "67-1_Fluo" sprayed the conidia suspension of *C. rosea* isolate 67-1 first and fluxapyroxad 24 h later. After 24 h, all of the above tomato seedlings treatments were inoculated with 5-mm-agar plugs of *Botrytis cinerea* isolate YN80 on the leaves. [z] Data are presented as the mean ± standard deviation (SD). The different lowercase letters indicate significant differences between different treatments in each repeat at the 5% level of probability. One-way analysis of variance (ANOVA) with a least significant difference (LSD) test in SPSS software (version 21.0; SPSS Inc.) was used to evaluate the significant differences between treatments.

3.4. qPCR for Specific Quantification of C. rosea and B. cinerea

The evaluation of the Ct values from the standard curve amplification for both *B. cinerea* and *C. rosea* revealed a linear dynamic range from 10^2 to 10^6 target copies, corresponding to a Ct range of 39~14 for *B. cinerea* and 32~14 for *C. rosea* (Figure 4). The lower limit of detection of *C. rosea* was determined around one target copy per reaction as 35 cycles were set to be the cutoff value for the method. Similarly, 40 cycles were set to be the cutoff value for *B. cinerea*. Linear regressions between the log-transformed number of target copies and the corresponding Ct values revealed R^2 values > 0.99 for both *B. cinerea* and *C. rosea* reactions. No PCR inhibition was observed when different amounts of plant DNA isolated from tomato plants were added to the qPCR, increasing concentrations from 1, 10, 25, 50, to 100 ng (data not shown).

3.4. qPCR for Specific Quantification of C. rosea and B. cinerea

The evaluation of the Ct values from the standard curve amplification for both B. cinerea and C. rosea revealed a linear dynamic range from 10^2 to 10^6 target copies, corresponding to a Ct range of 39~14 for B. cinerea and 32~14 for C. rosea (Figure 4). The lower limit of detection of C. rosea was determined around one target copy per reaction as 35 cycles were set to be the cutoff value for the method. Similarly, 40 cycles were set to be the cutoff value for B. cinerea. Linear regressions between the log-transformed number of target copies and the corresponding Ct values revealed R^2 values > 0.99 for both B. cinerea and C. rosea reactions. No PCR inhibition was observed when different amounts of plant DNA isolated from tomato plants were added to the qPCR, increasing concentrations from 1, 10, 25, 50, to 100 ng (data not shown).

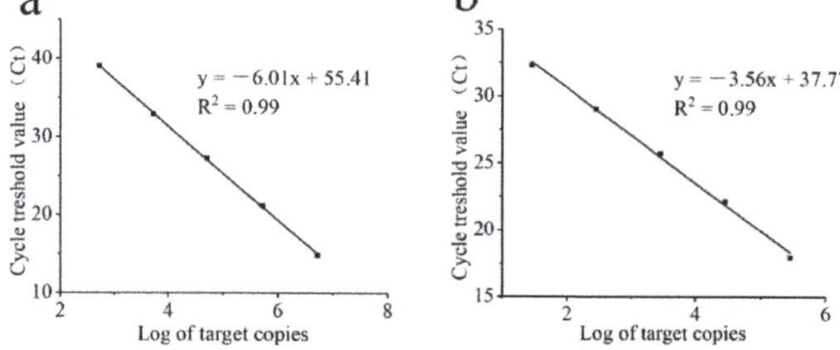

Figure 4. Standard curve of the qPCR for quantification of *Botrytis cinerea* (**a**) and *Clonostachys rosea* (**b**). qPCR standard regression was obtained from the log of the copy number of *B. cinerea* (**a**) and *C. rosea* (**b**) against the corresponding cycle threshold (Ct) values. Target range was from 5.25×10^2 to 5.25×10^6 copies per reaction for *B. cinerea*, and 2.91×10 to 2.91×10^5 per reaction for *C. rosea*. The number of target copies on a log-scaled X-axis were plotted against Ct values from 14 to 40 for *B. cinerea* isolate YN80 and 14 to 32 for *C. rosea* isolate 67-1 on the Y-axis. Linear regression equation of the *B. cinerea* standard curve was $Y = -6.01x + 55.41$ at $R^2 = 0.99$. Linear regression equation of the *C. rosea* standard curve was $Y = -3.56x + 37.77$ at $R^2 = 0.99$.

Thus, the qPCR method was applied to determine the survival of *B. cinerea* and *C. rosea* on tomato plants. In sample sets, *B. cinerea* and *C. rosea* were always detected when applied and not in the negative control samples. For *B. cinerea*, the trend of detection level for different treatments was inconsistent with those in the pot experiments. Take repeat 1, for example: two of the lowest levels of detection, with 1.66×10^4 copies and 1.08×10^4 copies, reflecting the lowest survival of *B. cinerea*, were found when treated with *C. rosea* combined with fluxapyroxad and fluopyram, respectively. When treated with *C. rosea*, the detection levels (with copies of 6.37×10^4) were higher than those that were treated with fluopyram or fluxapyroxad (with copies of 3.72×10^4 and 3.69×10^4, respectively) but lower than those that were treated with distilled water (with copies of 8.55×10^4). For *C. rosea*, the qPCR results showed that *C. rosea* could still be detected on tomato plants when mixed with fluopyram and fluxapyroxad (Figure 5).

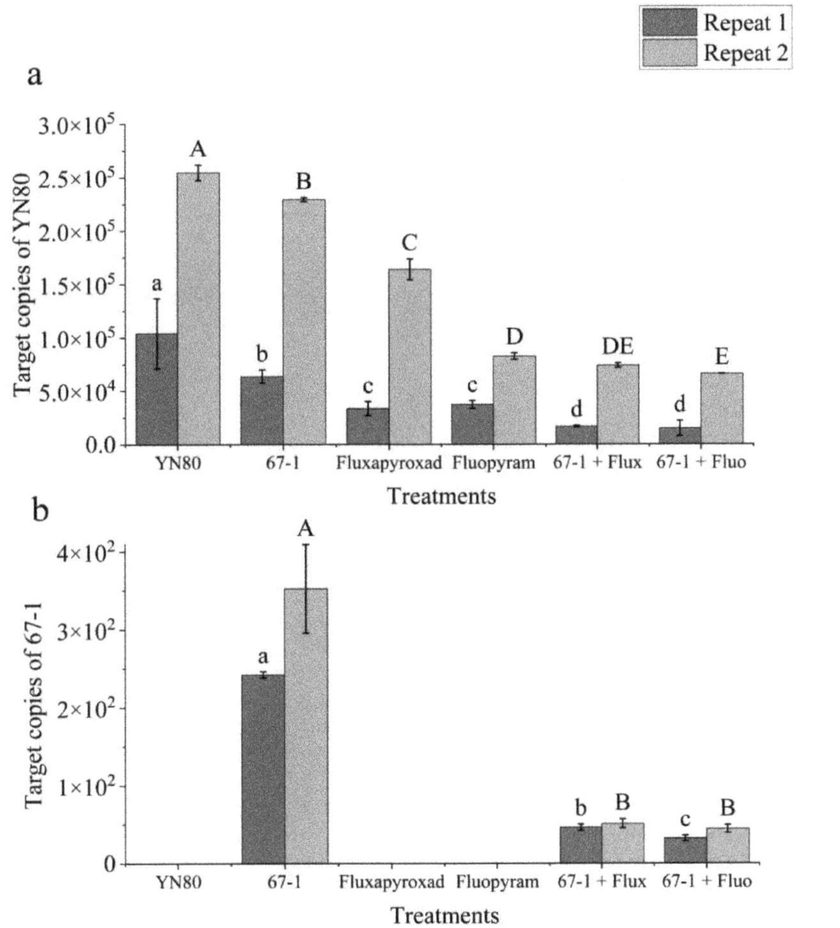

Figure 5. Quantification of *Botrytis cinerea* (**a**) and *Clonostachys rosea* (**b**) by TaqMan qPCR in treated tomato plants after 10 days under greenhouse conditions. Data are presented as the mean ± standard deviation (SD). Uppercase letters and lowercase letters represent two independent repeated tests. The different letters indicate significant differences between different treatments in each repeat ($\alpha = 0.05$). "YN80", only inoculated with mycelial plugs of *B. cinerea* isolate YN80; "67-1", sprayed with conidia suspension of *C. rosea* isolate 67-1; "Fluxapyroxad", sprayed with fluxapyroxad; "Fluopyram", sprayed with fluopyram; "67-1+Flux", sprayed with the mixture of 67-1 conidia suspension and fluxapyroxad; "67-1+Fluo" sprayed with the mixture of 67-1 conidia suspension and fluopyram.

4. Discussion

A combination of synthetic fungicides with BCA or a combination of different BCAs has been reported to reduce chemical application rates. Several combinations of BCA with fungicides have shown greater efficacy than the individual treatments. For example, combining *B. amyloliquefaciens* SDTB009 with difenoconazole is an effective strategy for tomato *Fusarium wilt* management [8]. Synergistic effects have been observed in the combined application of *Bacillus subtilis* H158 and strobilurins for rice sheath blight control [36]. The combination of *Trichoderma* and hymexazol enhanced antagonistic effects towards F. oxwith fungicides have shown greater efficacy than the individual treatments. For example, combining *B. amyloliquefaciens* SDTB009 with difenoconazole is an effective strategy for tomato *Fusarium wilt* management [8]. Synergistic effects have been observed in the combined application of *Bacillus subtilis* H158 and strobilurins for rice sheath blight control [36]. The combination of *Trichoderma* and hymexazol enhanced antagonistic effects towards *F. oxysporum* [37]. Besides, the combination of *Metarhizium robertsii* and *Trichoderma asperellum* reduced the malathion doses in controlling ambrosia beetles [38]. However, few studies showed the combination of *C. rosea* and fungicides or other BCAs in the control of plant disease. In this study, the compatibility of SDHIs fungicides was evaluated and the synergistic effect of the combined use of *C. rosea* and SDHI fungicides against tomato gray mold was investigated.

The action targets of fungicides against pathogenic fungi include cell membrane integrity, cell mitosis, nucleic acid metabolism, respiration, signal transduction, and protein synthesis [24]. However, some active ingredients of fungicides also act on non-target or beneficial microorganisms such as BCAs, which reduce the growth and population size of BCAs and limit the biocontrol effect [39]. Therefore, knowledge of the compatibility of BCAs and fungicides is essential to allow combined applications. Generally, fungal BCAs resistant to specific fungicides or bacterial BCAs have good compatibility. Compared with the biocontrol fungus, biocontrol bacteria, such as *B. amyloliquefaciens* and *B. subtillis* have been reported to tolerate many fungicides and exhibit synergistic effects when applied in combination [40–43]. The combination of hymexazol-resistant *Trichoderma* isolate with hymexazol also showed good compatibility and enhanced antagonistic potential [37]. Potential additive or synergistic effects of *C. rosea* and fungicides depend first on the biological compatibility between the biocontrol agent and the synthetic chemical. In this study, we screened several different categories of fungicides to identify their compatibility with *C. rosea*. Four FRAC code fungicides that are frequently used for the control of gray mold have been selected. *C. rosea* isolates were quite sensitive to carbendazim, pyraclostrobin, and tebuconazole in vitro. Fortunately, we found that *C. rosea* could tolerate SDHI fungicides, including boscalid, fluxapyroxad, fluopimomide, and fluopyram. Even when treated with 3000 μg mL^{-1} of SDHIs, the mycelium could grow quite well. The natural resistance of fungus to SDHI fungicides are not uncommon. The insensitivity of plant pathogens *Colletotrichum* species to boscalid, fluxapyroxad, and fluopyram have been confirmed on media and on plants [44]. Penflupen, a novel SDHI fungicide, exhibited good bioactivity against *F. fujikuroi*, but weak activity against other *Fusarium* spp. [45]. So far, the inherent resistance mechanisms in the above plant pathogens have remained unknown. As for *C. rosea*, the natural resistance to SDHIs allows them to be mixed with fungicides.

C. rosea 67-1 isolate has been reported to be a highly efficient biocontrol fungus targeting many plant pathogenic fungi, including *B. cinerea* [30,31]. Therefore, isolate 67-1 was selected for the following pot experiment. According to our data, the control effect of *C. rosea* alone was only slightly lower than the application of fungicides, which further proved that *C. rosea* 67-1 isolate is a promising BCA against *B. cinerea*. As *C. rosea* acts by competing for space and nutrients in wounded tissues [46], its efficacy in colonizing the host may depend on the amount of conidia applied. According to Borges et al., who compared the conidial concentration and disease control, the best results for control were obtained at a concentration above 10^6 conidia mL^{-1} one day before or simultaneously with the pathogens on tomato plants [15]. Thus, we applied *C. rosea* at 10^7 conidia mL^{-1} concentration for the control of *B. cinerea* in our pot experiments and halved the concentration of *C. rosea* to 5×10^6 conidia mL^{-1} when combined with the fungicides. Based on Chatterton and Punja's research, environmental factors such as temperature and pH were major factors that influenced population levels of *C. rosea* [14,47]. The optimum temperature for leaf colonization was 20–25 °C, and maximum population densities on the leaves required at least 12 h of continuous leaf wetness [14]. Hence, greenhouse environmental conditions were maintained at 90% relative humidity and 25 °C room temperature for the pot experiment to obtain a stable and efficient control effect.

Our study showed a significant synergistic effect of *C. rosea* with SDHIs. The control effect of the combination of *C. rosea* with fluxapyroxad or fluopyram against tomato gray mold was significantly increased compared to that of BCA or SDHI fungicide alone in combination treatment and rotation treatment; the combination allows a two-fold reduction of both the fungicide and BCA dose. Several possible mechanisms for the synergistic effects were observed upon the combined application of *C. rosea* and SDHIs. Firstly, as the primary biological control mechanism, *C. rosea* could secrete cell-wall-degrading enzymes (CWDEs) to degrade the cell wall of the host fungus [48–50]. Thus, with the lack of an essential barrier for cell protection, the gray mold might become more vulnerable to the fungicides treated. Second, *C. rosea* produced secondary metabolites such as antibiotics and toxins [51,52], and the combined application of these antibiotics or toxins with SDHIs may show the same synergistic effects as the synergistic effect shown in a combination of fungicides with one another. Third, treating *B. cinerea* infection with *C. rosea* has been reported to induce several defense mechanisms in tomatoes, including fortifying the plant cell wall and stimulating the expression of several signaling molecules [16,19,53]. In this way, the resistance of tomato plants to gray mold is enhanced when inoculated with *C. rosea*. After the fungicide treatment, the plants are less susceptible to gray mold, showing a synergistic effect.

Whether the BCAs survive on plants or colonize the plants successfully after the application is a crucial step for the biological control activity of many BCAs. Rapid activity loss is thought to be the main reason some BCAs are not successful in the field but show excellent performance in the lab [54]. It is reported that *B. subtilis* was rapidly lost 3 days after application on rice by using real-time qPCR detection [36]. This result is in accordance with a study of *B. subtilis* on a strawberry based on next-generation sequencing [55]. In terms of *C. rosea*, it was confirmed that *C. rosea* could successfully colonize the foliage of geraniums and the roots of cucumbers by using a GUS-transformed isolate, demonstrating the endophytic ability of *C. rosea* in foliar and root tissues [14,47]. In this study, DNA of *C. rosea* was directly extracted from tomato plants, and the fungal dynamics were analyzed by real-time qPCR to quantify *C. rosea* DNA. Although DNA extraction included dead and inactive fungi and may result in a higher gene expression level, it was believed to be the most available method because of its convenience and accuracy [36]. In the qPCR assays, though there were variations between the replicates, the replicates showed a similar trend (Figure 5). Because the absolute quantifications of *B. cinerea* and *C. rosea* were tested, it was very hard to repeat the absolute copy number from the two independent experiments. The environment and the status of the microorganisms can be slightly different from the two replicates, which ultimately influence the colonization. The qPCR test of *C. rosea* demonstrated that *C. rosea* could still be detected on tomatoes when used alone and mixed with fungicides. The qPCR test of *B. cinerea* showed that *C. rosea* and SDHI fungicide significantly reduced the biomass of *B. cinerea*. Compared to the control, the biomass of *B. cinerea* was the lowest in the combination treatment of *C. rosea* and SDHI fungicide, which is consistent with the control efficacy in the greenhouse.

In conclusion, our study showed that *C. rosea* isolates could tolerate high concentrations of SDHIs with no adverse growth effects, suggesting that they were fully compatible with these fungicides. Pot experiment and qPCR assays showed a significant synergistic effect of *C. rosea* with SDHIs in controlling tomato gray mold. These results showed that combining BCA with SDHIs may meet the demands of the Chinese government's "low fertilizer and low pesticides" campaign. Additional field trials and investigations to monitor the behavior of *C. rosea* in the field can help to determine the optimal timing and the method of this BCA application to control gray mold in tomato production.

Author Contributions: Writing—original draft, J.S., S.C.; Data curation, T.L., X.H., W.S.; Conceptualization, writing-review and editing, H.Y., S.C.; Funding acquisition, X.H., S.C.; Methodology S.C. All authors have read and agreed to the published version of the manuscript.

Funding: This material is based upon work supported by National Key Research Development Program of China (No. 2019YFD1002000) and Key Research and Development Program of Shanxi Province (201903D211001-1).

Data Availability Statement: The data used to support the findings of this study are available from the corresponding authors upon request.

Acknowledgments: We thank Manhong Sun of institution of plant protection, CAAS, for providing *C. rosea* isolates for this study.

Conflicts of Interest: The authors declare no conflict of interest.

References

1. Elad, Y.; Pertot, I.; Cotes Prado, A.M.; Stewart, A. Plant Hosts of *Botrytis* spp. In *Botrytis–the Fungus, the Pathogen and Its Management in Agricultural Systems*; Fillinger, S., Elad, Y., Eds.; Springer International Publishing: Cham, Switzerland, 2016; pp. 413–486.
2. Romanazzi, G.; Smilanick, J.L.; Feliziani, E.; Droby, S. Integrated management of postharvest gray mold on fruit crops. *Postharvest Biol. Technol.* **2016**, *113*, 69–76. [CrossRef]
3. Leroux, P.; Fritz, R.; Debieu, D.; Albertini, C.; Lanen, C.; Bach, J.; Gredt, M.; Chapeland, F. Mechanisms of resistance to fungicides in field strains of *Botrytis cinerea*. *Pest Manage. Sci.* **2002**, *58*, 876–888. [CrossRef]
4. Veloukas, T.; Kalogeropoulou, P.; Markoglou, A.N.; Karaoglanidis, G.S. Fitness and competitive ability of *Botrytis cinerea* field isolates with dual resistance to SDHI and QoI fungicides, associated with several sdhB and the cytb G143A mutations. *Phytopathology* **2013**, *104*, 347–356. [CrossRef] [PubMed]
5. Grimmer, M.K.; van den Bosch, F.; Powers, S.J.; Paveley, N.D. Fungicide resistance risk assessment based on traits associated with the rate of pathogen evolution. *Pest Manage. Sci.* **2015**, *71*, 207–215. [CrossRef] [PubMed]
6. Shao, W.; Zhao, Y.; Ma, Z. Advances in understanding fungicide resistance in *Botrytis cinerea* in China. *Phytopathology* **2021**, *111*, 455–463. [CrossRef] [PubMed]
7. McGovern, R.J. Management of tomato diseases caused by *Fusarium oxysporum*. *Crop Prot.* **2015**, *73*, 78–92. [CrossRef]
8. Xu, X.; Wang, Y.; Lei, T.; Sohail, M.A.; Wang, J.; Wang, H. Synergistic effects of *Bacillus amyloliquefaciens* SDTB009 and difenoconazole on Fusarium wilt of tomato. *Plant Dis.* **2022**, *106*, 2165–2171. [CrossRef]
9. Bi, Y.; Yu, Z. Diterpenoids from *Streptomyces* sp. SN194 and their antifungal activity against *Botrytis cinerea*. *J. Agric. Food Chem.* **2016**, *64*, 8525–8529. [CrossRef]
10. Jassbi, A.R.; Zare, S.; Asadollahi, M.; Schuman, M.C. Ecological roles and biological activities of specialized metabolites from the genus *Nicotiana*. *Chem. Rev.* **2017**, *117*, 12227–12280. [CrossRef]
11. Ji, X.; Li, J.; Meng, Z.; Zhang, S.; Dong, B.; Qiao, K. Synergistic effect of combined application of a new fungicide fluopimomide with a biocontrol agent *Bacillus methylotrophicus* TA-1 for management of gray mold in tomato. *Plant Dis.* **2019**, *103*, 1991–1997. [CrossRef]
12. Cota, L.V.; Maffia, L.A.; Mizubuti, E.S.G.; Macedo, P.E.F.; Antunes, R.F. Biological control of strawberry gray mold by *Clonostachys rosea* under field conditions. *Biol. Control* **2008**, *46*, 515–522. [CrossRef]
13. Zaldúa, S.; Sanfuentes, E. Control of *Botrytis cinerea* in *Eucalyptus globulus* mini-cuttings using *Clonostachys* and *Trichoderma* Strains. *Chil. J. Agric. Res.* **2010**, *70*, 576–582. [CrossRef]
14. Chatterton, S.; Punja, Z.K. Colonization of geranium foliage by *Clonostachys rosea* f. *catenulata*, a biological control agent of botrytis grey mould. *Botany* **2012**, *90*, 1–10. [CrossRef]
15. Borges, Á.V.; Saraiva, R.M.; Maffia, L.A. Biocontrol of gray mold in tomato plants by *Clonostachys rosea*. *Trop. Plant Pathol.* **2015**, *40*, 71–76. [CrossRef]
16. Mouekouba, L.D.O.; Zhang, L.L.; Guan, X.; Chen, X.L.; Chen, H.Y.; Zhang, J.; Zhang, J.F.; Li, J.F.; Yang, Y.J.; Wang, A. Analysis of *Clonostachys rosea*-induced resistance to tomato gray mold disease in tomato leaves. *PLoS ONE* **2014**, *9*, e102690. [CrossRef]
17. Mai, V.C.; Drzewiecka, K.; Jeleń, H.; Narożna, D.; Rucińska-Sobkowiak, R.; Kęsy, J.; Floryszak-Wieczorek, J.; Gabryś, B.; Morkunas, I. Differential induction of *Pisum sativum* defense signaling molecules in response to pea aphid infestation. *Plant Sci.* **2014**, *221–222*, 1–12. [CrossRef]
18. Quazi, S.A.J.; Meon, S.; Jaafar, H.; Ahmad, Z.A.B.M. The role of phytohormones in relation to bakanae disease development and symptoms expression. *Physiol. Mol. Plant Pathol.* **2015**, *90*, 27–38. [CrossRef]
19. Wang, Q.; Chen, X.; Chai, X.; Xue, D.; Zheng, W.; Shi, Y.; Wang, A. The involvement of jasmonic acid, ethylene, and salicylic acid in the signaling pathway of *Clonostachys rosea*-induced resistance to gray mold disease in tomato. *Phytopathology* **2019**, *109*, 1102–1114. [CrossRef]
20. Jacobsen, B.J.; Zidack, N.K.; Larson, B.J. The role of *Bacillus*-based biological control agents in integrated pest management systems: Plant diseases. *Phytopathology* **2004**, *94*, 1272–1275. [CrossRef]

21. Hu, M.-J.; Fernández-Ortuño, D.; Schnabel, G. Monitoring resistance to SDHI fungicides in *Botrytis cinerea* from strawberry fields. *Plant Dis.* **2016**, *100*, 959–965. [CrossRef]
22. Hägerhäll, C. Succinate: Quinone oxidoreductases. Variations on a conserved theme. *Biochim. Biophys. Acta. Bioenerg.* **1997**, *1320*, 107–141. [CrossRef] [PubMed]
23. Amiri, A.; Heath, S.M.; Peres, N.A. Resistance to fluopyram, fluxapyroxad, and penthiopyrad in *Botrytis cinerea* from strawberry. *Plant Dis.* **2013**, *98*, 532–539. [CrossRef] [PubMed]
24. FRAC. Fungicide Resistance Management. 2022. Available online: https://www.frac.info/fungicide-resistance-management/background (accessed on 20 December 2022).
25. Avenot, H.F.; Sellam, A.; Karaoglanidis, G.; Michailides, T.J. Characterization of mutations in the iron-sulphur subunit of succinate dehydrogenase correlating with boscalid resistance in *Alternaria alternata* from California pistachio. *Phytopathology* **2008**, *98*, 736–742. [CrossRef] [PubMed]
26. Broomfield, P.L.E.; Hargreaves, J.A. A single amino-acid change in the iron-sulphur protein subunit of succinate dehydrogenase confers resistance to carboxin in *Ustilago maydis*. *Curr. Genet.* **1992**, *22*, 117–121. [CrossRef]
27. Ito, Y.; Muraguchi, H.; Seshime, Y.; Oita, S.; Yanagi, S.O. Flutolanil and carboxin resistance in *Coprinus cinereus* conferred by a mutation in the cytochrome b 560 subunit of succinate dehydrogenase complex (Complex II). *Mol. Genet. Genomics* **2004**, *272*, 328–335. [CrossRef]
28. Matsson, M.; Ackrell, B.A.C.; Cochran, B.; Hederstedt, L. Carboxin resistance in *Paracoccus denitrificans* conferred by a mutation in the membrane-anchor domain of succinate:quinone reductase (complex II). *Arch. Microbiol.* **1998**, *170*, 27–37. [CrossRef]
29. Skinner, W.; Bailey, A.; Renwick, A.; Keon, J.; Gurr, S.; Hargreaves, J. A single amino-acid substitution in the iron-sulphur protein subunit of succinate dehydrogenase determines resistance to carboxin in *Mycosphaerella graminicola*. *Curr. Genet.* **1998**, *34*, 393–398. [CrossRef]
30. Sun, Z.B.; Li, S.D.; Ren, Q.; Xu, J.L.; Lu, X.; Sun, M.H. Biology and applications of *Clonostachys rosea*. *J. Appl. Microbiol.* **2020**, *129*, 486–495. [CrossRef]
31. Hasan, R.; Lv, B.; Uddin, M.J.; Chen, Y.; Fan, L.; Sun, Z.; Sun, M.; Li, S. Monitoring mycoparasitism of *Clonostachys rosea* against *Botrytis cinerea* using GFP. *J. Fungi* **2022**, *8*, 567. [CrossRef]
32. Myresiotis, C.K.; Bardas, G.A.; Karaoglanidis, G.S. Baseline sensitivity of *Botrytis cinerea* to pyraclostrobin and boscalid and control of anilinopyrimidine- and benzimidazole-resistant strains by these fungicides. *Plant Dis.* **2008**, *92*, 1427–1431. [CrossRef]
33. Wong, F.P.; Wilcox, W.F. Sensitivity to azoxystrobin among isolates of *Uncinula necator*: Baseline distribution and relationship to myclobutanil sensitivity. *Plant Dis.* **2002**, *86*, 394–404. [CrossRef] [PubMed]
34. Karlsson, M.; Durling, M.B.; Choi, J.; Kosawang, C.; Lackner, G.; Tzelepis, G.D.; Nygren, K.; Dubey, M.K.; Kamou, N.; Levasseur, A.; et al. Insights on the evolution of mycoparasitism from the genome of *Clonostachys rosea*. *Genome Biol. Evol.* **2015**, *7*, 465–480. [CrossRef] [PubMed]
35. Duan, Y.-B.; Ge, C.-Y.; Zhang, X.-K.; Wang, J.-X.; Zhou, M.-G. Development and evaluation of a novel and rapid detection assay for *Botrytis cinerea* based on loop-mediated isothermal amplification. *PLoS ONE* **2014**, *9*, e111094. [CrossRef] [PubMed]
36. Liu, L.; Liang, M.; Li, L.; Sun, L.; Xu, Y.; Gao, J.; Wang, L.; Hou, Y.; Huang, S. Synergistic effects of the combined application of *Bacillus subtilis* H158 and strobilurins for rice sheath blight control. *Biol. Control* **2018**, *117*, 182–187. [CrossRef]
37. Zhang, C.; Wang, W.; Xue, M.; Liu, Z.; Zhang, Q.; Hou, J.; Xing, M.; Wang, R.; Liu, T. The combination of a biocontrol agent *Trichoderma asperellum* SC012 and hymexazol reduces the effective fungicide dose to control Fusarium wilt in cowpea. *J. Fungi* **2021**, *7*, 685. [CrossRef]
38. Reynoso-López, E.A.; Méndez-Hernández, J.E.; Ek-Ramos, J.; Montesinos-Matías, R.; Loera, O. *Metarhizium robertsii* in combination with *Trichoderma asperellum* reduce the malathion doses used to control ambrosia beetles: The case of *Xyleborus affinis*. *Biocontrol Sci. Technol.* **2021**, *31*, 1080–1097. [CrossRef]
39. Yang, C.; Hamel, C.; Vujanovic, V.; Gan, Y. Fungicide: Modes of Action and Possible Impact on Nontarget Microorganisms. *ISRN Ecology* **2011**, *2011*, 130289. [CrossRef]
40. Peng, D.; Li, S.; Chen, C.; Zhou, M. Combined application of *Bacillus subtilis* NJ-18 with fungicides for control of sharp eyespot of wheat. *Biol. Control* **2014**, *70*, 28–34. [CrossRef]
41. Peng, D.; Li, S.; Wang, J.; Chen, C.; Zhou, M. Integrated biological and chemical control of rice sheath blight by *Bacillus subtilis* NJ-18 and jinggangmycin. *Pest Manage. Sci.* **2014**, *70*, 258–263. [CrossRef]
42. Kiewnick, S.; Jacobsen, B.J.; Braun-Kiewnick, A.; Eckhoff, J.L.A.; Bergman, J.W. Integrated control of Rhizoctonia crown and root rot of sugar beet with fungicides and antagonistic bacteria. *Plant Dis.* **2001**, *85*, 718–722. [CrossRef]
43. Kondoh, M.; Hirai, M.; Shoda, M. Integrated biological and chemical control of damping-off caused by *Rhizoctonia solani* using *Bacillus subtilis* RB14-C and flutolanil. *J. Biosci. Bioeng.* **2001**, *91*, 173–177. [CrossRef] [PubMed]
44. Ishii, H.; Zhen, F.; Hu, M.; Li, X.; Schnabel, G. Efficacy of SDHI fungicides, including benzovindiflupyr, against *Colletotrichum* species. *Pest Manage. Sci.* **2016**, *72*, 1844–1853. [CrossRef] [PubMed]
45. Sun, Y.; Shi, H.; Mao, C.; Wu, J.; Zhang, C. Activity of a SDHI fungicide penflufen and the characterization of natural-resistance in *Fusarium fujikuroi*. *Pestic. Biochem. Physiol.* **2021**, *179*, 104960. [CrossRef] [PubMed]
46. Morandi, M.A.B.; Maffia, L.A.; Mizubuti, E.S.G.; Alfenas, A.C.; Barbosa, J.G. Suppression of *Botrytis cinerea* sporulation by *Clonostachys rosea* on rose debris: A valuable component in Botrytis blight management in commercial greenhouses. *Biol. Control* **2003**, *26*, 311–317. [CrossRef]

47. Chatterton, S.; Punja, Z.K. Factors influencing colonization of cucumber roots by *Clonostachys rosea* f. *catenulata*, a biological disease control agent. *Biocontrol Sci. Technol.* **2010**, *20*, 37–55. [CrossRef]
48. Chatterton, S.; Punja, Z.K. Chitinase and β-1,3-glucanase enzyme production by the mycoparasite *Clonostachys rosea* f. *catenulata* against fungal plant pathogens. *Can. J. Microbiol.* **2009**, *55*, 356–367. [CrossRef]
49. Seidl, V. Chitinases of filamentous fungi: A large group of diverse proteins with multiple physiological functions. *Fungal Biol. Rev.* **2008**, *22*, 36–42. [CrossRef]
50. Fatema, U.; Broberg, A.; Jensen, D.F.; Karlsson, M.; Dubey, M. Functional analysis of polyketide synthase genes in the biocontrol fungus *Clonostachys rosea*. *Sci. Rep.* **2018**, *8*, 15009. [CrossRef]
51. Rodríguez, M.A.; Cabrera, G.; Gozzo, F.C.; Eberlin, M.N.; Godeas, A. *Clonostachys rosea* BAFC3874 as a *Sclerotinia sclerotiorum* antagonist: Mechanisms involved and potential as a biocontrol agent. *J. Appl. Microbiol.* **2011**, *110*, 1177–1186. [CrossRef]
52. Zhai, M.-M.; Qi, F.-M.; Li, J.; Jiang, C.-X.; Hou, Y.; Shi, Y.-P.; Di, D.-L.; Zhang, J.-W.; Wu, Q.-X. Isolation of secondary metabolites from the soil-derived fungus *Clonostachys rosea* YRS-06, a biological control agent, and evaluation of antibacterial activity. *J. Agric. Food Chem.* **2016**, *64*, 2298–2306. [CrossRef]
53. Gong, C.; Liu, Y.; Liu, S.-Y.; Cheng, M.-Z.; Zhang, Y.; Wang, R.-H.; Chen, H.-Y.; Li, J.-F.; Chen, X.-l.; Wang, A.-X. Analysis of *Clonostachys rosea*-induced resistance to grey mould disease and identification of the key proteins induced in tomato fruit. *Postharvest Biol. Technol.* **2017**, *123*, 83–93. [CrossRef]
54. Crane, J.M.; Bergstrom, G.C. Spatial distribution and antifungal interactions of a *Bacillus* biological control agent on wheat surfaces. *Biol. Control* **2014**, *78*, 23–32. [CrossRef]
55. Wei, F.; Hu, X.; Xu, X. Dispersal of *Bacillus subtilis* and its effect on strawberry phyllosphere microbiota under open field and protection coditions. *Sci. Rep.* **2016**, *6*, 22611. [CrossRef] [PubMed]

microorganisms

MDPI

Article

A Lipopolysaccharide Synthesis Gene *rfaD* from *Mesorhizobium huakuii* Is Involved in Nodule Development and Symbiotic Nitrogen Fixation

Yuan Liu [1], Ye Lin [1], Ning Guan [1], Yuting Song [1], Youguo Li [1,*] and Xianan Xie [2,*]

1 State Key Laboratory of Agricultural Microbiology, College of Life Science and Technology, Huazhong Agricultural University, Wuhan 430070, China
2 State Key Laboratory of Conservation and Utilization of Subtropical Agro-Bioresources, Guangdong Laboratory for Lingnan Modern Agriculture, Guangdong Key Laboratory for Innovative Development and Utilization of Forest Plant Germplasm, College of Forestry and Landscape Architecture, South China Agricultural University, Guangzhou 510642, China
* Correspondence: youguoli@mail.hzau.edu.cn (Y.L.); xiexianan8834203@126.com (X.X.); Tel.: +86-127-8728-1685 (Y.L.); +86-159-1855-2425 (X.X.)

Abstract: Rhizobium lipopolysaccharide (LPS) is an important component of the cell wall of gram-negative bacteria and serves as a signal molecule on the surface of rhizobia, participating in the symbiosis during rhizobia–legume interaction. In this study, we constructed a deletion mutant of ADP-L-glycerol-D-mannoheptosyl-6-exoisomerase (*rfaD*) of *Mesorhizobium huakuii* 7653R and a functional complementary strain. The results showed that the deletion of *rfaD* did not affect the free-living growth rate of 7653R, but that it did affect the LPS synthesis and that it increased sensitivity to abiotic stresses. The *rfaD* promoter-GUS reporter assay showed that the gene was mainly expressed in the infection zone of the mature nodules. The root nodules formation of the *rfaD* mutant was delayed during symbiosis with the host plant of *Astragalus sinicus*. The symbiotic phenotype analyses showed that the nodules of *A. sinicus* lost symbiotic nitrogen fixation ability, when inoculated with the *rfaD* mutant strain. In conclusion, our results reveal that the 7653R *rfaD* gene plays a crucial role in the LPS synthesis involved in the symbiotic interaction between rhizobia and *A. sinicus*. This study also provides new insights into the molecular mechanisms by which the rhizobia regulate their own gene expression and cell wall components enabling nodulation in legumes.

Keywords: symbiotic nitrogen fixation; gram-negative bacteria; *Mesorhizobium huakuii* 7653R; lipopolysaccharide; *rfaD*

Citation: Liu, Y.; Lin, Y.; Guan, N.; Song, Y.; Li, Y.; Xie, X. A Lipopolysaccharide Synthesis Gene *rfaD* from *Mesorhizobium huakuii* Is Involved in Nodule Development and Symbiotic Nitrogen Fixation. *Microorganisms* **2023**, *11*, 59. https://doi.org/10.3390/microorganisms11010059

Academic Editors: Beibei Ge, Tomislav Cernava and Kyungseok Park

Received: 28 November 2022
Revised: 21 December 2022
Accepted: 21 December 2022
Published: 25 December 2022

1. Introduction

The successful establishment of the mutualistic endosymbiosis between rhizobia and legumes results from mutual recognition and exchange of signal molecules [1,2]. Flavonoids secreted by the roots of legumes can activate the nodulation genes of rhizobia to synthesize lipochitooligosaccharides (Nod factors; NFs) [3,4]. Subsequently, NFs are sensed by the cell surface LysM receptor heteromer of leguminous plants, resulting in the root hair curling, plasma membrane invagination, and cell division to form the infection threads and root nodules [5–7]. The rhizobia type III secretion system (T3SS) enables rhizobia symbiosis through the NFs-independent signaling pathway [8]. The type III effectors secreted by T3SS, also known as nodulation outer proteins (nops), play an important role in determining symbiosis specificity [9]. Meanwhile, the establishment of a rhizobia symbiosis requires an intimate connection between two symbionts, and cell surface structures will be involved in the early interaction between the rhizobia and legumes. Rhizobia secrete a variety of glycans on their cell surfaces, including lipopolysaccharides (LPS), exopolysaccharides (EPS), capsular polysaccharides (KPS), and cycloglucans (CG) [10,11]. These polysaccharides and glucans are necessary for the successful establishment of rhizobia symbiosis, especially at the early infection stages, and also determine the symbiotic specificity in the rhizobia–legumes interactions [11–15].

LPS forms the outer leaflet of the outer membrane of gram-negative bacteria, and maintains the cell membrane permeability of bacteria, enabling them to survive in harsh environments [16,17]. The LPS of rhizobia is essential for the formation of infection threads and symbionts, but inhibits the defense responses of legumes to accommodate the rhizobia within the host cells [18–20]. In *Rhizobium* sp. NGR234, the *lpsB* encodes a glycosyl transferase acting early in synthesis with the core part of the LPS; the LPS synthesis of *lpsB* mutant was affected, which seriously affected the establishment of its symbiotic relationship with *Vigna unguiculata*. [21]. Moreover, the *lpsB* and *lpsE* are two key genes in *Sinorhizobium fredii* HH103 involved in the LPS biosynthesis; however, the corresponding mutants cause the changes in the LPS structure and early nodule senescence, thereby impairing the symbiosis between *S. fredii* HH103 and soybean [22]. In addition, two genes, *lpxA* (encoding acyl-ACP-UDP-N-acetylglucosamine O-acyltransferase) and *lpxE* (encoding lipid A 1-phosphatase) in *Rhizobium* sp. PRF81, participate in the lipid A biosynthesis and LPS modification, and the expression of these two genes can be induced by the exudates released from common bean seeds [23]. Moreover, the purified LPS of *Sinorhizobium meliloti* inhibits the alkalization and reactive oxygen species burst in the alfalfa cells suspension, but not in non-host tobacco, which may suppress pathogenic responses in the host alfalfa to promote the rhizobia symbiosis [24]. The role of some genes regulating the LPS synthesis in rhizobia–legume symbiosis has been investigated on the bacterial side; however, the function of rhizobia LPS during symbiosis is not fully understood.

The LPS of rhizobia is composed of lipid A, core oligosaccharides, and O-antigen polysaccharide chains [16]. It has a general structure similar to that of animal gram-negative pathogens [25]. The core oligosaccharides of the LPS can be further divided into the inner and outer cores [26]. The part of the inner core is composed of 3-deoxy-D-manno-oct-2-ulosonic acid (Kdo) and ADP-heptose residues, while the outer core is composed of hexoses and 2-acetoamido-2-deoxy-hexoses and offers the attachment site for the O-polysaccharide chain. Furthermore, the ADP-heptose is an essential part of the inner core of the LPS, which links the exterior of the LPS to the Kdo between Kdo2-lipidA and O antigen. The ADP-heptose is also required for maintaining outer membrane integrity and limiting permeability [16,26,27]. The biosynthesis of the LPS has been extensively studied in *Escherichia coli* [28]. Studies have shown that the synthesis process includes five catalytic steps required to generate ADP-L-glycero-D-mannose-heptose, a molecule necessary for the first heptose that can be transferred to Kdo [27,28]. The chromosomal *rfa* in *E. coli* operably encodes the enzymes required for the stepwise assembly of major core oligosaccharides [16]. Among these genes, the *rfaD* gene encodes an ADP-L-*glycero*-D-manno-heptose-6-epimerase, which catalyzes the conversion of ADP-D-*glycero*-D-mannose-heptose to ADP-L-glycero-D-mannose-heptose, and this is the last step in the synthesis pathway of ADP-L-glycerol-D-mannose-heptose [27]. However, the function of *rfaD* genes in the symbiosis between rhizobia and legumes remains elusive.

Chinese milk vetch (*Astragalus sinicus*) is a traditional green manure legume widely grown in southern China. It can be used for improving the fertility of paddy fields, and can also be used as animal forage and nectar source plants [29]. *Mesorhizobium huakuii* 7653R is a Gram-negative bacterium which was isolated from the indeterminate root nodules of *A. sinicus* [30]. The symbiotic relationships of *A. sinicus* and *M. huakuii* 7653R have been extensively investigated [31–33]; moreover, the *M. huakuii* 7653R genome contains a single copy of the *rfaD* gene [34]. However, the function of *rfaD* in the symbiosis between *M. huakuii* 7653R and *A. sinicus* remains unknown.

In this study, we characterized the roles of the *rfaD* gene from *M. huakuii* 7653R. By constructing the *rfaD* deletion mutant, the phenotypes showed that the *rfaD* mutation affected the LPS synthesis and the host plants lost nitrogen fixation ability. The in situ gene expression analysis showed that the *rfaD* was mainly expressed in the infection zone of nodules of *A. sinicus*. Further observation of early infection events found that the nodule formation in the *rfaD* mutant was relatively delayed when compared with that in the wild type. Taken together, our findings revealed that the *rfaD* gene is required for rhizobia infection and nodulation in *A. sinicus*, and additionally provides new insights into the molecular mechanism by which the rhizobia orchestrate their own genes and cell wall components in the establishment of endosymbiosis in the roots of legumes.

2. Materials and Methods

2.1. Bacterial Strains and Plant Material and Growth Conditions

The bacterial strains used in this study are listed in Supplemental Table S1. Wild-type *M. huakuii* 7653R, and strains of 7653RΔ*rfaD*, 7653RΔ*rfaD*-C, 7653RΔ*rfaD*-GFP and 7653R pRG960-*rfaD* derived from 7653R were grown on tryptone yeast (TY) medium at 28 °C. *Escherichia coli* DH5a and S17-1 were cultured on Luria–Bertani (LB) medium at 37 °C. The following concentrations of antibiotics were used for the above strains: streptomycin, 100 μg mL^{-1}; kanamycin, 50 μg mL^{-1}; gentamycin, 25 μg mL^{-1}; ampicillin, 100 μg mL^{-1}; spectinomycin, 100 μg mL^{-1}; tetracycline, 12.5 μg mL^{-1}.

2.2. Symbiotic Phenotype Evaluation

Astragalus sinicus was used for symbiotic phenotype analysis. The seeds were surface-sterilized in 75% (*v*/*v*) ethanol for 10 min, washed six times with sterile water for 1 min each, and then treated with 5% sodium hypochlorite for 10 min, followed by 5–7 additional washes in sterile water [31,35]. To synchronize the germination, the sterilized seeds were immersed in distilled water for 18–24 h, and then placed on the sterile plates containing 0.5% sucrose and 0.55% agar. Seedlings were grown under a 16-h/8-h light/dark cycle at 25 °C/18 °C. After 5 days, the plantlets were transferred into sterile sand, and inoculated with 1 mL corresponding rhizobial culture at OD_{600} = 0.2 and watered with Fahraeus nitrogen-free plant nutrient solution [35]. At 28 days post-inoculation (dpi), the symbiotic plants were harvested, and the symbiotic phenotypes were examined.

2.3. Construction of the Deletion Mutants and Functional Complement Strains

In order to explore the biological function of *rfaD* in rhizobia symbiosis, the *cre-lox* system was used to construct the in-frame 7653R *rfaD* gene deletion mutant [36]. The fragments containing the flanking sequences of the 5′- and 3′-terminal coding regions of *rfaD* were amplified and cloned into the *Acc65*I/*Nde*Iand *Apa*I/*Age*Imultiple cloning sites of pCM351. The recombinant plasmid was verified by PCR and Sanger sequencing, and transformed into *E. coli* S17-1, then conjugated into *M. huakuii* 7653R by bi-parental mating. To eliminate the gentamicin resistance gene flanked by the *loxP* site, the plasmid pCM157 carrying the *Cre* gene was introduced into the 7653RΔ*rfaD*::Gm by bi-parental conjugation. The transconjugants sensitive to gentamicin were screened, and the strains sensitive to tetracycline were then selected. The candidate strains were further confirmed by PCR and sequencing. To validate the correction of the obtained mutant strain, the genomic DNAs from wild-type and mutant strains were extracted, and amplified by PCR with the *rfaD*-MAP-F/R primers; the results showed that a large DNA fragment (3.1 Kb) was amplified using the genomic DNA from wild-type *M. huakuii* 7653R as a template, while a small DNA fragment (2.1 Kb) was specifically amplified using the *rfaD* mutant genomic DNA as a template (Figure 1). Meanwhile, these two genomic DNAs were also detected using the *rfaD*-ORF-F/R primers, and results showed that a 526 bp internal fragment of the *rfaD* ORF fragment was amplified using the genomic DNA from wild-type *M. huakuii* 7653R as a template, whereas no fragment was obtained using the *rfaD* mutant genomic DNA as a template (Figure S1). All the fragments obtained were subsequently sequenced, and the results showed the absence of the *rfaD* ORF in the 7653RΔ*rfaD* mutant strain. The strain with the *rfaD* deletion mutant was named as 7653RΔ*rfaD*.

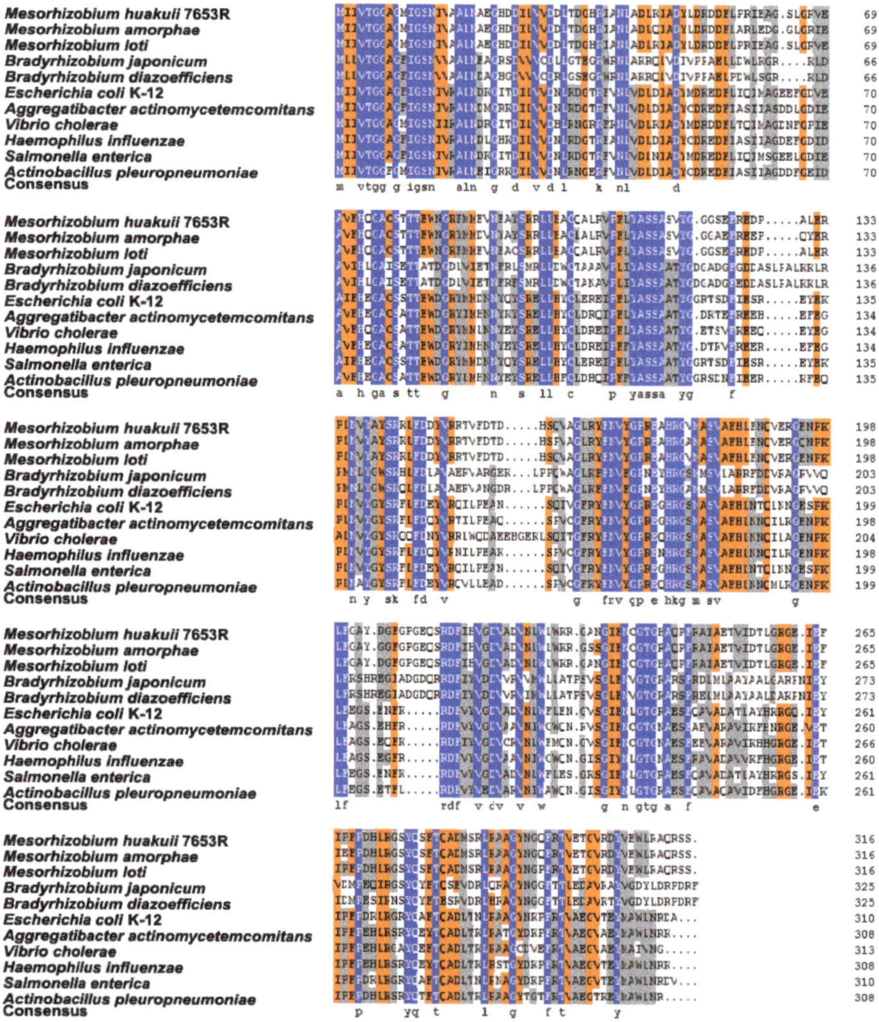

Figure 1. Multiple amino acids sequence alignments of the RfaD proteins from various bacteria species. Sequence alignments of the representatives of RfaD proteins from *Mesorhizobium*, *Bradyrhizobium* and some pathogens. Conserved residues are highlighted in blue.

To construct the recombinant plasmid carrying *rfaD*, the fragment containing the *rfaD* coding sequence and its promoter region (500 bp) was amplified and cloned into the *Sma*I/*EcoR*Isites of the broad-host-range vector pBBR1MCS-5 [37]. The resulting construct pBBR1MCS-5-rfaD was transformed into the strain 7653RΔ*rfaD* by bi-parental mating. The strain obtained was named as 7653RΔ*rfaD*-C. The complementary strain was verified by PCR with the *rfaD*-MAP-F/R and *rfaD*-ORF-F/R primers (Figure S1); the DNA fragments represent the absence of the *rfaD* gene from the genome (2.1 kb) of *M. huakuii* 7653R and the presence of the *rfaD* gene in the complement plasmid pBBR1MCS-5-*rfaD* (526 bp), respectively. The gene-specific primers used for gene amplification in this study are listed in Supplemental Table S2.

2.4. LPS Extraction and Silver Staining

The LPS from the *M. huakuii* 7653R and 7653RΔ*rfaD* strains were extracted with an LPS extraction kit (iNtRON 17141) according to the manufactural procedures. The prepared LPS were added with 5 × SDS loading buffer and boiled, then separated by SDS-PAGE. The gel was fixed in 30% ethanol, 10% glacial acetic acid and 7 g/L periodate at 22 °C for 20 min, followed by being washed three times in sterile water. Subsequently, 1 g/L $AgNO_3$ at 30 °C was added for 30 min for silver staining, with washing in sterile water for 10 s. Subsequently, 30 g/L Na_2CO_3 (4 °C), 0.02% formaldehyde color development for 20 min (fresh preparation before use) was added. Finally 10% glacial acetic acid was added to terminate the color reaction.

2.5. Plate Inhibition Test

The *M. huakuii* 7653R, 7653RΔ*rfaD*, and 7653RΔ*rfaD*-C strains were grown in liquid TY medium overnight to OD_{600} = 1.0. The cultures were mixed with the heated and melted TY solid medium sufficiently. The mixture was then poured into sterile plates. A sterile filter paper with a diameter of 5 mm was placed in the center of each plate. SDS (detergent), HCl (acid) and NaClO (oxidizing) were then dropped onto the filter paper. The plates were incubated at 28 °C for 3 days.

2.6. Construction of Promoter-GUS Reporter System and GUS Activity Assay

A 500-bp of upstream fragment of the *rfaD* was chosen as the putative promoter region, and then obtained from *M. huakuii* 7653R genomic DNA by PCR using the primers *rfaD-P-F* and *rfaD-P-R*. The amplicon was cloned into the *Sma*I and *Pst*I sites of pGR960 to create the pGR960-rfaD-GUS construct [38]. The recombinant plasmid was transformed into the wild-type *M. huakuii* 7653R. The recombinant strain was then inoculated to the roots of *A. sinicus*.

The roots and nodules of the *A. sinicus* plants were harvested and washed with distilled water to enable us to detect the *GUS* expression. The histochemical staining of GUS activity in the roots and nodules was performed according to the previous descriptions [39]. Roots and nodules were stained with the GUS buffer (0.1 M sodium phosphate buffer pH 7.0, 2 mM K_4Fe $(CN)_6$, 2 mM K_4Fe $(CN)_6$, 0.1% Triton X-100, 10 mM EDTA, 0.5 mg mL^{-1} X-Gluc dissolved in Dimethylformamide). Samples were incubated at 37 °C for 10 h in darkness and then washed with 70% (*v*/*v*) ethanol. The root and nodule sections were observed with a light microscope (BX51, Olympus, Japan) to visualize the spatial GUS activity.

2.7. Observation and Quantification of Early Infection Events

The wild-type 7653R and mutant 7653RΔ*rfaD* were labeled with a GFP reporter gene from vector pMP2463 [40]. The *A. sinicus* seedlings were inoculated with the strain 7653RΔ*rfaD*-GFP or 7653R-GFP. The early infection events, including root hair curling, infection threads, and nodular primordia, were observed and quantified at 3, 5, and 7days post-inoculation using a fluorescence microscope (BX51, Olympus, Tokyo, Japan).

2.8. Nitrogenase Activity Measurement

The nitrogenase activity of nodules was determined using the acetylene reduction activity method as described previously [33,41]. Briefly, the nodules of *A. sinicus* plants were harvested at 28 dpi for nitrogenase activity analysis. For each group, nine *A. sinicus* plants were randomly selected for the measurement. The root systems of each three plants, including roots and nodules, were incubated with 2 mL of acetylene for 2 h at 28 °C. The amount of ethylene was detected by the gas chromatograph (East & West Analytical Instrument GC 4000A, Dongxi, Beijing, China).

2.9. Analysis of Histological Nodule Cross-Sections

The microscopic analysis was performed according to methods previously described [31]. For light microscopic analysis, some nodules were fixed in formaldehyde-acetic acid buffer and dehydrated. The treated nodules were embedded in paraffin and cut longitudinally. The slides were stained with the Toluidine blue O and observed with a light microscope (BX51, Olympus, Tokyo, Japan). For electron microscopic analysis, the other nodules were first fixed in 2.5% glutaraldehyde for 4 h, then immersed in 1% osmium tetroxide for 3 h, dehydrated, and finally embedded in the London resin white [42]. The ultrathin sections were observed with an electron microscope (H-7650; Hitachi, Tokyo, Japan).

2.10. Phylogenetic Analysis and Conserved Sequence Alignment

The phylogenetic analysis of the rfaD proteins from bacteria was performed with MEGA11 software [43]. The amino acid sequences of rfaD were searched through the BLASTp program (https://blast.ncbi.nlm.nih.gov/Blast.cgi, accessed on 8 May 2022). The multiple protein sequences were aligned using the ClustalW, and the aligned sequences were loaded into the MEGA11 to generate an unrooted phylogenetic tree. The evolutionary history was inferred using the neighbor-joining method. The conserved sequence alignment was completed with DNAMAN software.

2.11. Accession Numbers

The sequence data of this article can be found in the GeneBank according to the following accession numbers: *Mesorhizobium huakuii* 7653R rfaD (WP_038648977.1). The rfaD homologous sequences for phylogenetic analysis were provided as below: *Bradyrhizobium japonicum* (WP_014493906.1), *Bradyrhizobium diazoefficiens* (WP_200519224.1), *Bradyrhizobium sp.* WSM1743 (WP_027576844.1), *Bradyrhizobium shewense* (WP_091959050.1), *Bradyrhizobium elkanii* (WP_028163631.1), *Bradyrhizobium lablabi* (WP_079543334.1), *Escherichia coli* K-12 (HBP1552200.1), *Mesorhizobium amorphae* (WP_192183840.1), *Mesorhizobium loti* (WP_019860550.1), *Mesorhizobium muleiense* (WP_091594992.1), *Mesorhizobium intechi* (WP_143977314.1), *Mesorhizobium comanense* (WP_137934890.1), *Mesorhizobium australicum* (WP_015316155.1), *Aggregatibacter actinomycetemcomitans* (WP_205772981.1), *Vibrio cholerae* (BCN19765.1), *Haemophilus influenzae* (WP_005657299.1), *Salmonella enterica* (WP_079802599.1), *Actinobacillus pleuropneumoniae* (WP_005605737.1).

2.12. Statistical Analysis

Statistical significance was determined by an independent-sample t test and single-factor analysis of the variance method with SPSS16 software (SPSS, Chicago, IL, USA). Three biological replicates were set up for each experiment. The error bars in the figures represent the standard deviation of three biological replicates. The asterisks indicate means that are statistically different at * $p < 0.05$, ** $p < 0.01$ and *** $p < 0.001$.

3. Results

3.1. Identification of rfaD in the Genome of Mesorhizobium huakuii 7653R

In the genome of *M. huakuii* 7653R, we found a single copy of the *rfaD* gene encoding an ADP-L-glycero-D-manno-heptose-6-epimerase [34]. Bioinformatics analysis showed that the *rfaD* gene had a 951 bp length of the open reading frame encoding 316 amino acids, with a predicted molecular weight of 35.01 kD and a theoretical pI of 5.69. Amino acid sequence alignment results showed that the rfaD proteins were relatively conserved across bacteria species, and the homologous proteins were conserved at the N-terminal regions (Figure 1). Phylogenetic analysis showed that the rfaD protein exists in a variety of Gram-negative bacteria, mainly in *Bradyrhizobium*, *Mesorhizobium* and pathogenic bacteria, but not in *Sinorhizobium* (Figure 2). This result indicated that the function of the *rfaD* in rhizobia-legume symbiosis may be species–specific.

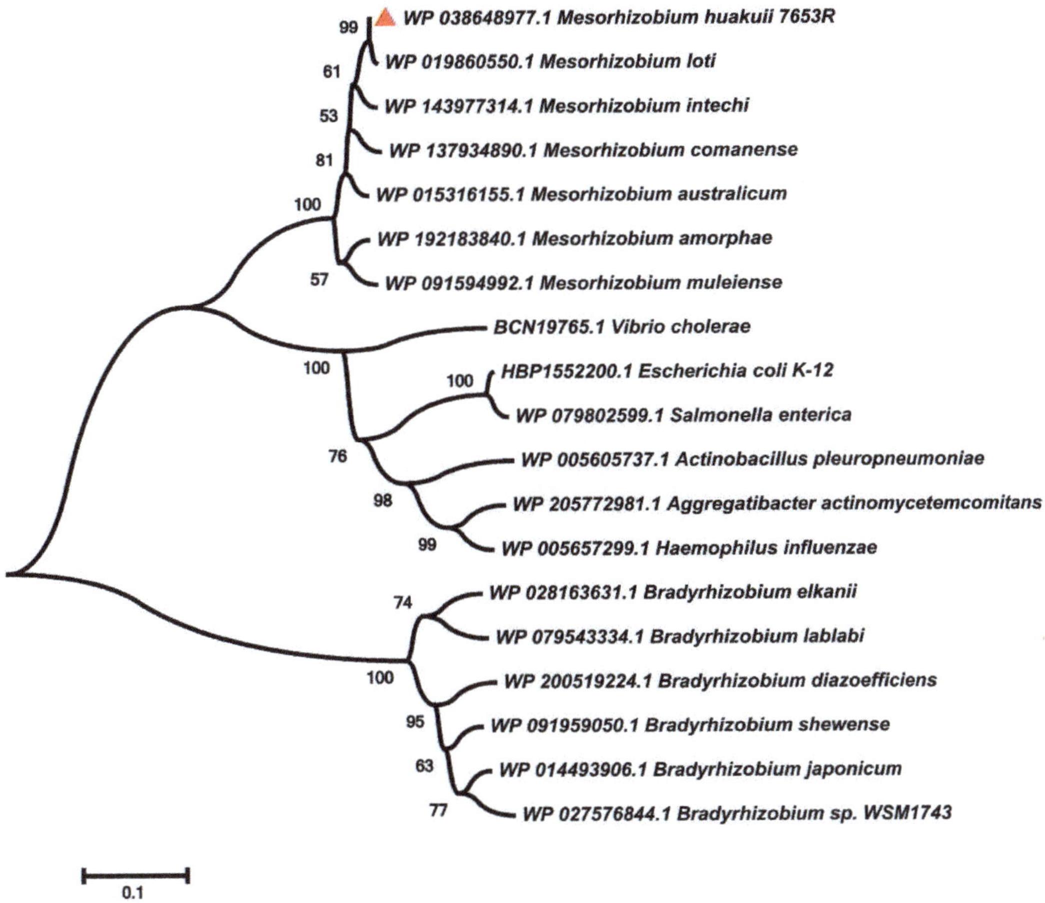

Figure 2. Phylogenetic analysis of RfaD homologues proteins among bacteria species. The phylogenetic relationship was inferred by MEGA11 software using the neighbor-joining method [43]. The evolutionary distance was computed using the *p*-distance method and presented in the number of amino acid differences per site. The branch of *M. huakuii* 7653R RfaD is marked with a red triangle. Tree scale, 0.1.

3.2. The 7653RΔrfaD Mutant Was Sensitive to Abiotic Stresses and Antibiotics

We initially examined the free-living growth of 7653R, 7653RΔ*rfaD*, and 7653RΔ*rfaD*-C strains. The result showed that the deletion of the *rfaD* gene had no significant effect on the growth of the 7653RΔ*rfaD* strain under free-living conditions (Figure S2). Subsequently, we examined the LPS synthesis in the wild-type 7653R and 7653RΔrfaD strains. The results showed that the LPS content of the *rfaD* mutant strain was clearly reduced when compared with that of the wild-type strain (Figure S2), indicating that the deletion of the *rfaD* affected the LPS synthesis of the 7653R strain.

In order to further test whether the deletion of the *rfaD* would affect the stress resistance of the 7653R strain, we selected the reagents SDS (detergent), HCl (acidic) and NaClO (oxidant) to perform the plate inhibition zone experiments. The results showed that the mutation of the *rfaD* in 7653R led to a significant decrease in the stress resistance of rhizobia, whereas the functional complementary strain in response to abiotic stresses was comparable to that of the wild type (Figure 3). The previous report has shown that LPS is also resistant further detected the growth of *M. huakuii* 7653R, 7653RΔ*rfaD* and 7653RΔ*rfaD*-C strains on the TY plates in response to the polymyxin B addition. It was found that the 7653RΔ*rfaD* strain was more sensitive to the polymyxin B (50 μg/mL) than the wild-type strain (Figure S4). Collectively, these results suggest that the rhizobia LPS producing the gene *rfaD* plays a crucial role in the bacterial cell resistance to external stresses and antibiotics.

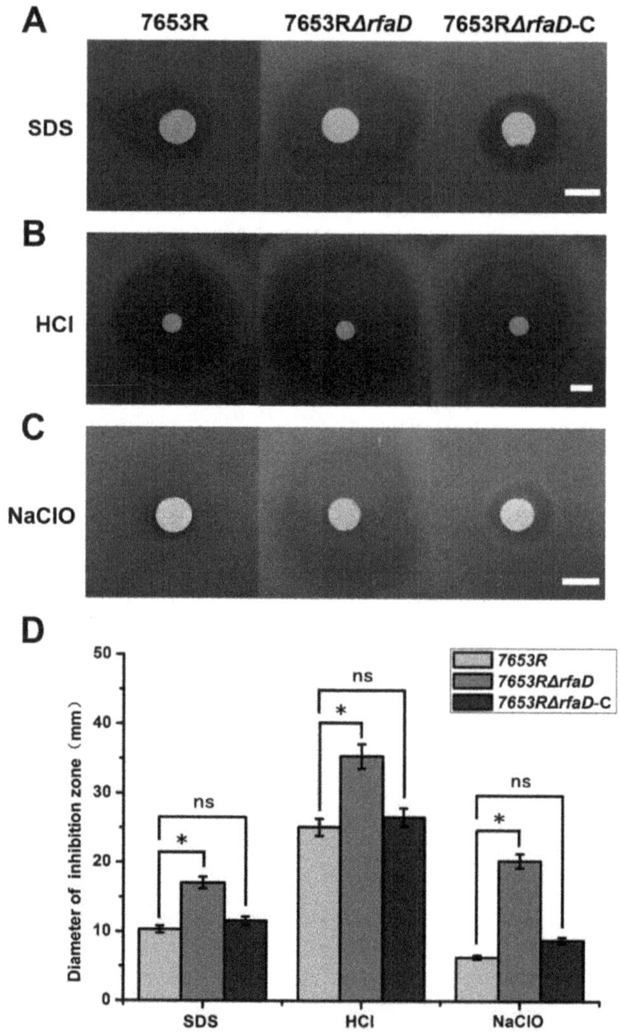

Figure 3. Inhibition zone determination of the wild-type 7653R, 7653RΔ*rfaD* and 7653RΔ*rfaD*-C strains. Two percent SDS (**A**), 5M HCl (**B**), NaClO (**C**) were used to detect the inhibition zone of wild-type 7653R, 7653RΔ*rfaD* and 7653RΔ*rfaD*-C. (**D**), the diameters of inhibition zones of the different indicated strains. Scale bars: 5 mm. Significance: * $p < 0.05$; ns, not significant; Student's *t*-test. Data are shown as mean ± SD.

3.3. The In-Situ Expression of the rfaD Gene in Nodules during Symbiosis

To investigate the in-situ spatial expression of the *rfaD* gene during nodulation in *A. sinicus*, we cloned the *rfaD* promoter, a 500 bp genomic DNA upstream fragment. This promoter fragment was fused with the GUS reporter in the pRG960 vector. The recombinant plasmid pRG960-rfaD was subsequently introduced into the wild-type *M. huakuii* 7653R, which was then inoculated into *A. sinicus* seedlings. The samples including roots and nodules were harvested separately at 3, 5, 9, 14 and 28 dpi for GUS staining analysis (Figure 4). At 3 dpi and 5 dpi, no GUS activity signals were detected at the curled root hairs and nodule primordia (Figure 4A,B). At 9 dpi, strong GUS activity was detected at the base of young nodules (Figure 4C). At 14 dpi and 28 dpi, GUS activity signals were found in the infection zone of mature nodules (Figure 4D,E). In addition, we also detected high GUS activity in the free-living bacterial cells of the 7653R strain carrying pRG960-rfaD; however, no GUS activity was detected in the *M. huakuii* 7653R strain with the empty vector pRG960 (Figure 4F). The spatiotemporal expression results showed that the *rfaD* gene was highly expressed in the infection zone of the mature nodules, suggesting that it may play a role in the infection process during symbiosis.

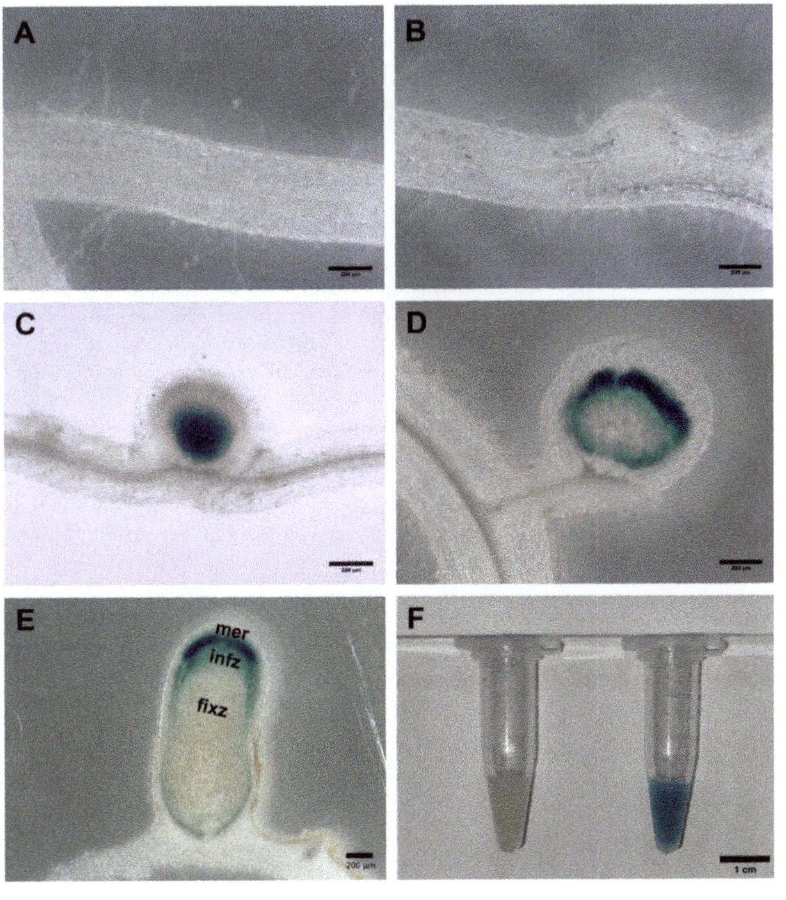

Figure 4. The in-situ expression pattern of the 7653R *rfaD* in *A. sinicus* roots during early infection and nodulation. (**A–E**), GUS staining of *A. sinicus* roots and nodules induced by *M. huakuii* 7653R carrying P_{rfaD}-GUS. The *A. sinicus* roots were harvested at 3 dpi (**A**), 5 dpi (**B**), 9 dpi (**C**), 14 dpi (**D**) and 28 dpi (**E**). (**F**), GUS activity in the 7653R carrying empty plasmid pRG960 (left) and P_{rfaD}-GUS fusions (right) in free-living conditions. Scale bars: 200 μm (**A–E**), 1 cm (**F**). mer, meristem zone; infz, infection zone; fixz, nitrogen fixation zone.

3.4. *M. huakuii 7653R rfaD Gene Engages in the Early Infection Events during Rhizobia-A. sinicus Interaction*

To further explore the effect of the *rfaD* gene on the infection process during *M. huakuii* 7653R-*A. sinicus* interaction, we introduced the pMP2463-GFP plasmid into the *rfaD* mutant and 7653R strains. Therefore, the 7653RΔ*rfaD* mutant strain was marked by the GFP reporter gene, and could constitutively express the GFP protein. This transgenic strain was named 7653RΔ*rfaD*-GFP. Subsequently, we inoculated 7653RΔ*rfaD*-GFP into the roots of *A. sinicus* seedlings, while the wild-type 7653R-GFP was used as a control line. By visualizing the GPF signals during infection processes, we observed and quantified the rhizobia infection events during *M. huakuii* 7653R-*A. sinicus* interaction. In the *A. sinicus* roots with 7653R-GFP, the root hair curling was formed at the tips of root hairs, and infection threads were present in the curled root hairs, while the infection threads extended along the root hairs to the cortex, thereby forming a nodule primordium to develop the root nodules (Figure 5A,C). The early infection events in *A. sinicus* roots carrying the 7653RΔ*rfaD*-GFP were also observed and quantified at 7 dpi. The results showed that the number of curled root hairs and infection threads was significantly increased in the roots of *A. sinicus* inoculated with the 7653RΔ*rfaD*-GFP, while the number of nodule primordia was obviously decreased in this *rfaD* mutant when compared with the control strain 7653R-GFP (Figure 5B,D). Collectively, these findings suggest that the *M. huakuii* 7653R *rfaD* gene is involved in the early infection events during rhizobia-*A. sinicus* interaction and nodule formation.

3.5. *M. huakuii 7653R rfaD Mutant Results in the Defective Nodules in A. sinicus*

To further investigate the role of the *rfaD* gene from *M. huakuii* 7653R in the processes of the rhizobia–legume symbiosis, we inoculated the strains of 7653R, 7653RΔ*rfaD*, or 7653RΔ*rfaD*-C into the roots of the host legume *A. sinicus*, while the plants without inoculation were used as the control lines. At 28 dpi, we harvested all the inoculated and non-inoculated plants. The *A. sinicus* plants inoculated with 7653RΔ*rfaD* exhibited a defective symbiotic phenotype: the biomass of the shoots of the plants decreased, the leaves turned yellow, and the morphology and color of the formed nodules were abnormal, being small white round nodules. However, the *A. sinicus* plants inoculated with the functional complementary strain exhibited normal growth and were almost similar to the 7653R wild-type plants; the shoot biomass was larger, the leaves were green, and the morphology and color of root nodules were normal with a pink color (Figure 6A–F).

The quantitative analysis showed that, compared with the 7653R wild-type, the symbiotic phenotype of the *M. huakuii* 7653RΔ*rfaD* strain, the fresh weight of the shoot biomass and nodules were significantly reduced, whereas the number of root small nodules was significantly increased, and the nitrogenase activity of root nodules after being inoculated with 7653RΔrfaD was not detectable (Figure 6G–J). Additionally, combined with the results of early infection events, it is suggested that the lack of *rfaD* results there is a delay in the plant–*rfaD* mutant symbiosis, and that they finally form ineffective root nodules. On the other hand, there was no significant difference in the shoot biomass, root nodules, nodules number, or the nitrogenase activity of nodules between the 7653RΔ*rfaD*-C and the wild-type 7653R strains. These results indicated that the 7653RΔ*rfaD*-C strain restored the defective nitrogen fixation ability of the 7653RΔ*rfaD* mutant.

To next dissect the structures of root nodules of the *A. sinicus*, we performed a histological analysis of the nodules from the 7653R, 7653RΔ*rfaD*, and 7653RΔ*rfaD*-C strains. The results showed that the nodules from the 7653RΔ*rfaD* mutant contained a considerable number of uninfected root cells, while only a small number of rhizobia cells were found in the infective zone of the nodules. In contrast, the nodules of plants inoculated with the 7653R wild-type or complement strain 7653RΔ*rfaD*-C contained large numbers of symbiotic cells with bacteroid (Figure 7). In order to further examine the effect of the *rfaD* deletion on the development of bacteroid, we observed the ultrastructure of wild-type and Δ*rfaD* mutant nodules by transmission electron microscopy. The results showed that the rhizobia in the wild-type nodules differentiated into normal elongated bacteroid surrounded by symbiosome membranes (Figure 8A,B). On the contrary, in the Δ*rfaD* mutant nodules, the differentiation in the bacteroid was abnormal; they were small spheres (Figure 8C,D).

Taken together, these findings reveal that the *rfaD* deletion impairs the normal differentiation in the bacteroid, resulting in the formation of defective nodules.

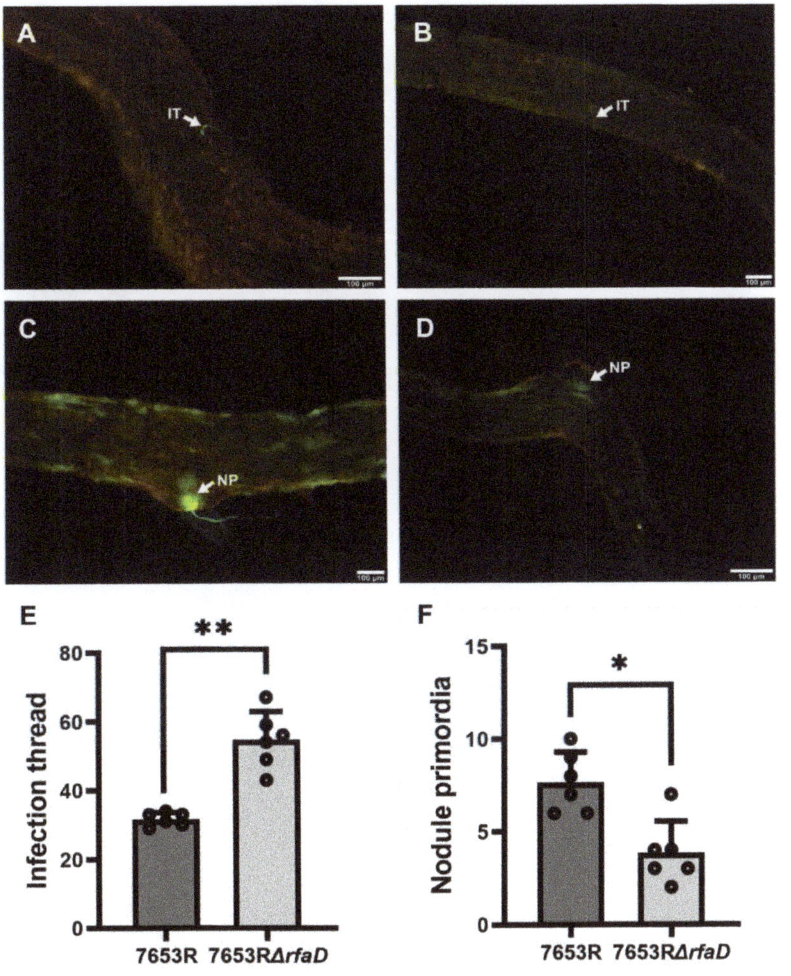

Figure 5. Observation and quantification of early infection events in *A. sinicus* roots inoculated with 7653R or 7653RΔ*rfaD*. (**A**–**D**), images of infection events of GFP-labeled 7653R (**A**,**C**) and 7653RΔ*rfaD* (**B**,**D**) at 7 dpi. A–B, infection thread; C–D, nodule primordium. E–F, Frequencies of infection threads (**E**) and nodule primordia (**F**) per root. The white arrows refer to infection thread and nodule primordium. IT, infection thread; NP, nodule primordium. Scale bars: 100 μm. Significance: * $p < 0.05$; ** $p < 0.01$; Student's *t*-test. Data are shown as means ± SD.

type plants; the shoot biomass was larger, the leaves were green, and the morphology and color of root nodules were normal with a pink color (Figure 6A–F).

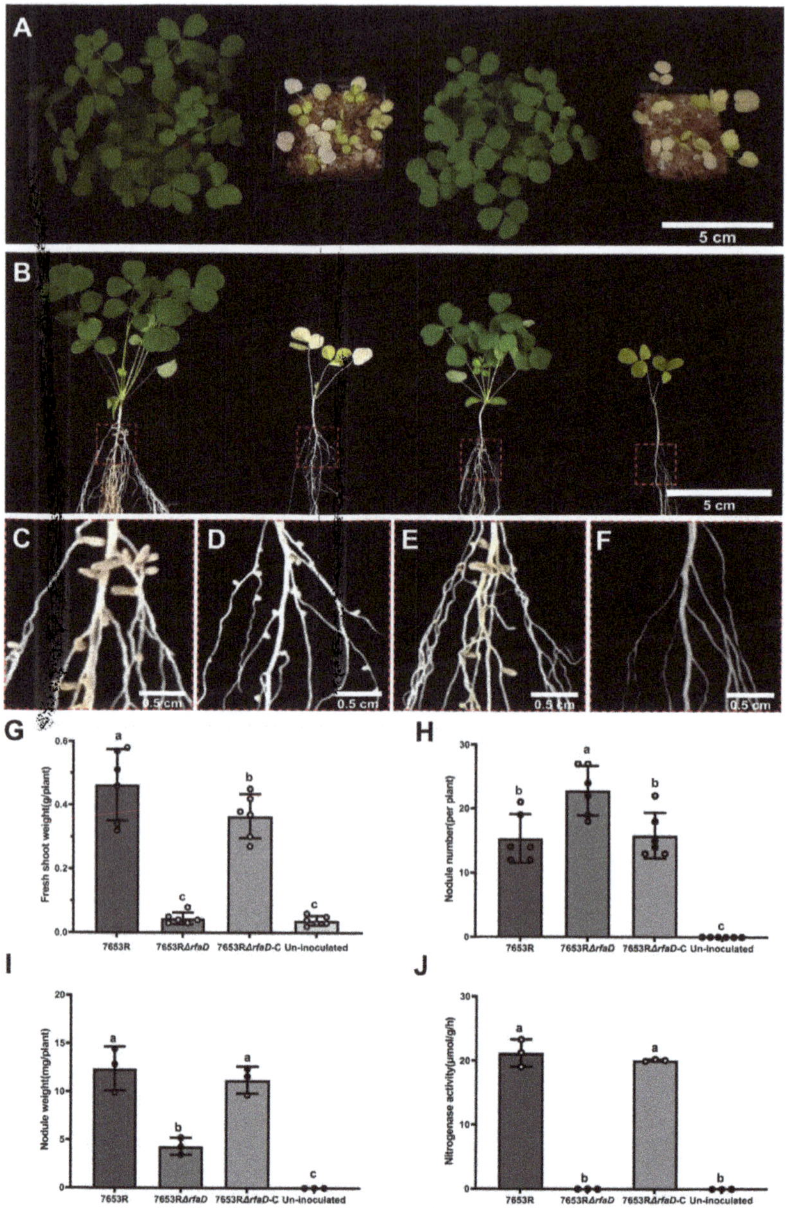

Figure 6. Symbiotic phenotypes of the *A. sinicus* plants inoculated with *M. huakuii* 7653R, 7653RΔ*rfaD*, or 7653RΔ*rfaD*-C strains. *A. sinicus* plants were inoculated with the indicated strains and harvested at 28 dpi. A to F, the *A. sinicus* plants presented from the left to the right panels were inoculated with strain 7653R, 7653RΔ*rfaD*, 7653RΔ*rfaD*-C and without inoculation, respectively. (**A**), shoot growth performance; (**B**), whole-plant performance; (**C–F**), root nodule phenotypes, enlarged view of the red square; (**G–J**), fresh shoot weight, nodule numbers, nodule weights and the nitrogenase activity of plants inoculated with the strain 7653R, 7653RΔ*rfaD*, 7653RΔ*rfaD*-C, or not (un-inoculated), respectively. Scale bars = 5 cm (**A**,**B**), 2 mm (**C**,**E**) and 1 mm (**D**,**F**). Different letters are significantly different ($p < 0.05$) by one-way analysis of variance multiple comparisons.

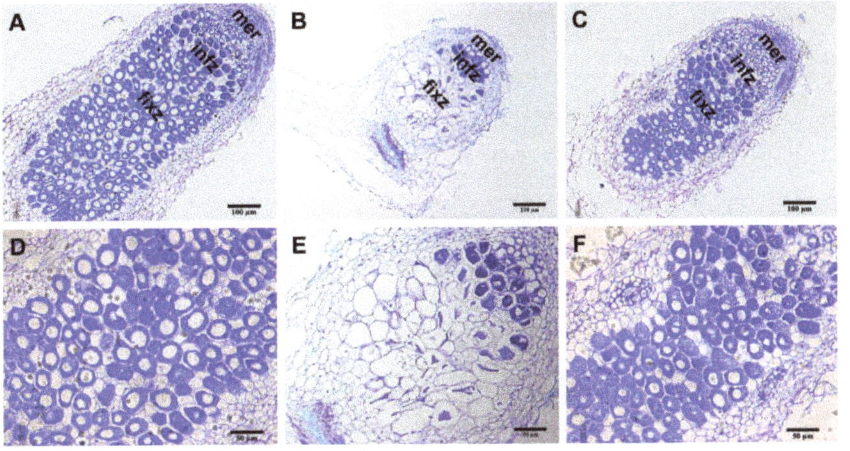

Figure 7. Observation of the nodule structures from *A. sinicus* roots inoculated with *M. huakuii* 7653R, 7653RΔ*rfaD*, or 7653RΔrfaD-C strains. (**A–F**), the paraffin sections of nodules from *A. sinicus* roots after staining with Toluidine blue O. A and D, inoculation with wild-type 7653R; (**B**,**E**), inoculation with 7653RΔ*rfaD* mutant; (**C**,**F**), inoculation with 7653RΔ*rfaD*-C. mer, meristem zone; infz, infection zone; fixz, nitrogen fixation zone. Scale bars: 100 μm (**A–C**) and 50 μm (**D–F**).

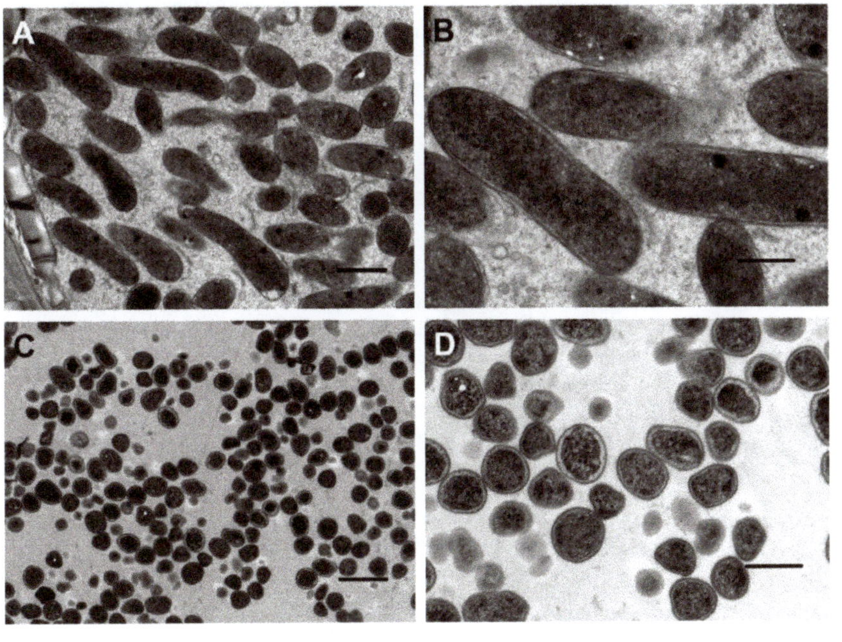

Figure 8. Observation of the ultrastructures of *A. sinicus* nodules generated by the *M. huakuii* 7653R and 7653RΔ*rfaD*. Transmission electron microscope images of the nodules from *A. sinicus* roots at 28 dpi. (**A**,**B**), inoculation with wild-type 7653R; (**C**,**D**), inoculation with 7653RΔ*rfaD*. Scale bars: 2 μm (**A**,**C**) and 1 μm (**B**,**D**).

4. Discussion

The symbiotic nitrogen fixation between rhizobia and legumes is a delicate and complex process, which is involved in a large number of genes and chemical components from both symbionts [2,45]. Among them, the lipopolysaccharide (LPS) is an important component of the cell wall of gram-negative bacteria and plays an important role in the symbiosis between rhizobia and leguminous plants [10,18]. The LPS in *E. coli* was extensively studied [27,46–48], and some rhizobia also have the gene information of the LPS synthetic pathway, but the precise functions of LPS synthesis-related genes in the process of symbiotic nitrogen fixation remain elusive. It has been reported that in some Gram-negative bacteria, the deletion of the *rfaD* gene (encoding an ADP-L-glycero-D-manno-heptose-6-epimerase) will affect the biosynthesis of the inner core of the LPS [16,27]. Here, we focused on the *rfaD* gene in *M. huakuii* 7653R, by investigating its roles in LPS synthesis and stress resistance, its expression patterns during symbiosis and its roles in the early infection and nodulation.

In this study, we constructed a *rfaD* gene deletion mutant using the *cre-loxp* system and explored its functions in free-living growth and during symbiosis. Quantitative detection of the LPS extracted from 7653RΔ*rfaD* and wild-type 7653R strains showed that the deletion of the *rfaD* gene did directly lead to a lower level of LPS synthesis. Furthermore, the ability of the 7653RΔ*rfaD* strain to resist external stresses decreased significantly under free-living conditions. It is reasonably proposed that this strain sensitivity to stresses may have resulted from the lower LPS production of this Δ*rfaD* mutant. Because the LPS is an important component of the cell wall of Gram-negative bacteria, the very low amounts of LPS may change the permeability of the cell wall, thereby affecting the ability of bacteria to resist external stress. These results further indicate that the epimerase encoded by the *rfaD* converts the ADP-D-glycerol-β-D-mannoheptose into the ADP-L-glycerol-β-D-mannoheptose; this reaction is necessary for LPS synthesis [27]. Therefore, the ADP-L-glycerol-β-D-mannoheptose can participate in the synthesis of the inner core of the LPS. The core polysaccharide is connected to the other two important components of the LPS: lipid A and O antigen; thus, the defective synthesis of the core polysaccharide has an impact on the structure of the LPS.

LPS is also considered as one of the essential components in the symbiosis between rhizobia and legumes [19,49]. The mutant of the *rfaD* gene of *Azorhizobium caulinodans* ORS571, which forms nodules on the stems and roots of *Sesbania rostrata*, was more advanced than the wild-type strain, but the mutant nodules were ineffective [50]. In this work, the inoculated 7653RΔ*rfaD* into the roots of *A. sinicus*, the symbiotic phenotype analysis showed that the shoots of the symbiotic plants grew poorly, and the leaves were yellow in color (an indicator of lacking N). Furthermore, the development of root nodules was abnormal and defective: in the Δ*rfaD* mutant roots, nodules were small and white, and the nitrogenase activity of nodules was not detectable. This led us to think about and analyze the possible causes of these symbiotic phenotypes. Because of the deletion mutation of the *rfaD*, the structure or synthetic quantity of the LPS changed in the mutant strain, which further interfered with the balance of symbiosis and immunity during the interactions between the rhizobia and the host, thus affecting the normal release and survival of the rhizobia from the infection threads (ITs), and significantly reducing the number of infected plant cells in the root nodules; the mutant forms were therefore ineffective and white nodules. Furthermore, the enhanced sensitivity of the outer membrane of the rfaD mutant may affect the normal differentiation of the bacteroid, which leads to a loss of nitrogen fixation ability of the mutant nodules. The results indicated that the *rfaD* is an important gene in the process of symbiotic nitrogen fixation, suggestive of the activation of symbiotic nitrogen fixation by regulating the LPS synthesis. The previous studies reported that the LPS could regulate symbiotic nitrogen fixation at the multiple infection stages, such as the inhibition of plant defense responses during infection, and participation in the formation of infection threads [18–20]. Core polysaccharide and O-antigen of the LPS may change during development and symbiotic interaction [20]. The rhizobia LPS would become more hydrophobic during bacterial development, and would also change due to plant exudate [51–53]. Therefore, the LPS synthesis-related genes may play different roles in the autogenesis growth and symbiotic interaction. The results from the in-situ expression experiment showed that the *rfaD* gene was highly expressed in the infection zone of root nodules, but not in the early infection process. This result may be due to the low expression of the gene in the early stage of infection. Alternatively, it is predicted that the *rfaD* gene also functions in the infected zone at the late stage of nodulation, such as protecting bacterial cells in the nitrogen fixation zone as a barrier, or participating in the molecular information exchange at the symbiotic interface. The researches on *rfaD* in some pathogenic bacteria reported that the deletion of this gene would lead to a short-of-LPS-sugar chain, thereby reducing the adhesion of bacteria to the host cells and reducing the pathogenicity of bacteria [26,27]. Moreover, the observation on the 7653RΔ*rfaD* inoculated into *A. sinicus* roots found that the number of infection threads increased at the early stage of infection; this result may be caused by the decrease in the adhesion ability of the rhizobia 7653RΔ*rfaD* (due to the lower LPS synthesis) during the infection processes, resulting in a defective nodule development. In the process of symbiosis with the rhizobia, plants increase their nitrogen fixation ability by regulating the number of nodules [54,55]. However, the low number of nodule primordia in the 7653RΔ*rfaD* strain may be due to the lagging process of nodule formation. The above findings indicate that the *rfaD* gene plays a role in the infected zone after nodule formation, regulating nodule development and nitrogen fixation, although it is not highly expressed at the early stage of nodule formation.

In conclusion, the present study showed an ADP-L-glycerol-D-mannoheptosyl-6-exoisomerase coding gene, *rfaD* from *M. huakuii* 7653R. The expression, localization and function experiments revealed the *rfaD* engaged in the LPS synthesis required for nodulation during the *M. huakuii* 7653R-*A. sinicus* interaction. These new findings advance our understanding of the function of LPS synthesis-related genes from rhizobia in the establishment of endosymbiosis in legumes.

Supplementary Materials: The following supporting information can be downloaded at: https://www.mdpi.com/article/10.3390/microorganisms11010059/s1, Figure S1: Verification of the constructed M. huakuii 7653R *rfaD* deletion mutant (7653RΔ*rfaD*) [56]; Figure S2: Effects of 7653R *rfaD* on the rhizobia growth under free-living conditions; Figure S3: The LPS produced by wild-type 7653R and 7653RΔ*rfaD* mutant; Figure S4: Sensitivity of the 7653R, 7653RΔ*rfaD* and 7653RΔrfaD-C strains to the polymyxin B. Table S1: The strains and plasmids used in this study; Table S2: Primers used in this study.

Author Contributions: Conceptualization, X.X. and Y.L. (Youguo Li); methodology, Y.L. (Yuan Liu), Y.S., Y.L. (Ye Lin); software, N.G., Y.L. (Yuan Liu); validation, Y.L. (Yuan Liu), X.X., Y.L. (Youguo Li); formal analysis, Y.L. (Youguo Li); investigation, Y.L. (Yuan Liu)., Y.S.; resources, Y.L. (Youguo Li); data curation, Y.L. (Yuan Liu), Y.L. (Youguo Li); writing—original draft preparation, Y.L. (Yuan Liu)., N.G.; writing—review and editing, Y.L. (Youguo Li); visualization, Y.L. (Yuan Liu); supervision, Y.L. (Youguo Li), X.X.; project administration, Y.L. (Youguo Li); funding acquisition, Y.L. (Youguo Li). All authors have read and agreed to the published version of the manuscript.

Funding: National Key Research and Development Program of China (grant no. 2022YFA0912100, 2021YFD1700201-05), the National Natural Science Foundation of China (grant no. 31970267, 32270259) and Hubei Natural Science Foundation (grant no. 2020CFA008).

Data Availability Statement: National Center for Biotechnology Information (NCBI) at https://www.ncbi.nlm.nih.gov/.

Acknowledgments: We thank Li Luo of Shanghai University for the gift of the pRG960 plasmid.

Conflicts of Interest: The authors declare no competing interests.

References

1. Oldroyd, G.E.; Murray, J.D.; Poole, P.S.; Downie, J.A. The rules of engagement in the legume-rhizobial symbiosis. *Annu. Rev. Genet.* **2011**, *45*, 119–144. [CrossRef]
2. Via, V.D.; Zanetti, M.E.; Blanco, F. How legumes recognize rhizobia. *Plant Signal. Behav.* **2016**, *11*, e1120396. [CrossRef]
3. Hassan, S.; Mathesius, U. The role of flavonoids in root-rhizosphere signalling: Opportunities and challenges for improving plant-microbe interactions. *J. Exp. Bot.* **2012**, *63*, 3429–3444. [CrossRef]
4. Wang, D.; Yang, S.; Tang, F.; Zhu, H. Symbiosis specificity in the legume: Rhizobial mutualism. *Cell. Microbiol.* **2012**, *14*, 334–342. [CrossRef] [PubMed]
5. Madsen, E.B.; Madsen, L.H.; Radutoiu, S.; Olbryt, M.; Rakwalska, M.; Szczyglowski, K.; Sato, S.; Kaneko, T.; Tabata, S.; Sandal, N.; et al. A receptor kinase gene of the LysM type is involved in legume perception of rhizobial signals. *Nature* **2003**, *425*, 637–640. [CrossRef]
6. Madsen, L.H.; Tirichine, L.; Jurkiewicz, A.; Sullivan, J.T.; Heckmann, A.B.; Bek, A.S.; Ronson, C.W.; James, E.K.; Stougaard, J. The molecular network governing nodule organogenesis and infection in the model legume *Lotus japonicus*. *Nat. Commun.* **2010**, *1*, 10. [CrossRef]
7. Oldroyd, G.E. Speak, friend, and enter: Signalling systems that promote beneficial symbiotic associations in plants. *Nat. Rev. Microbiol.* **2013**, *11*, 252–263. [CrossRef]
8. Okazaki, S.; Kaneko, T.; Sato, S.; Saeki, K. Hijacking of leguminous nodulation signaling by the rhizobial type III secretion system. *Proc. Natl. Acad. Sci. USA* **2013**, *110*, 17131–17136. [CrossRef]
9. Sugawara, M.; Takahashi, S.; Umehara, Y.; Iwano, H.; Tsurumaru, H.; Odake, H.; Suzuki, Y.; Kondo, H.; Konno, Y.; Yamakawa, T.; et al. Variation in bradyrhizobial NopP effector determines symbiotic incompatibility with Rj2-soybeans via effector-triggered immunity. *Nat. Commun.* **2018**, *9*, 3139. [CrossRef]
10. Fraysse, N.; Couderc, F.; Poinsot, V. Surface polysaccharide involvement in establishing the rhizobium-legume symbiosis. *Eur. J. Biochem.* **2003**, *270*, 1365–1380. [CrossRef]
11. Kawaharada, Y.; Kelly, S.; Nielsen, M.W.; Hjuler, C.T.; Gysel, K.; Muszyński, A.; Carlson, R.W.; Thygesen, M.B.; Sandal, N.; Asmussen, M.H.; et al. Receptor-mediated exopolysaccharide perception controls bacterial infection. *Nature* **2015**, *523*, 308–312. [CrossRef] [PubMed]
12. Becker, A.; Fraysse, N.; Sharypova, L. Recent advances in studies on structure and symbiosis-related function of rhizobial K-antigens and lipopolysaccharides. *Mol. Plant-Microbe Interact.* **2005**, *18*, 899–905. [CrossRef] [PubMed]
13. Jones, K.M.; Kobayashi, H.; Davies, B.W.; Taga, M.E.; Walker, G.C. How rhizobial symbionts invade plants: The Sinorhizobium-Medicago model. *Nat. Rev. Microbiol.* **2007**, *5*, 619–633. [CrossRef] [PubMed]
14. Janczarek, M.; Rachwał, K.; Marzec, A.; Grządziel, J.; Palusińska-Szysz, M. Signal molecules and cell-surface components involved in early stages of the legume–rhizobium interactions. *Appl. Soil Ecol.* **2015**, *85*, 94–113. [CrossRef]
15. Chu, H.; Mazmanian, S.K. Innate immune recognition of the microbiota promotes host-microbial symbiosis. *Nat. Immunol.* **2013**, *14*, 668–675. [CrossRef]
16. Raetz, C.; Whitfield, C. Lipopolysaccharide endotoxins. *Annu. Rev. Biochem.* **2002**, *71*, 365–700. [CrossRef]
17. Maldonado, R.F.; Sá-Correia, I.; Valvano, M.A. Lipopolysaccharide modification in Gram-negative bacteria during chronic infection. *FEMS Microbiol. Rev.* **2016**, *40*, 480–493. [CrossRef]
18. Lerouge, I.; Vanderleyden, J. O-antigen structural variation: Mechanisms and possible roles in animal/plant-microbe interactions. *FEMS Microbiol. Rev.* **2002**, *26*, 17–47. [CrossRef]
19. Scheidle, H.; Gross, A.; Niehaus, K. The Lipid A substructure of the Sinorhizobium meliloti lipopolysaccharides is sufficient to suppress the oxidative burst in host plants. *New Phytol.* **2005**, *165*, 559–565. [CrossRef]
20. Xiang, Q.; Wang, J.; Qin, P.; Adil, B.; Xu, K.; Gu, Y.; Yu, X.; Zhao, K.; Zhang, X.; Ma, M.; et al. Effect of common bean seed exudates on growth, lipopolysaccharide production, and lipopolysaccharide transport gene expression of Rhizobium anhuiense. *Can. J. Microbiol.* **2020**, *66*, 186–193. [CrossRef]
21. Ardissone, S.; Noel, K.D.; Klement, M.; Broughton, W.J.; Deakin, W.J. Synthesis of the flavonoid-induced lipopolysaccharide of Rhizobium Sp. *strain NGR*234 requires rhamnosyl transferases encoded by genes rgpF and wbgA. *Mol. Plant-Microbe Interact.* **2011**, *24*, 1513–1521. [CrossRef]
22. Margaret, I.; Lucas, M.M.; Acosta-Jurado, S.; Buendía-Clavería, A.M.; Fedorova, E.; Hidalgo, Á.; Rodríguez-Carvajal, M.A.; Rodriguez-Navarro, D.N.; Ruiz-Sainz, J.E.; Vinardell, J.M. The Sinorhizobium fredii HH103 lipopolysaccharide is not only relevant at early soybean nodulation stages but also for symbiosome stability in mature nodules. *PloS ONE* **2013**, *8*, e74717. [CrossRef]
23. Oliveira, L.R.; Rodrigues, E.P.; Marcelino-Guimarães, F.C.; Oliveira, A.L.; Hungria, M. Fast induction of biosynthetic polysaccharide genes lpxA, lpxE, and rkpI of Rhizobium sp. strain PRF 81 by common bean seed exudates is indicative of a key role in symbiosis. *Funct. Integr. Genom.* **2013**, *13*, 275–283. [CrossRef]
24. Albus, U.; Baier, R.; Holst, O.; Pühler, A.; Niehaus, K. Suppression of an elicitor-induced oxidative burst reaction in Medicago sativa cell cultures by *Sinorhizobium meliloti* lipopolysaccharides. *New Phytol.* **2001**, *151*, 597–606. [CrossRef]
25. Carlson, R.W.; Forsberg, L.S.; Kannenberg, E.L. Lipopolysaccharides in Rhizobium-legume symbioses. *Sub-Cell. Biochem.* **2010**, *53*, 339–386. [CrossRef]

26. Chang, P.C.; Wang, C.J.; You, C.K.; Kao, M.C. Effects of a HP0859 (rfaD) knockout mutation on lipopolysaccharide structure of Helicobacter pylori 26695 and the bacterial adhesion on AGS cells. *Biochem. Biophys. Res. Commun.* **2011**, *405*, 497–502. [CrossRef]
27. Kuo, C.J.; Chen, J.W.; Chiu, H.C.; Teng, C.H.; Hsu, T.I.; Lu, P.J.; Syu, W.J.; Wang, S.T.; Chou, T.C.; Chen, C.S. Mutation of the enterohemorrhagic *Escherichia coli* core LPS biosynthesis enzyme RfaD confers hypersusceptibility to host intestinal innate immunity in vivo. *Front. Cell. Infect. Microbiol.* **2016**, *6*, 82. [CrossRef]
28. Valvano, M.A.; Messner, P.; Kosma, P. Novel pathways for biosynthesis of nucleotide-activated glycero-manno-heptose precursors of bacterial glycoproteins and cell surface polysaccharides. *Microbiology* **2002**, *148*, 1979–1989. [CrossRef]
29. Cho, H.J.; Yong, X.; Murooka, Y. Formation of adventitious shoots and plant regeneration by culture of cotyledon segment in Astragalus sinicus (Chinese Milk Vetch). *Plant Tissue Cult. Lett.* **1995**, *12*, 87–90. [CrossRef]
30. Chen, H.K.; Shu, M.K. Note on the root-nodule bacteria of Astragalus sinicus L. *Soil Sci.* **1944**, *58*, 291–294. [CrossRef]
31. Lei, L.; Chen, L.; Shi, X.; Li, Y.; Wang, J.; Chen, D.; Xie, F.; Li, Y. A nodule-specific lipid transfer protein AsE246 participates in transport of plant-synthesized lipids to symbiosome membrane and is essential for nodule organogenesis in Chinese milk vetch. *Plant Physiol.* **2014**, *164*, 1045–1058. [CrossRef]
32. Si, Z.; Guan, N.; Zhou, Y.; Mei, L.; Li, Y.; Li, Y. A methionine sulfoxide reductase B is required for the establishment of Astragalus sinicus-Mesorhizobium symbiosis. *Plant Cell Physiol.* **2020**, *61*, 1631–1645. [CrossRef]
33. Zhou, D.; Li, Y.; Wang, X.; Xie, F.; Chen, D. *Mesorhizobium huakuii HtpG Interaction with nsLTP AsE*246 Is Required for Symbiotic Nitrogen Fixation. *Plant Physiol.* **2019**, *180*, 509–528. [CrossRef]
34. Wang, S.; Hao, B.; Li, J.; Gu, H.; Peng, J.; Xie, F.; Zhao, X.; Frech, C.; Chen, N.; Ma, B.; et al. Whole-genome sequencing of Mesorhizobium huakuii 7653R provides molecular insights into host specificity and symbiosis island dynamics. *BMC Genom.* **2014**, *15*, 440. [CrossRef]
35. Li, Y.; Zhou, L.; Li, Y.; Chen, D.; Tan, X.; Lei, L.; Zhou, J. A nodule-specific plant cysteine proteinase, AsNODF32, is involved in nodule senescence and nitrogen fixation activity of the green manure legume *Astragalus sinicus*. *New Phytol.* **2008**, *180*, 185–192. [CrossRef]
36. Marx, C.J.; Lidstrom, M.E. Broad-host-range cre-lox system for antibiotic marker recycling in gram-negative bacteria. *BioTechniques* **2002**, *33*, 1062–1067. [CrossRef]
37. Kovach, M.E.; Elzer, P.H.; Hill, D.S.; Robertson, G.T.; Farris, M.A.; Roop, R.M., 2nd; Peterson, K.M. Four new derivatives of the broad-host-range cloning vector pBBR1MCS, carrying different antibiotic-resistance cassettes. *Gene* **1995**, *166*, 175–176. [CrossRef]
38. Van den Eede, G.; Deblaere, R.; Goethals, K.; Van Montagu, M.; Holsters, M. Broad host range and promoter selection vectors for bacteria that interact with plants. *Mol. Plant-Microbe Interact.* **1992**, *5*, 228–234. [CrossRef]
39. Wilson, K.J.; Sessitsch, A.; Corbo, J.C.; Giller, K.E.; Akkermans, A.D.; Jefferson, R.A. beta-Glucuronidase (GUS) transposons for ecological and genetic studies of rhizobia and other gram-negative bacteria. *Microbiology* **1995**, *141 Pt 7*, 1691–1705. [CrossRef]
40. Stuurman, N.; Pacios Bras, C.; Schlaman, H.R.; Wijfjes, A.H.; Bloemberg, G.; Spaink, H.P. Use of green fluorescent protein color variants expressed on stable broad-host-range vectors to visualize rhizobia interacting with plants. *Mol. Plant-Microbe Interact.* **2000**, *13*, 1163–1169. [CrossRef]
41. Hardy, R.; Burns, R.C.; Holsten, R.D. Applications of the acetylene-ethylene assay for measurement of nitrogen fixation. *Soil Biol. Biochem.* **1973**, *5*, 47–81. [CrossRef]
42. Si, Z.; Yang, Q.; Liang, R.; Chen, L.; Chen, D.; Li, Y. Digalactosyldiacylglycerol synthase gene MtDGD1 plays an essential role in nodule development and nitrogen fixation. *Mol. Plant-Microbe Interact.* **2019**, *32*, 1196–1209. [CrossRef]
43. Tamura, K.; Stecher, G.; Kumar, S. MEGA11: Molecular Evolutionary Genetics Analysis Version 11. *Mol. Biol. Evol.* **2021**, *38*, 3022–3027. [CrossRef]
44. Pegues, J.C.; Chen, L.S.; Gordon, A.W.; Ding, L.; Coleman, W.G., Jr. Cloning, expression, and characterization of the *Escherichia coli* K-12 rfaD gene. *J. Bacteriol.* **1990**, *172*, 4652–4660. [CrossRef] [PubMed]
45. Loutet, S.A.; Flannagan, R.S.; Kooi, C.; Sokol, P.A.; Valvano, M.A. A complete lipopolysaccharide inner core oligosaccharide is required for resistance of *Burkholderia cenocepacia* to antimicrobial peptides and bacterial survival in vivo. *J. Bacteriol.* **2006**, *188*, 2073–2080. [CrossRef]
46. Walker, L.; Lagunas, B.; Gifford, M.L. Determinants of host range specificity in legume-rhizobia symbiosis. *Front. Microbiol.* **2020**, *11*, 585749. [CrossRef]
47. Kneidinger, B.; Marolda, C.; Graninger, M.; Zamyatina, A.; Mcarthur, F.; Kosma, P.; Valvano, M.A.; Messner, P. Biosynthesis Pathway of ADP-L-glycero-beta-D-manno-Heptose in *Escherichia coli*. *J. Bacteriol.* **2002**, *184*, 363–369. [CrossRef]
48. Wang, J.; Ma, W.; Wang, Z.; Li, Y.; Wang, X. Construction and characterization of an *Escherichia coli* mutant producing Kdo_2-lipid A. *Mar. Drugs* **2014**, *12*, 1495–1511. [CrossRef]
49. Tang, G.; Wang, Y.; Luo, L. Transcriptional regulator LsrB of Sinorhizobium meliloti positively regulates the expression of genes involved in lipopolysaccharide biosynthesis. *Appl. Environ. Microbiol.* **2014**, *80*, 5265–5273. [CrossRef]
50. Wakao, S.; Siarot, L.; Aono, T.; Oyaizu, H. Effects of alteration in LPS structure in Azorhizobium caulinodans on nodule development. *J. Gen. Appl. Microbiol.* **2015**, *61*, 248–254. [CrossRef]
51. Kannenberg, E.L.; Carlson, R.W. Lipid A and O-chain modifications cause Rhizobium lipopolysaccharides to become hydrophobic during bacteroid development. *Mol. Microbiol.* **2001**, *39*, 379–391. [CrossRef] [PubMed]

52. Noel, K.D.; Box, J.M.; Bonne, V.J. 2-O-methylation of fucosyl residues of a rhizobial lipopolysaccharide is increased in response to host exudate and is eliminated in a symbiotically defective mutant. *Appl. Environ. Microbiol.* **2004**, *70*, 1537–1544. [CrossRef] [PubMed]
53. D'Haeze, W.; Leoff, C.; Freshour, G.; Noel, K.D.; Carlson, R.W. Rhizobium etli CE3 bacteroid lipopolysaccharides are structurally similar but not identical to those produced by cultured CE3 bacteria. *J. Biol. Chem.* **2007**, *282*, 17101–17113. [CrossRef] [PubMed]
54. Ferguson, B.J.; Indrasumunar, A.; Hayashi, S.; Lin, M.H.; Lin, Y.H.; Reid, D.E.; Gresshoff, P.M. Molecular analysis of legume nodule development and autoregulation. *J. Integr. Plant Biol.* **2010**, *52*, 61–76. [CrossRef] [PubMed]
55. Hastwell, A.H.; Corcilius, L.; Williams, J.T.; Gresshoff, P.M.; Payne, R.J.; Ferguson, B.J. Triarabinosylation is required for nodulation-suppressive CLE peptides to systemically inhibit nodulation in Pisum sativum. *Plant Cell Environ.* **2019**, *42*, 188–197. [CrossRef]
56. Simon, R.; Priefer, U.; Puhler, A. A broad host mobilization system for in vivo genetic engineering: Transposon mutagenesis in Gram-negative bacteria. *Bio/Technolgy* **1983**, *1*, 37–45. [CrossRef]

microorganisms

Article

The Improved Biocontrol Agent, F1-35, Protects Watermelon against Fusarium Wilt by Triggering Jasmonic Acid and Ethylene Pathways

Xiao-Min Dong †, Qing-Gui Lian †, Jing Chen, Rui-Min Jia, Zhao-Feng Zong, Qing Ma and Yang Wang *

College of Plant Protection, Northwest A&F University, Yangling, Xianyang 712100, China
* Correspondence: wangyang2006@nwsuaf.edu.cn
† These authors contributed equally to this work.

Abstract: Watermelon Fusarium wilt, caused by *Fusarium oxysporum* f. sp. *niveum* (FON), is one of the most important diseases, and has become a major limiting factor to watermelon production worldwide. Previous research has found that the improved biocontrol agent, F1-35, had a high control efficiency to watermelon Fusarium wilt. In this study, the control efficiency of F1-35 to watermelon Fusarium wilt was firstly tested, and the control efficiency was 61.7%. Then, we investigated the mode of action of F1-35 in controlling watermelon Fusarium wilt. Using a pairing assay, we found that F1-35 did not inhibit the normal growth of FON. To know more about the interaction between F1-35 and watermelon root, the protein expressions of roots after 12, 24, and 48 h post-inoculation were examined. A total of 1109 differentially expressed proteins were obtained. KEGG analysis found that the most differentially expressed proteins occurred in alpha-linolenic acid metabolism, cysteine and methionine metabolism, plant–pathogen interaction, and the MAPK signaling pathway to the plant. A further analysis of differentially expressed proteins showed that F1-35 triggered the jasmonic acid and ethylene pathways in watermelon. To validate our results, the qRT-PCR was used to analyze the gene expression levels of *PAL*, *LOX1*, and *CTR1*. The gene expression results showed that those genes, which were positive correlated with the JA pathway, were up-expressed, including *PAL* and *LOX1*, and the negative associated gene, *CTR1*, was down-expressed. In conclusion, the improved biocontrol agent, F1-35, improves the resistance of watermelons to FON by triggering the JA and ET pathways.

Keywords: watermelon Fusarium wilt; biological control; JA and ET pathway; proteomes; qRT-PCR

Citation: Dong, X.-M.; Lian, Q.-G.; Chen, J.; Jia, R.-M.; Zong, Z.-F.; Ma, Q.; Wang, Y. The Improved Biocontrol Agent, F1-35, Protects Watermelon against Fusarium Wilt by Triggering Jasmonic Acid and Ethylene Pathways. *Microorganisms* **2022**, *10*, 1710. https://doi.org/10.3390/microorganisms10091710

Academic Editors: Beibei Ge, Tomislav Cernava and Kyungseok Park

Received: 17 July 2022
Accepted: 22 August 2022
Published: 25 August 2022

Publisher's Note: MDPI stays neutral with regard to jurisdictional claims in published maps and institutional affiliations.

1. Introduction

Fusarium wilt of watermelons (*Citrullus lanatus*), caused by *Fusarium oxysporum* f. sp. *niveum* (FON), is one of the most devastating diseases of watermelons around the world [1,2]. The pathogen which causes Fusarium wilt of watermelons exists in many different habitats, including saline soil, dry-tropical rainforests, and hospital sewage [3–6], and produces microspores, macrospores, and chlamydospores [7].

Fungicides, especially some fumigants, were once the most efficient way to control Fusarium wilt, but those fungicides were banned because of their defects, including being unfriendly to the environment and hardly degradable [8,9]. According to the Montreal Protocols, some alternative soil fumigants are used to control Fusarium wilt, but they are not as effective as MeBr at controlling FON or weeds, they are expensive, and they also negatively impact the environment [10]. Grafting is another method to control watermelon Fusariun wilt, which was first used commercially on watermelons in Japan and Korea [11,12]. Watermelons are primarily grafted onto *Cucurbita* rootstocks, which are non-hosts to FON [13–15]. However, as Mahamed [16] reported, grafting watermelons influences the growth, productivity, and quality of the fruit. Consequently, it is necessary to find an environmentally friendly, high-efficiency method to control Fusarium wilt.

Some researchers reported that *F. oxysporum*, as a hemi-biotrophic pathogen, could be inhibited by jasmonic acid and ethylene: the interaction between *F. oxysporum* and *Arabidopsis thaliana* reveals that salicylic acid (SA), which connects to systemic acquired resistance (SAR) [17,18], and JA/ET, which are induced systemic resistance (ISR) response hormones to beneficial microorganisms [19], also take part in the resistance process to *Fusarium* [20–22]. In addition, Kasote et al. [23] also found that ME-JA and JA-Ile may play an important role in the watermelon defense response against FON pathogens. Interestingly, some *F. oxysporum* species can enter into a plant as a beneficial microorganism and stimulate the JA pathway in the early stages [18,24]. As such, JA and ET play a role of resistance in the interaction between plant and *F. oxysporum*.

Many studies over the past decades have shown that certain microorganisms interact intimately with plants, suppressing plant pathogens and promoting plant growth [25–28]. Those microorganisms able to suppress plant pathogens are called biological control agents (BCA). The antagonism of BCAs includes: (1) diffusible antibiotics, volatile organic compounds, toxins, and biosurfactants inhibiting microbial growth; (2) competition space and nutrients; (3) degradation toxins and other pathogenicity factors; and (4) hyperparasitic, secreting cell-wall-degrading enzymes, such as chitinases and β-1,3-glucanase [29–31]. At present, some BCAs have been reported to have inhibitory effects on Fusarium wilt, including *Bacillus velezensis, Fluorescent pseudomonads, Trichoderma saturnisporum*, and non-pathogenic *F. oxysporum* [32–35].

Furthermore, some BCAs induce the expression of plant hormones, such as SA, ET, and JA [36–39]. Those plant hormones suppress infection by plant pathogens by plant hormone signal transduction. SA, JA, and ethylene (ET) are the major phytohormones in SAR and ISR, respectively. SAR and ISR are the important strategies by which plants respond to a wide range of biotic and abiotic stresses [40,41].

One of the BCAs, non-pathogenic *Fusarium oxysporum* Fo47, was isolated from soil in France [42]. Many researchers have reported that it has the ability to control many plant diseases worldwide, including Fusarium wilt, *Verticillium*, and *Phytophthora* in pepper and tomatoes [42–47]. In former research, protoplast fusion was carried out between Fo47 and actinomycetes 153 by inactivation and mutation in order to improve the characteristics of the Fo47 strain [48]. F1-35, one of the fusion products, improved the properties of inoculation, fungicide resistance, and growth promotion. We also found that the rates of promoting the growth of F1-35 depended on the dose of gibberellin. The recombinant strain, F1-35, produced a higher content of gibberellin (223.8 mg/L) than Fo47 (134.5 mg/L) after being separately grown in 200 mL Czapek broth for 3 days in a 250 mL flask at 25 °C, agitated at 150 rmp. F1-35 also had better growth-promoting than that of Fo47 in watermelons and cucumber seedlings. With regard to the control efficiency of FON, F1-35 reached 59.04%, which was higher than that of Fo47 (34.94%) [48].

The aim of this study was to determine the mechanism of F1-35 control of watermelon Fusarium wilt by using the pairing assay and isobaric tags for relative and absolute quantitation (iTRAQ) coupled with the LC-MS/MS method and qRT-PCR to study the interaction between F1-35 and watermelon plants.

2. Materials and Methods

2.1. *Pathogenic Fungus, Biocontrol Agent F1-35, and Plant Materials*

The watermelon Fusarium wilt pathogen, *Fusarium oxysporum* f. sp. *niveum* (FON), and the improved biocontrol strain, F1-35, were provided by the Vegetable Disease and Biocontrol Laboratory, College of Plant Protection, Northwest A&F University, Yangling, Shaanxi Province, China. FON and F1-35 were cultured on potato dextrose agar (PDA) plates at 28 °C for 7 d. A cultivated variety of watermelon, NW-5, which is sensitive to watermelon Fusarium wilt, was used in this study; this was provided by Associate Professor Jianxiang Ma, College of Horticulture, Northwest A&F University, Yangling, Shaanxi Province, China. The seeds were treated according to the procedures of Kong et al. [49] as follows: seeds were first sterilized with 1.5% sodium hypochlorite, soaked in distilled water for 4 h, and maintained at 30 °C for germination. The germinated seedlings were planted in a sterilized peat-perlite substrate (2:1 v/v) and cultured in the greenhouse under a 16-h diurnal light cycle at 28 °C with 80 to 85% relative humidity inside a controlled environment chamber. Seedlings with five true leaves were used for the experiment.

2.2. Inoculum Preparation and Plant Inoculation

Agar plugs (5-mm diameter) were aseptically cut out from the edge of the fungal colonies of FON and F1-35 after 7 d of growth in PDA plates. FON and F1-35 were then separately grown in 200 mL potato dextrose broth (PDB) in a 250 mL flask at 28 °C for 7 d, with agitation at 160 rmp. Spore suspensions of FON and F1-35 in PDB were filtered and prepared at concentrations of 10^6 cfu/mL and 10^8 cfu/mL, respectively, determined using a hemocytometer [50].

All seedling roots were inoculated by dipping in a spore suspension for 10 min. There were four treatment groups: control (CK) (treated with PDB), FON (treated with FON), F1-35 (treated with F1-35), and F1-35/FON (treated with FON and F1-35 simultaneously). Each treatment was conducted in triplicate, with 24 seedlings in each replicate. Samples of CK, FON, and F1-35 were separately collected at 0, 12, 24, 48, 72, and 96 h post-inoculation (hpi), with six seedlings for each time point, immediately frozen in liquid nitrogen and stored at −80 °C for further study.

2.3. Control Efficiency Test

Following the treatments described above, and with three replicates for each treatment, the disease index and control efficiency were counted using the methodology described by Pu et al. [50].

2.4. Pairing Assay

To determine the F1-35 inhibition of FON, dual cultures were used according to the method of Veloso and Díaz [43], including only FON and F1-35 + FON, with three replicates, each replicate comprising three Petri dishes. The mycelia of FON, which were located beside F1-35, were examined under a microscope.

2.5. Proteomic Analysis

2.5.1. Proteome Extraction and iTRAQ Labeling

Protein from the CK, and 12 hpi, 24 hpi, and 48 hpi root of F1-35 samples, comprising three biological replicates, was extracted using TCA/acetone precipitation and SDT lysis [51]. Proteins (20 μg) from each sample were then mixed with 5× loading buffer, and boiled for 5 min. The proteins were separated on 12.5% SDS-PAGE gel (constant current 14 mA for 90 min). Protein bands were visualized by Coomassie Blue R-250 staining. After filter-aided sample preparation (FASP digestion) using the Wisniewski method [52], 100 μg of peptide mixture from each sample was labeled using the iTRAQ reagent according to the manufacturer's instructions (Applied Biosystems).

2.5.2. Peptide Fractionation with Strong Cation Exchange (SCX) Chromatography

The labeled peptides in each group were homogenized and mixed with an AK-TA purifier system (GE Healthcare). As buffer A, we used 10 mM KH_2PO_4, 25% ACN, pH 3.0, whereas buffer B contained 10 mM KH_2PO_4, 500 mM KCl, 25% ACN, pH 3.0. The column was equilibrated with buffer A, and then, peptides were eluted at a flow rate of 1 Ml/min with a gradient of 0–8% buffer B (500 mM KCl, 10 mM KH_2PO_4 in 25% of ACN, pH 3.0) for 22 min, 8–52% buffer B for 22–47 min, 52–100% buffer B for 47–50 min, and 100% buffer B for 50–58 min; then, buffer B was reset to 0% after 58 min. The absorbance was set at 214 nm for monitoring elution, and fractions were collected every 1 min. After collecting fractions, the fractions were collected on C18 Cartridges (Empore™ SPE Cartridges C18, standard density, bed I.D. 7 mm, 3-mL volume; Sigma, St. Louis, MO, USA) and vacuum-centrifuged for concentration.

2.5.3. LC-MS/MS Analysis

The specific steps are as follows: (1) buffer A (0.1% formic acid) and buffer B (84% acetonitrile and 0.1% formic acid) were prepared. (2) The peptide mixture was loaded onto a reverse phase trap column (Thermo Scientific Acclaim PepMap100, 100 μm × 2 cm, nanoViper C18, Boston, MA, USA) and connected to a C18-reverse phase analytical column (Thermo Scientific Easy Column, 10 cm long, 75 μm inner diameter, 3μm resin). Buffer A and buffer B were used for elution at a flow rate of 300 nL/min, which was controlled by IntelliFlow technology. (3) The samples were analyzed by a Q Exactive mass spectrometer (Thermo Scientific), and the detection mode was set to positive ion for 60/120/240 min. The parent scan was in the range of 300 to 1800 m/z with a resolution of 70,000 at 200 m/z. The automatic gain control (AGC) target was set to 3e6 at a maximum injection time of 10 ms. The duration of dynamic exclusion was 40.0 s. The parent ions with the top 10 ionic strength in the full scan were selected and broken by high-energy collisional dissociation (HCD) fragmentation at a resolution of 17,500 at 200 m/z. The collision energy was normalized to 30 eV, and the underfill ratio was 0.1%.

2.5.4. MS Data Analysis

MS/MS spectra were searched using the MASCOT engine (Matrix Science, London, UK; version 2.2) embedded into Proteome Discoverer 1.4 [53]. Protein data, uniprot_Beninc aseae_27316_20170228.fasta, were downloaded on 28 February 2017 at http://www.uniprot.org. The following parameters were set: enzyme as trypsin, max missed cleavages at 2, fixed modification as carbamidomethyl (C), iTRAQ 4/8 plex (N-term) and iTRAQ 4/8 plex (K), variable modifications as oxidation (M) and iTRAQ 4/8 plex (Y), peptide mass tolerance at ±20 ppm level, fragment mass tolerance as 0.1 Da, Benincaseae as database, decoy as database pattern, peptide FDR at ≤0.01 level, and the protein ratios calculated as the median of only the unique peptides of the protein. All peptide ratios were normalized by the median protein ratio; the median protein ratio should, therefore, be 1 after normalization.

2.5.5. Analysis of Differentially Expressed Proteins

Compared to CK, proteins with a differential expression greater than ±1.2-fold and a significant p-value ≤ 0.05 were annotated by gene ontology (GO) according to [54] and [55], and KEGG pathway annotation according to Kanehisa et al. [56].

2.6. *qRT-PCR Analysis*

2.6.1. Screening of Candidate Reference Genes and Primer Design

A total of 4 candidate reference genes were evaluated from http://www.icugi.org/ (accessed on 9 April 2017). These genes were chosen based on previous work by Guo et al. [57] and included: PAL, LOX1, CTR1, and beta-actin. All gene primers for qRT-PCR, except beta-actin, which was referenced by Kong et al. [49], were designed by Beacon Designer 7. All primers were synthesized by Quintarabio, Wuhan, China. The specificity of the PCR amplification product for each primer pair was further studied by melting curve analysis and amplification efficiency (E). All reactions were performed in a 20-μL volume containing 10 μL 2 × UltraSYBR Mixture (Cwbio Biotech Co., Ltd., Beijing, China), 1μL 10Mm forward and reverse primers, and 500 ng of cDNA. Melting curve analysis and amplification efficiency were conducted using Bio-Rad CFX Manager™ Software on Bio-Rad iQ5 real-time PCR instruments (Bio-Rad Laboratories, Hercules, CA, USA). The PCR program was as follows: 95 °C for 1 min, 40 cycles of 95 °C for 10 s, 60 °C for 10 s, and 72 °C for 40 s. Finally, dissociation curves were generated by increasing the temperature from 65 to 95 °C.

2.6.2. Total RNA Extraction and cDNA Synthesis

Total RNA was extracted from root samples using the Biozol method (Biomiga, Shanghai, China). Complementary DNA (cDNA) synthesis was then performed using a PrimeScript™ RT Reagent Kit with gDNA Eraser (Takara Biotechnology Co., Ltd., Dalian, China) according to the manufacturer's instructions. The primer pairs for quantitative RT-PCR were designed using Beacon Designer (Premier Biosoft, Palo Alto, CA, USA). The PCR reaction consisted of 10 μL of 2× Ultra SYBR Mixture (Cwbio, Beijing, China), 40 nM primers, and 2 μL of 1:10 diluted template cDNA in a total volume of 20 μL. No template controls were set for each primer pair. Quantitative RT-PCR was performed employing the Bio-Rad CF X96 System and Opticon Monitor software (Bio-Rad, USA). CT values were determined by averaging three technical replicates and three biological replicates. For each gene of interest, standard curves were generated. We converted Ct values to relative amounts of cDNA and calculated gene expression using the $2^{-\Delta\Delta CT}$ method [58].

2.6.3. Quantitative Determination and Statistical Analysis

The expression levels of the tested reference genes were determined by CT values. The value of E for each reference gene was calculated according to the following equation: E (%) = $(10^{-1/\text{slope}} - 1) \times 100$, where the slope is the standard curve slope determined using Excel 2016. The calculations and comparisons of treatment means for each experiment were conducted using analysis of variance (ANOVA) and SPSS 20.0 (IBM, Armonk, NY, USA); means were analyzed with Tukey's honestly significant difference (HSD) test at $p = 0.05$.

2.7. Data Analysis

All data were analyzed using SPSS 20.0 (IBM, Armonk, NY, USA) and Excel 2016 (Microsoft, Redmond, WA, USA). The results of qRT-PCR were calculated by REST using the $2^{-\Delta\Delta CT}$ method, and all figures were generated using GraphPad Prism 6 (GraphPad Software, San Diego, CA, USA).

2.8. MS/MS Data Submission

The raw data used in the present study for proteomics were submitted to the Integrated Proteome Resources database under accession number, IPX0001180000.

3. Results

3.1. The Control Efficiency of F1-35 against Watermelon Fusarium Wilt

To test the efficiency of F1-35 in controlling watermelon Fusarium wilt, F1-35, F1-35, and FON were simultaneously inoculated, and the incidence of watermelon seedling infections was recorded. The resulting control efficiency of F1-35 on watermelon Fusarium wilt was 61.7% (Table 1).

Table 1. Control efficiency of F1-35 on watermelon Fusarium wilt.

Treatment	Disease Index (%)	Control Efficiency (%)
FON	58.46 ± 4.48	-
F1-35 and FON	22.39 ± 5.01	61.7 ± 3.79

3.2. Pairing Assay

FON grew synchronously in the presence of F1-35, with no reduction compared to FON only. The radius of the FON colony was 28.5 mm and 27.7 mm, respectively, with no significant difference by the *t*-test. Therefore, F1-35 did not change the growth rates of FON. There was a boundary between F1-35 and FON. Subsequently, the mycelia of FON, which was located beside F1-35 colony, were observed with a microscope. There were no changes in either the mycelia or the spores. As such, no direct interactions between FON and F1-35 were observed under the microscope.

3.3. Influence of the Root Proteome Expressed after F1-35 Inoculation

To investigate the mechanism of Fusarium wilt reduction under F1-35, watermelon root proteomes during treatment with F1-35 (samples collected at 12, 24, and 48 hpi) were compared to CK. A total of 31,440 peptides were identified, including 27,618 unique peptides, and 6243 proteins were blasted to Benincaseae. To establish authenticity and accuracy, the MASCOT ion score was set as the index, which was carried with MASCOT 2.2 to form a blast in the database. The result of MASCOT showed that 70% of total peptides ion scores were higher than 20, with the middle ion score being 30. Comparing the FDR < 0.01 and ion score distribution, the MS result appears to be reliable and reproducible.

Proteomic results were summarized in the Venn diagram in Figure 1. This clearly demonstrated the differently expressed proteins in watermelon response to F1-35 at 12, 24, and 48 hpi. For the 12-hpi root proteins compared to CK, there were 184 differentially expressed proteins, 110 with up-regulated expression and 83 with down-regulated expression. For the 24-hpi root proteins, the differentially expressed proteins increased to 412, with 221 up-regulated and 191 down-regulated. The 48-hpi root differentially expressed proteins were similar to the 24-hpi root proteins, where 413 proteins were differentially expressed, 186 up-regulated and 227 down-regulated.

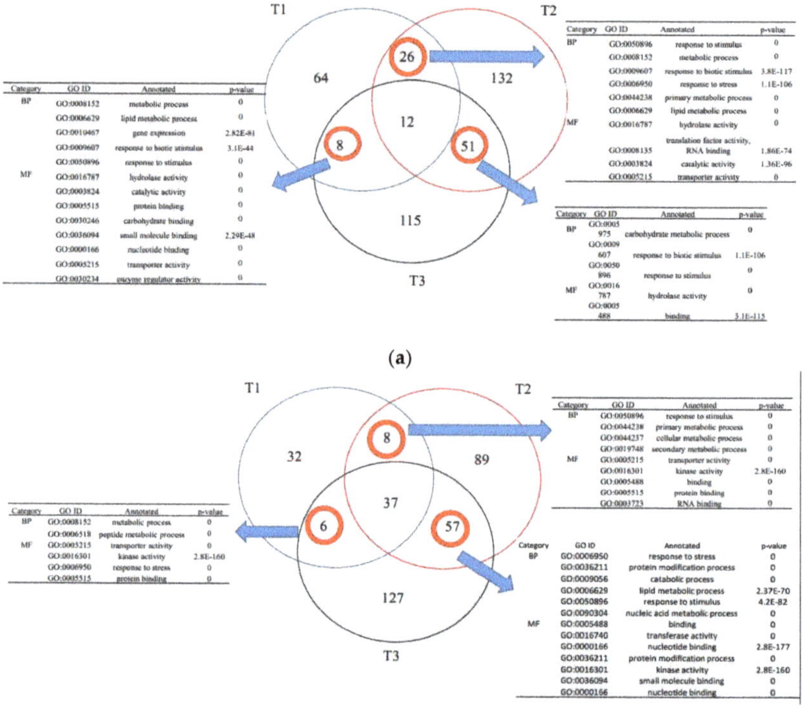

Figure 1. Distribution of differentially expressed proteins. (**a**): The up-regulated expression in watermelon response to F1-35 at 12, 24, and 48 hpi. (**b**): The down-regulated expression in watermelon response to F1-35 at 12, 24, and 48 hpi. **T1:** CK treatment at 12 hpi; **T2**: CK treatment at 24 hpi; **T3**: CK treatment at 48 hpi.

To investigate the interaction between F1-35 and watermelons, the identified proteins were analyzed. Compared to CK at 12 hpi, there were three up-regulated proteins and three down-regulated proteins. Having blasted the function of those proteins, the three up-regulated proteins related to soil stress or drought stress, and the three down-regulated proteins were Gip1-like proteins, which respond to some stress. Auxin-repressed protein (Arp), induced by JA, could increase resistance after being silenced. Compared to CK at 24 hpi, the highly expressed protein after treatment for 24 h with F1-35 had not only GIP, Arp, and the negative-regulated resistance protein, but also some JA-associated proteins,

PR5 and glyceraldehyde-3-phosphate dehydrogenase. The down-regulated proteins were similar to the 12-hpi samples. After treatment for 48 h with F1-35, the major differentially expressed proteins were similar to 12 h and 24 h. The down-regulated proteins were WRKY 17, a negative protein to JA, and a negative-regulated resistance protein (Table 2).

3.4. Identification and Functional Classification of Differentially Expressed Proteins

GO analysis revealed that most differentially expressed proteins were involved in metabolic processes, and single organism and cellular processes, among others. Many differentially expressed proteins were associated with molecular functions, such as binding, catalytic activity, and the binding of small molecules. Moreover, many differentially expressed proteins were associated with the cellular component, such as the organelle, membrane, and cell part, as seen in (Figure 2).

The differentially expressed proteins were also grouped into main functional classes based on KEGG [59]. The KEGG pathway analysis showed that most differentially expressed proteins occurred in plant–pathogen interaction, phenylpropanoid biosynthesis, oxidative phosphorylation, carbon metabolism, alpha-linolenic acid metabolism, cysteine and methionine metabolism, and the MAPK signaling pathway to the plant (Figure 3). In the root proteins at 12 and 24 hpi, the key enzymes were identified in alpha-linolenic acid metabolism and cysteine/methionine metabolism pathways. The plant hormones, JA and ET, are the product of those pathways. In the 24- and 36-hpi root proteins, JAR1 and ChiB were identified.

Table 2. Function analysis of differentially expressed proteins.

12 hpi /CK				24 hpi /CK				48 hpi /CK			
Up	**Function**	**Down**	**Function**	**Up**	**Function**	**Down**	**Function**	**Up**	**Function**	**Down**	**Function**
Q6I673	Drought stress	A0A0A0LG20	Drought stress	V5LF72	Broad-spectrum resistance	D1MWZ6	Arp	Q6I673	Broad-spectrum resistance	E7CEW6	WRKY17
A3F570	Drought stress	D1MWZ6	Drought stress	A0A097BU00	Alcohol dehydrogenase	A0A0A0L4M8	Susceptibility	A0A0A0LVN5	Alcohol dehydrogenase	A0A0A0L4M8	Susceptibility
P0DI61	Soil stress	A0A0A0L4M8	Soil stress	Q6I673	Drought stress			A0A0A0LD49	Drought stress	A0A0A0LT35	Susceptibility
				A0A0A0LWG8	Related to JA					D1MWZ6	Arp
				A0A1D8RFV5	PR5					A0A0A0LW64	Negative correlation to heat shock
				A0A0A0LW70	Physiological stress						
				A0A0A0K1Y0	Ca						
				H6TB43	Osmotic stress						
				A0A0A0KGB3	Related to JA						
				A5X4I4	glycerol 3-phosphate degradation						
				A0A0A0KBB7	Precursors of JA						

GO analysis revealed that most differentially expressed proteins were involved in metabolic processes, and single organism and cellular processes, among others. Many differentially expressed proteins were associated with molecular functions, such as binding, catalytic activity, and the binding of small molecules. Moreover, many differentially expressed proteins were associated with the cellular component, such as the organelle, membrane, and cell part, as seen in (Figure 2).

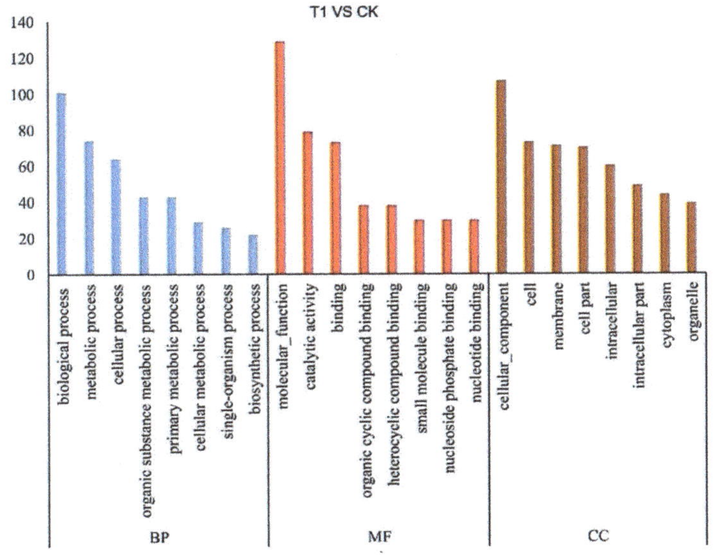

(a)

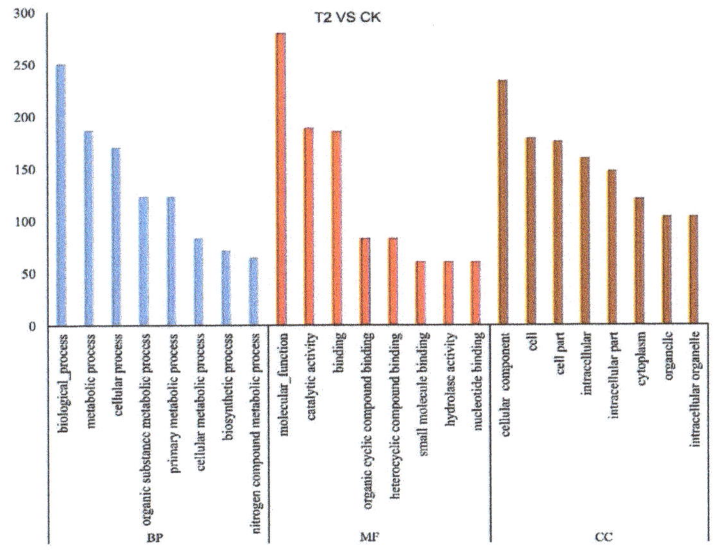

(b)

Figure 2. *Cont.*

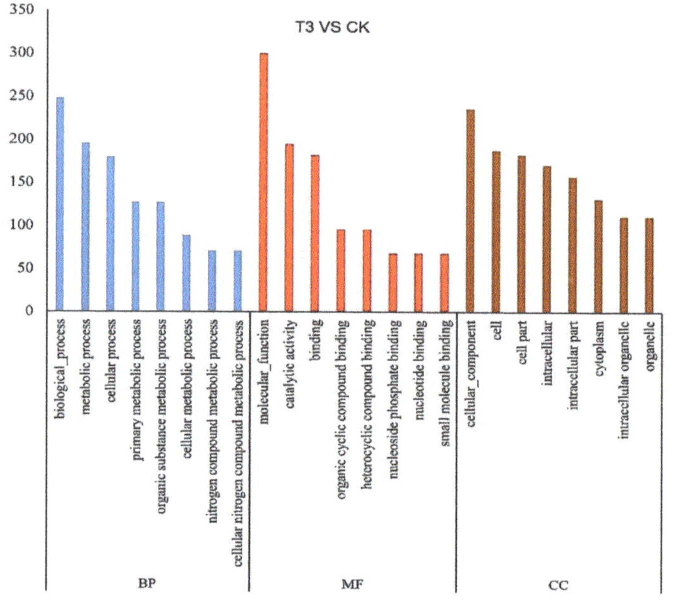

(c)

Figure 2. Gene ontology annotation. (**a**): CK treatment at 12 hpi; (**b**): CK treatment at 24 hpi; (**c**): CK treatment at 48 hpi. **BP:** biological process; **MF:** molecular function; **CC:** cellular component.

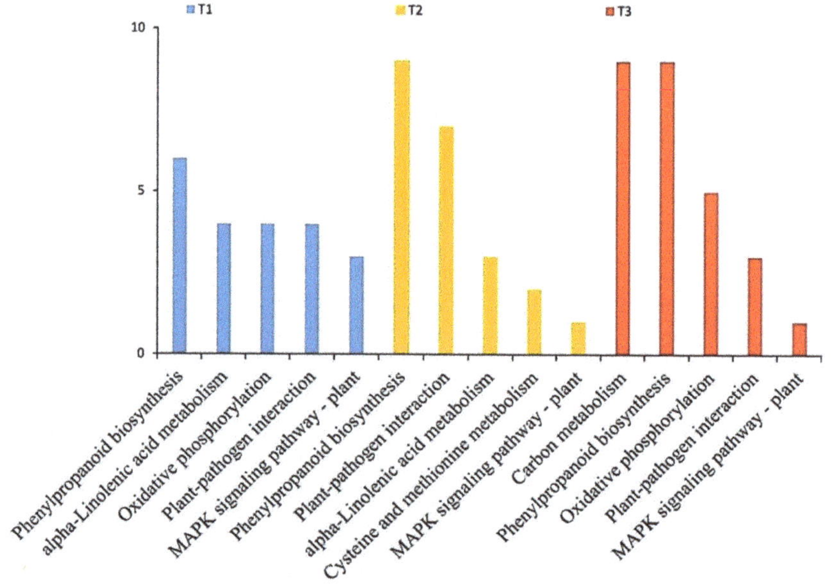

Figure 3. KEGG annotation. T1 is CK treatment at 12 hpi; T2 is CK treatment at 24 hpi; and T3 is CK treatment at 48 hpi.

Based on the above results, we consider that the mechanism of F1-35 protection watermelon against Fusarium wilt involves triggering the JA and ET pathways (Figure 4).

Figure 3. KEGG annotation. T1 is CK treatment at 12 hpi; T2 is CK treatment at 24 hpi; and T3 is CK treatment at 48 hpi.

Based on the above results, we consider that the mechanism of F1-35 protection watermelon against Fusarium wilt involves triggering the JA and ET pathways (Figure 4).

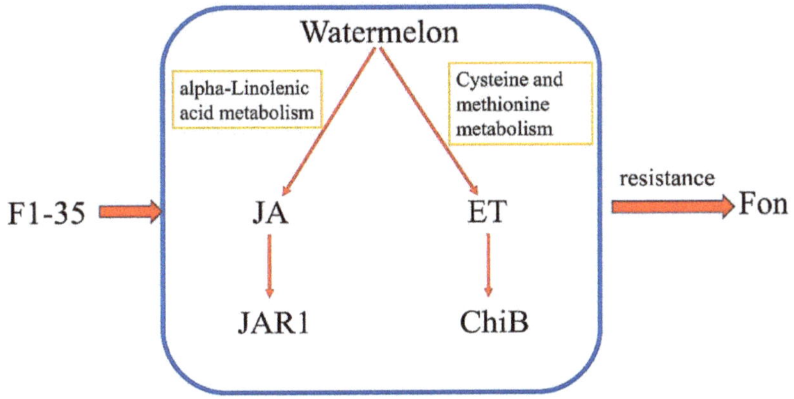

Figure 4. A summary of the mechanism of F1-35 protection of watermelons against FON.

Figure 4. A summary of the mechanism of F1-35 protection of watermelons against FON.

3.5. qRT-PCR Analysis

3.5. qRT-PCR Analysis

To validate the above hypothesis, qRT-PCR was used. The primers were designed with Beacon Designer 7 according to the cDNA sequence, which can be downloaded from http://www.icugi.org/ (accessed on 9 April 2017), with the exception of beta-actin (Table 3). To determine a suitable temperature for qRT-PCR, a gradient PCR was carried out, with results shown for 60 °C; each primer had a well-defined peak. The results of E show that all primers had amplification efficiency values of 96.28 to 224.35%, and all R^2 values were greater than 0.96, all of which indicated that those primers could be used in qRT-PCR and $2^{-\Delta\Delta CT}$ [58].

To validate the above hypothesis, qRT-PCR was used. The primers were designed with Beacon Designer 7 according to the cDNA sequence, which can be downloaded from http://www.icugi.org/ (accessed on 9 April 2017), with the exception of beta-actin (Table 3). To determine a suitable temperature for qRT-PCR, a gradient PCR was carried out, with results shown for 60 °C; each primer had a well-defined peak. The results of E show that all primers had amplification efficiency values of 96.28 to 224.35%, and all R^2 values were greater than 0.96, all of which indicated that those primers could be used in qRT-PCR and $2^{-\Delta\Delta CT}$ [58].

Table 3. The primers for qRT-PCR.

Gene Name	Gene ID		Primers	Production Size (bp)	Amplification Efficiency %	R^2
β-actin	MU51303	F R	CCTGGTATCGCTGACCGTAT TACTGAGCGATGCAAGGATG	133	96.7	
PAL	Cla018297	F R	TGCTATGGCTTCCTATT ATGTCAATGGCTTCTTC	141	114.35	0.9912
LOX1	Cla019905	F R	AATGCTTGCTGGAGTGA TGCTATGTGTTCTTCTGTTATG	121	99.25	0.9946
CTR1	Cla017731	F R	GAAGTTGCTGTGAAGAT TAGGATGTCGTAAGGATT	100	96.28	0.9656

Note: beta-actin data taken from Kong et al. [49].

Together, these results suggest that all primers were suitable for qRT-PCR. Therefore, samples collected after treatment for 0, 12, 24, 48, 72, and 96 h were used to analyze gene expression, including the PAL, LOX1, and CTR1 genes. In F1-35 treated roots, PAL gene transcriptional levels were also induced after 48 hpi, whereas in FON-treated roots, the PAL gene was highly expressed before 48 hpi. In F1-35-treated stems, the PAL gene was induced at 48 hpi, and peaked at 72 hpi, which was higher than in the CK or FON treatments. In F1-35-treated leaves, the PAL gene was induced at 24 hpi. The peak of the PAL gene was 5.34 at 24 hpi, and then it slowly returned to normal (1.34). With FON treatment, PAL gene expression was increased after an initial decrease. The LOX1 gene was induced after treatment with F1-35 at 12 hpi, and peaked at 24 hpi in the roots. In stems and leaves, LOX1 was also induced. CTR1, a negative key gene connected with the ET pathway, was down-regulated after treatment with F1-35 and even FON (Figure 5).

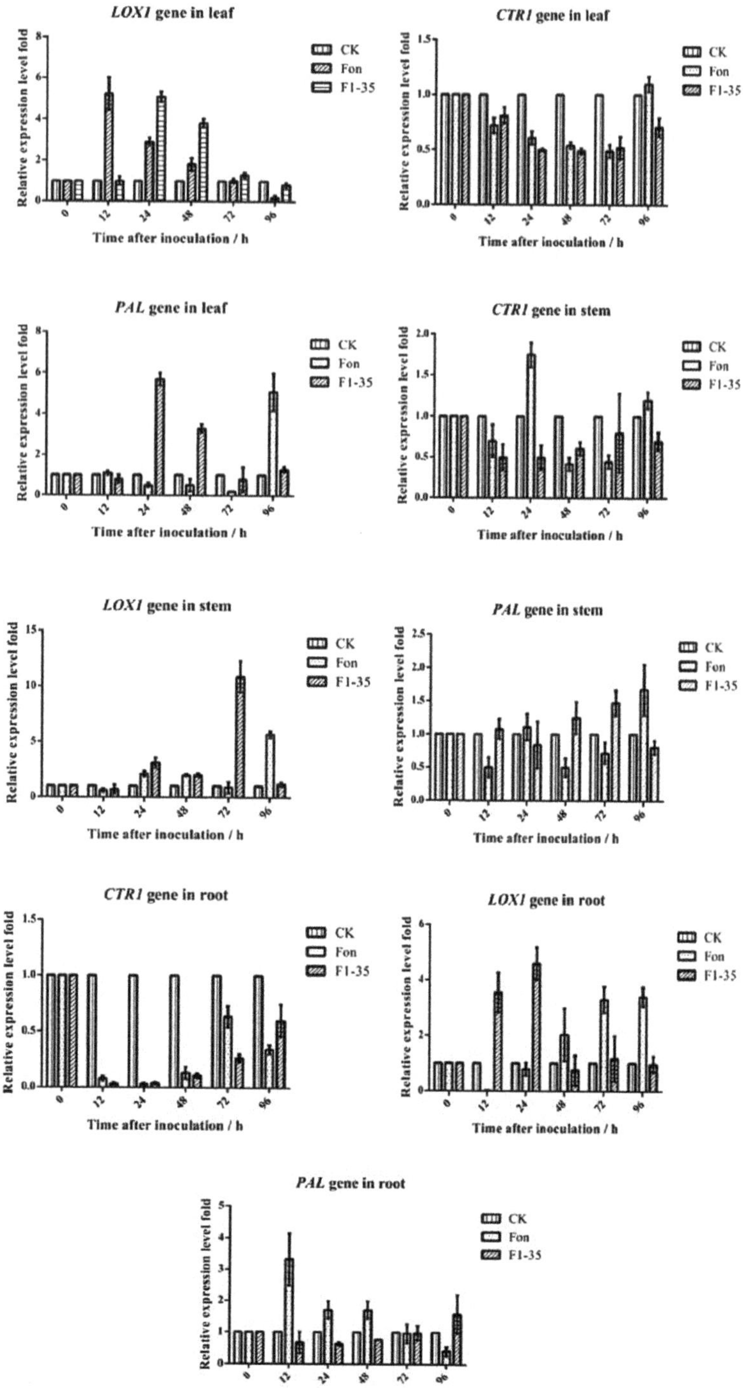

Figure 5. Target gene expression level in the roots, stems, or leaves of plants treated with F1-35 or FON.

4. Discussion

Fusarium oxysporum f. sp. *niveum* (FON), the pathogen causing watermelon Fusarium wilt, is wildly distributed around the world, and *F. oxysporum* is a hemi-biotrophic pathogen that can infect many plants, including tomatoes, bananas, and *Arabidopsis* [60]. FON poses a serious threat to the yield and quality of watermelons. A biocontrol agent is an effective natural means of controlling plant disease. All BCAs have been isolated from the rhizosphere and plants, but there are some obstacles in the application of BCAs. One of the important factors is the loss of anti-fungal activity, a result of a longtime subculture on artificial media [61–64]. Several methods can resolve this loss, including protoplast fusion and molecular genetics. As a Fo47-actinomycetes-153 fusion strain, previous research found that F1-35 produce more gibberellin and have a higher control efficiency of FON in watermelon plants, up to 59%, which was higher than Fo47 [48]. Our control efficiency experiment demonstrated a 61.7% control effect of FON in watermelon seedlings. Indeed, Fo47, improved with actinomycetes 153 by protoplast fusion, had stable biocontrol activity.

Further, the mechanism of F1-35-controlled Fusarium wilt in watermelon seedlings was explored. Previous studies reported that Fo47 controls plant disease by nutrition and ecological niche competition, inducing the expression of some defense genes. Actinomycetes 153 has activity against some fungi and bacteria [48]. As such, after protoplast fusion, did the Fo47-actinomycetes-153 fusion strain have those abilities? In this study, pair assaying was used to explore the antifungal ability of Fo47 after improvement with the protoplast of actinomycetes 153. The pair assay results showed that F1-35 could not cause the inhibition of FON growth because both the mycelia growth rates and spore morphology were normal.

In plants, *F. oxysporum* is controlled by manipulating defense pathways: some research has indicated that *Arabidopsis thaliana* responds against *F. oxysporum* via the JA/ET pathway [65]. Some studies have reported that SA and JA biosynthesis genes can be induced to expression at the early stage of FON infection of watermelons, and a lower concentration of SA may enhance watermelon resistance to FON. In addition, in resistant watermelon varieties, the crosstalk net between SA, JA, and ABA may help the JAR, NPRs, and PYLs family genes trigger the plant immune system against FON infection [22,66]. Other studies reported that *F. oxysporum* resistance is independent of SA-dependent defense gene expression in impaired JA-signaling mutants, such as myc2, coi1, and pft1/med25 [67–69]. In our study, the improved control agent, F1-35, increased some key enzymes of JA and ET pathways, which is similar to other BCAs, such as *Bacillus pumilus* SE34 [70,71]. Moreover, our results also indicate that F1-35 not only stimulates the JA and ET pathways, but also triggers plant PAMP defense and many broad-spectrum resistance proteins. Compared to CK treatment at 12 hpi, the up-regulated proteins were Q6I673, A3F570, and P0DI61. Former research has shown that these are a response to a poor soil environment and drought stress [72,73]. The down-regulated proteins were A0A0A0LG20, D1MWZ6, and A0A0A0L4M8. A0A0A0LG20 is a gibberellin-binding protein. D1MWZ6 is an auxin-repressing protein and A0A0A0L4M8 is a negative-regulated resistance protein [74]. Compared to CK treatment at 24 hpi, the up-regulated proteins were V5LF72, A0A097BU00, Q6I673, A0A0A0LWG8, A0A1D8RFV5, A0A0A0LW70, A0A0A0K1Y0, H6TB43, A0A0A0KGB3, A5X4I4, and A0A0A0KBB7. The main function of these proteins is drought resistance, including the PR5 protein or connecting with the JA pathway. The down-regulated proteins were D1MWZ6 and A0A0A0L4M8, which is similar to the result of 12-hpi vs. CK treatment. Compared to CK treatment at 48 hpi, the up-regulated proteins were Q6I673, A0A0A0LVN5, and A0A0A0LD49. These are connected to the JA pathway and some resistance proteins. The down-regulated proteins were E7CEW6, A0A0A0L4M8, A0A0A0LT35, D1MWZ6, and A0A0A0LW64. These proteins are negative-regulated resistance proteins or have a negative correlation with JA.

Furthermore, plant hormones were systemically analyzed. JA and ET, as plant hormones, play important roles in plant life, including fruit-ripening, and responding to pathogen and environment stress [75]. The signal transduction pathways occur in the induced responses of a plant to the pathogen, wounding, and herbivory. Jasmonic acid is involved in inducible defenses against pathogens and insects, and ET is important in induced plant defense [76,77].

In this study, lipoxygenase (LOX) and allene oxide synthase (AOS), key enzymes in the JA-producing pathway [78], were identified. Indeed, jasmonic acid-amido synthetase (JAR1), a JA-induced protein [79], could also be identified. In addition, chitinase B (ChiB), a protein related to downstream ethylene signal transduction [80], was elevated. Therefore, the JA and ET pathways were induced after the inoculation of F1-35 in the watermelons. To validate our results, qRT-PCR was used to analyze *PAL*, *LOX1*, and *CTR1* gene expression levels. The gene expression results show that *PAL* and *LOX1*, which have a positive correlation to JA, have up-expression, and the negative-associated gene, *CTR1*, was down-expressed. In conclusion, the improved biocontrol agent, F1-35, can improve the resistance of watermelons to FON by triggering the JA and ET pathways.

Author Contributions: Conceptualization, Y.W., Q.M., X.-M.D. and Q.-G.L.; Investigation, Q.-G.L., X.-M.D. and J.C.; Methodology, Q.-G.L. and R.-M.J.; Validation, X.-M.D. and J.C.; Interpretation of the data, Y.W. and R.-M.J.; Writing—original draft, Q.-G.L. and X.-M.D.; Writing—review and editing, Z.-F.Z., Q.M. and Y.W. All authors have read and agreed to the published version of the manuscript.

Funding: This research was funded by National Natural Science Foundation of China (31371978); The Key Industry Chain Innovation Project of Shaanxi province (2020ZDLNY07-02, 2021ZDLNY03-05).

Data Availability Statement: The raw data used in the present study for proteomics were submitted to the Integrated Proteome Resources database under accession number IPX0001180000.

Acknowledgments: We are grateful to John Richard Schrock (Emporia State University, USA) for proofreading the manuscript.

Conflicts of Interest: The authors have no conflict of interest to declare.

References

1. Ren, Y.; Jiao, D.; Gong, G.Y.; Zhang, H.Y.; Guo, S.G.; Zhang, J.; Xu, Y. Genetic analysis and chromosome mapping of resistance to *Fusarium oxysporum* f. sp. *niveum* (FON) race 1 and race 2 in watermelon (*Citrullus lanatus* L.). *Mol. Breed.* **2015**, *35*, 9. [CrossRef] [PubMed]
2. Hudson, O.; Fulton, J.C.; Dong, A.K.; Dufault, N.S.; Ali, M.E. *Fusarium oxysporum* f. sp. *niveum* molecular diagnostics past, present and future. *Int. J. Mol. Sci.* **2021**, *22*, 9735. [CrossRef]
3. Massaccesi, G.; Romero, M.C.; Cazau, M.C.; Bucsinszky, A.M. Cadmium removal capacities of filamentous soil fungi isolated from industrially polluted sediments, in La Plata (Argentina). *World J. Microbiol. Biot.* **2002**, *18*, 817–820. [CrossRef]
4. Mandeel, Q.A. Biodiversity of the genus *Fusarium* in saline soil habitats. *J. Basic Microbiol.* **2006**, *46*, 480–494. [CrossRef] [PubMed]
5. Ortega-Larrocea, M.D.; Xoconostle-Cazares, B.; Maldonado-Mendoza, I.E.; Carrillo-Gonzalez, R.; Hernandez-Hernandez, J.; Garduno, M.D.; Lopez-Meyer, M.; Gomez-Flores, L.; Gonzalez-Chavez, M.D.A. Plant and fungal biodiversity from metal mine wastes under remediation at Zimapan, Hidalgo, Mexico. *Environ. Pollut.* **2010**, *158*, 1922–1931. [CrossRef] [PubMed]
6. Babic, M.N.; Zalar, L.; Zenko, B.; Schroers, H.J.; Deroski, S.; Gunde-Cimerman, N. *Candida* and *Fusarium* species known as opportunistic human pathogens from customer-accessible parts of residential washing machines. *Fungal Biol.* **2015**, *119*, 95–113. [CrossRef]
7. Rahman, M.Z.; Ahmad, K.; Siddiqui, Y.; Saad, N.; Hun, T.G.; Mohd Hata, E.; Rashed, O.; Hossain, M.I.; Kutawa, A.B. First report of *Fusarium* wilt disease on watermelon caused by *Fusarium oxysporum* f. sp. *niveum* (FON) in Malaysia. *Plant Dis.* **2021**, *105*, 4169. [CrossRef]
8. Everts, K.L.; Himmelstein, J.C. *Fusarium* wilt of watermelon: Towards sustainable management of a re-emerging plant disease. *Crop Prot.* **2015**, *73*, 93–99. [CrossRef]
9. Zhang, X.; Xue, C.; Fang, D.; He, X.H.; Wei, M.Y.; Zhuo, C.J.; Jin, J.Y.; Shen, B.A.; Li, R.; Ling, N.; et al. Manipulating the soil microbiomes during a community recovery process with plant beneficial species for the suppression of *Fusarium* wilt of watermelon. *Amb. Express.* **2021**, *11*, 1. [CrossRef]
10. Sande, D.; Mullen, J.; Wetzstein, M.; Houston, J. Environmental Impacts from Pesticide Use: A case study of Soil fumigation in Florida tomato production. *Int. J. Environ. Res. Public Health* **2011**, *8*, 4649–4661. [CrossRef]
11. Davis, A.R.; Perkins-Veazie, P.; Sakata, Y.; Lopez-Galarza, S.; Maroto, J.V.; Lee, S.G.; Huh, Y.C.; Sun, Z.Y.; Miguel, A.; King, S.R.; et al. Cucurbit grafting. *Crit. Rev. Plant Sci.* **2008**, *27*, 50–74. [CrossRef]
12. King, S.R.; Davis, A.R.; Liu, W.G.; Levi, A. Grafting for disease resistance. *Hortscience* **2008**, *43*, 1673–1676. [CrossRef]
13. Cohen, R.; Tyutyunik, J.; Fallik, E.; Oka, Y.; Tadmor, Y.; Edelstein, M. Phytopathological evaluation of exotic watermelon germplasm as a basis for rootstock breeding. *Sci. Hortic.* **2014**, *165*, 203–210. [CrossRef]

14. Keinath, A.P.; Hassell, R.L. Control of *Fusarium* wilt of watermelon by grafting onto bottlegourd or interspecific hybrid squash despite colonization of rootstocks by *Fusarium*. *Plant Dis.* **2014**, *98*, 255–266. [CrossRef] [PubMed]
15. Keinath, A.P.; Wechter, W.P.; Rutter, W.B.; Agudelo, P.A. Cucurbit rootstocks resistant to *Fusarium oxysporum* f. sp. *niveum* remain resistant when coinfected by *Meloidogyne incognita* in the Field. *Plant Dis.* **2019**, *103*, 1383–1390. [CrossRef]
16. Mohamed, F.; El-Hamed, K.; Elwan, M.; Hussien, M.-A. Impact of grafting on watermelon growth, fruit yield and quality. *J. Fruit Ornam. Plant Res.* **2012**, *76*, 99–118. [CrossRef]
17. Ryals, J.A.; Neuenschwander, U.H.; Willits, M.G.; Molina, A.; Steiner, H.Y.; Hunt, M.D. Systemic acquired resistance. *Plant Cell* **1996**, *8*, 1809–1819. [CrossRef]
18. Chen, Y.C.; Kidd, B.N.; Carvalhais, L.C.; Schenk, P.M. Molecular defense responses in roots and the rhizosphere against *Fusarium oxysporum*. *Plant Signal. Behav.* **2014**, *9*, e977710. [CrossRef]
19. Van Loon, L.C.; Bakker, P.A.; Pieterse, C.M. Systemic resistance induced by rhizosphere bacteria. *Annu. Rev. Phytopathol.* **1998**, *36*, 453–483. [CrossRef]
20. Pieterse, C.M.; Zamioudis, C.; Berendsen, R.L.; Weller, D.M.; Van Wees, S.C.; Bakker, P.A. Induced systemic resistance by beneficial microbes. *Annu. Rev. Phytopathol.* **2014**, *52*, 347–375. [CrossRef]
21. Di, X.T.; Takken, F.L.W.; Tintor, N. How phytohormones shape interactions between plants and the soil-borne fungus *Fusarium oxysporum*. *Front. Plant Sci.* **2016**, *7*, 170. [CrossRef] [PubMed]
22. Zhu, F.; Wang, Z.; Fang, Y.; Tong, J.; Xiang, J.; Yang, K.; Wang, R. Study on the role of phytohormones in resistance to watermelon *Fusarium* wilt. *Plants* **2022**, *11*, 156. [CrossRef] [PubMed]
23. Kasote, D.M.; Jayaprakasha, G.K.; Ong, K.; Crosby, K.M.; Patil, B.S. Hormonal and metabolites responses in Fusarium wilt-susceptible and -resistant watermelon plants during plant-pathogen interactions. *BMC Plant Biol.* **2020**, *20*, 481. [CrossRef]
24. Fernandes, L.B.; Ghag, S.B. Molecular insights into the jasmonate signaling and associated defense responses against wilt caused by *Fusarium oxysporum*. *Plant Physiol. Biochem.* **2022**, *174*, 22–34. [CrossRef]
25. Alvarez, B.; Biosca, E.G. Bacteriophage-based bacterial wilt biocontrol for an environmentally sustainable agriculture. *Front. Plant Sci.* **2017**, *8*, 1218. [CrossRef]
26. Vurukonda, S.S.K.P.; Giovanardi, D.; Stefani, E. Plant growth promoting and biocontrol activity of *Streptomyces spp.* as endophytes. *Int. J. Mol. Sci.* **2018**, *19*, 952. [CrossRef] [PubMed]
27. Bubici, G.; Kaushal, M.; Prigigallo, M.I.; Gomez-Lama Cabanas, C.; Mercado-Blanco, J. Biological control agents against Fusarium wilt of Banana. *Front. Microbiol.* **2019**, *10*, 616. [CrossRef]
28. Zheng, Y.; Han, X.; Zhao, D.; Wei, K.; Yuan, Y.; Li, Y.; Liu, M.; Zhang, C.S. Exploring biocontrol agents from microbial keystone taxa associated to suppressive soil: A new attempt for a biocontrol strategy. *Front. Plant Sci.* **2021**, *12*, 655673. [CrossRef]
29. Raaijmakers, J.M.; de Bruijn, I.; de Kock, M.J.D. Cyclic lipopeptide production by plant-associated *Pseudomonas* spp.: Diversity, activity, biosynthesis, and regulation. *Mol. Plant-Microbe Interact.* **2006**, *19*, 699–710. [CrossRef]
30. Deketelaere, S.; Tyvaert, L.; Franca, S.C.; Hofte, M. Desirable traits of a good biocontrol agent against *Verticillium* wilt. *Front. Microbiol.* **2017**, *8*, 1186. [CrossRef]
31. Kohl, J.; Kolnaar, R.; Ravensberg, W.J. Mode of action of microbial biological control agents against plant diseases: Relevance beyond efficacy. *Front. Plant Sci.* **2019**, *10*, 845. [CrossRef] [PubMed]
32. Tziros, G.T.; Lagopodi, A.L.; Tzavella-Klonari, K. Reduction of Fusarium wilt in watermelon by *Pseudomonas chlororaphis* PCL1391 and *P. fluorescens* WCS365. *Phytopathol. Mediterr.* **2007**, *46*, 320–323.
33. Martinez, F.D.; Santos, M.; Carretero, F.; Marin, F. *Trichoderma saturnisporum*, a new biological control agent. *J. Sci. Food Agric.* **2016**, *96*, 1934–1944. [CrossRef]
34. Jiang, C.H.; Yao, X.F.; Mi, D.D.; Li, Z.J.; Yang, B.Y.; Zheng, Y.; Qi, Y.J.; Guo, J.H. Comparative transcriptome analysis reveals the biocontrol mechanism of *Bacillus velezensis* F21 against *Fusarium* wilt on watermelon. *Front. Microbiol.* **2019**, *10*, 652. [CrossRef]
35. Hua, G.K.H.; Wang, L.; Chen, J.; Ji, P.S. Biological control of *Fusarium* wilt on watermelon by *Fluorescent pseudomonads*. *Biocontrol Sci. Techn.* **2020**, *30*, 212–227. [CrossRef]
36. Yoshioka, Y.; Ichikawa, H.; Naznin, H.A.; Kogure, A.; Hyakumachi, M. Systemic resistance induced in *Arabidopsis thaliana* by *Trichoderma asperellum* SKT-1, a microbial pesticide of seedborne diseases of rice. *Pest Manag. Sci.* **2012**, *68*, 60–66. [CrossRef] [PubMed]
37. Nawrocka, J.; Malolepsza, U. Diversity in plant systemic resistance induced by *Trichoderma*. *Biol. Control* **2013**, *67*, 149–156. [CrossRef]
38. Angelopoulou, D.J.; Naska, E.J.; Paplomatas, E.J.; Tjamos, S.E. Biological control agents (BCAs) of *Verticillium* wilt: Influence of application rates and delivery method on plant protection, triggering of host defence mechanisms and rhizosphere populations of BCAs. *Plant Pathol.* **2014**, *63*, 1062–1069. [CrossRef]
39. Jiao, X.R.; Takishita, Y.; Zhou, G.S.; Smith, D.L. Plant associated rhizobacteria for biocontrol and plant growth enhancement. *Front. Plant Sci.* **2021**, *12*, 634796. [CrossRef]
40. Ku, Y.S.; Sintaha, M.; Cheung, M.Y.; Lam, H.M. Plant Hormone Signaling Crosstalks between Biotic and Abiotic Stress Responses. *Int. J. Mol. Sci.* **2018**, *19*, 3206. [CrossRef]
41. Vlot, A.C.; Sales, J.H.; Lenk, M.; Bauer, K.; Brambilla, A.; Sommer, A.; Chen, Y.Y.; Wenig, M.; Nayem, S. Systemic propagation of immunity in plants. *New Phytol.* **2021**, *229*, 1234–1250. [CrossRef] [PubMed]

42. Alabouvette, C.; Couteaudier, Y.; Lemanceau, P. Nature of intrageneric competition between pathogenic and non-pathogenic *Fusarium* in a wilt-suppressive soil. In *Iron, Siderophores, and Plant Diseases*; Springer: Berlin/Heidelberg, Germany, 1986; pp. 165–178. [CrossRef]
43. Veloso, J.; Díaz, J. *Fusarium oxysporum* Fo47 confers protection to pepper plants against *Verticillium dahliae* and *Phytophthora capsici*, and induces the expression of defence genes. *Plant Pathol.* **2012**, *61*, 281–288. [CrossRef]
44. Aimé, S.; Alabouvette, C.; Steinberg, C.; Olivain, C. The endophytic strain *Fusarium oxysporum* Fo47: A good candidate for priming the defense responses in tomato roots. *Mol. Plant-Microbe Interact.* **2013**, *26*, 918–926. [CrossRef] [PubMed]
45. Veloso, J.; Alabouvette, C.; Olivain, C.; Flors, V.; Pastor, V.; García, T.; Díaz, J. Modes of action of the protective strain Fo47 in controlling *Verticillium* wilt of pepper. *Plant Pathol.* **2016**, *65*, 997–1007. [CrossRef]
46. Constantin, M.E.; Vlieger, B.V.; Takken, F.L.W.; Rep, M. Diminished pathogen and enhanced endophyte colonization upon coInoculation of endophytic and pathogenic *Fusarium* strains. *Microorganisms* **2020**, *8*, 544. [CrossRef]
47. De Lamo, F.J.; Spijkers, S.B.; Takken, F.L.W. Protection to Tomato wilt disease conferred by the nonpathogen *Fusarium oxysporum* Fo47 is more effective than that conferred by avirulent strains. *Phytopathology* **2021**, *111*, 253–257. [CrossRef]
48. Chunling, L.; Zhaoyi, S.; Zhaofeng, Z. The improvement of *Fusarium oxysporum* Fo47 by protoplast fusion. *J. Northwest A F Univ. Nat. Sci. Ed.* **2010**, *38*, 126–130. (In Chinese) [CrossRef]
49. Kong, Q.S.; Yuan, J.X.; Niu, P.H.; Xie, J.J.; Jiang, W.; Huang, Y.; Bie, Z.L. Screening suitable reference genes for normalization in reverse transcription quantitative real-time PCR analysis in Melon. *PLoS ONE* **2014**, *9*, 1. [CrossRef]
50. Pu, X.M.; Xie, B.Y.; Li, P.Q.; Mao, Z.C.; Ling, J.; Shen, H.F.; Zhang, J.X.; Huang, N.; Lin, B.R. Analysis of the defence-related mechanism in cucumber seedlings in relation to root colonization by nonpathogenic *Fusarium oxysporum* CS-20. *FEMS Microbiol. Lett.* **2014**, *355*, 142–151. [CrossRef]
51. Jorrin-Novo, J.V. Plant proteomics methods and protocols. *Methods Mol. Biol.* **2014**, *1072*, 3–13. [CrossRef]
52. Wisniewski, J.R.; Zougman, A.; Nagaraj, N.; Mann, M. Universal sample preparation method for proteome analysis. *Nat. Methods* **2009**, *6*, 359–362. [CrossRef] [PubMed]
53. Colaert, N.; Barsnes, H.; Vaudel, M.; Helsens, K.; Timmerman, E.; Sickmann, A.; Gevaert, K.; Martens, L. Thermo-msf-parser: An open source Java library to parse and visualize Thermo Proteome Discoverer msf files. *J. Proteome Res.* **2011**, *10*, 3840–3843. [CrossRef] [PubMed]
54. Gotz, S.; Garcia-Gomez, J.M.; Terol, J.; Williams, T.D.; Nagaraj, S.H.; Nueda, M.J.; Robles, M.; Talon, M.; Dopazo, J.; Conesa, A. High-throughput functional annotation and data mining with the Blast2GO suite. *Nucleic Acids Res.* **2008**, *36*, 3420–3435. [CrossRef]
55. Quevillon, E.; Silventoinen, V.; Pillai, S.; Harte, N.; Mulder, N.; Apweiler, R.; Lopez, R. InterProScan: Protein domains identifier. *Nucleic Acids Res.* **2005**, *33*, W116–W120. [CrossRef] [PubMed]
56. Kanehisa, M.; Goto, S.; Sato, Y.; Furumichi, M.; Tanabe, M. KEGG for integration and interpretation of large-scale molecular data sets. *Nucleic Acids Res.* **2012**, *40*, D109–D114. [CrossRef] [PubMed]
57. Guo, S.; Zhang, J.; Sun, H.; Salse, J.; Lucas, W.J.; Zhang, H.; Zheng, Y.; Mao, L.; Ren, Y.; Wang, Z.; et al. The draft genome of watermelon (*Citrullus lanatus*) and resequencing of 20 diverse accessions. *Nat. Genet.* **2013**, *45*, 51–58. [CrossRef]
58. Schmittgen, T.D.; Livak, K.J. Analyzing real-time PCR data by the comparative CT method. *Nat. Protoc.* **2008**, *3*, 1101–1108. [CrossRef]
59. Moriya, Y.; Itoh, M.; Okuda, S.; Yoshizawa, A.C.; Kanehisa, M. KAAS: An automatic genome annotation and pathway reconstruction server. *Nucleic Acids Res.* **2007**, *35*, W182–W185. [CrossRef]
60. Berrocal-Lobo, M.; Molina, A. *Arabidopsis* defense response against *Fusarium oxysporum*. *Trends Plant Sci.* **2008**, *13*, 145–150. [CrossRef]
61. Whipps, J.M. Microbial interactions and biocontrol in the rhizosphere. *J. Exp. Bot.* **2001**, *52*, 487–511. [CrossRef]
62. Wang, L.Y.; Xie, Y.S.; Cui, Y.Y.; Xu, J.; He, W.; Chen, H.G.; Guo, J.H. Conjunctively screening of biocontrol agents (BCAs) against *Fusarium* root rot and *Fusarium* head blight caused by *Fusarium graminearum*. *Microbiol. Res.* **2015**, *177*, 34–42. [CrossRef] [PubMed]
63. Pellan, L.; Dieye, C.A.T.; Durand, N.; Fontana, A.; Strub, C.; Schorr-Galindo, S. Biocontrol agents: Toolbox for the screening of weapons against Mycotoxigenic Fusarium. *J. Fungi* **2021**, *7*, 446. [CrossRef] [PubMed]
64. Chen, J.; Hu, L.; Chen, N.; Jia, R.; Ma, Q.; Wang, Y. The biocontrol and plant growth-promoting properties of *Streptomyces alfalfae* XN-04 revealed by functional and genomic analysis. *Front. Microbiol.* **2021**, *12*, 745766. [CrossRef] [PubMed]
65. Thatcher, L.F.; Manners, J.M.; Kazan, K. *Fusarium oxysporum* hijacks COI1-mediated jasmonate signaling to promote disease development in *Arabidopsis*. *Plant J.* **2009**, *58*, 927–939. [CrossRef]
66. Xu, W.; Wang, Z.; Wu, F. Companion cropping with wheat increases resistance to *Fusarium* wilt in watermelon and the roles of root exudates in watermelon root growth. *Physiol. Mol. Plant Pathol.* **2015**, *90*, 12–20. [CrossRef]
67. Anderson, J.P.; Badruzsaufari, E.; Schenk, P.M.; Manners, J.M.; Desmond, O.J.; Ehlert, C.; Maclean, D.J.; Ebert, P.R.; Kazan, K. Antagonistic interaction between abscisic acid and jasmonate-ethylene signaling pathways modulates defense gene expression and disease resistance in *Arabidopsis*. *Plant Cell* **2004**, *16*, 3460–3479. [CrossRef]
68. Kidd, B.N.; Edgar, C.I.; Kumar, K.K.; Aitken, E.A.; Schenk, P.M.; Manners, J.M.; Kazan, K. The mediator complex subunit PFT1 is a key regulator of jasmonate-dependent defense in *Arabidopsis*. *Plant Cell* **2009**, *21*, 2237–2252. [CrossRef]
69. Cole, S.J.; Yoon, A.J.; Faull, K.F.; Diener, A.C. Host perception of jasmonates promotes infection by *Fusarium oxysporum* formae speciales that produce isoleucine- and leucine-conjugated jasmonates. *Mol. Plant Pathol.* **2014**, *15*, 589–600. [CrossRef]

70. Compant, S.; Duffy, B.; Nowak, J.; Clement, C.; Barka, E.A. Use of plant growth-promoting bacteria for biocontrol of plant diseases: Principles, mechanisms of action, and future prospects. *Appl. Environ. Microbiol.* **2005**, *71*, 4951–4959. [CrossRef]
71. Samaras, A.; Roumeliotis, E.; Ntasiou, P.; Karaoglanidis, G. *Bacillus subtilis* MBI600 promotes growth of tomato plants and induces systemic resistance contributing to the control of soilborne pathogens. *Plants* **2021**, *10*, 1113. [CrossRef]
72. Chaumont, F.; Moshelion, M.; Daniels, M.J. Regulation of plant aquaporin activity. *Biol. Cell* **2005**, *97*, 749–764. [CrossRef] [PubMed]
73. Nanasato, Y.; Akashi, K.; Yokota, A. Co-expression of cytochrome b(561) and ascorbate oxidase in leaves of wild watermelon under drought and high light conditions. *Plant Cell Physiol.* **2005**, *46*, 1515–1524. [CrossRef] [PubMed]
74. Milc, J.; Infantino, A.; Pecchioni, N.; Aragona, M. Identification of tomato genes differentially expressed during compatible interaction with *Pyrenochaeta lycopersici*. *J. Plant Pathol.* **2012**, *94*, 283–296.
75. Alvarez-Florez, F.; Lopez-Cristoffanini, C.; Jauregui, O.; Melgarejo, L.M.; Lopez-Carbonell, M. Changes in ABA, IAA and JA levels during calyx, fruit and leaves development in cape gooseberry plants (*Physalis peruviana* L.). *Plant Physiol. Biochem.* **2017**, *115*, 174–182. [CrossRef] [PubMed]
76. Shigenaga, A.M.; Argueso, C.T. No hormone to rule them all: Interactions of plant hormones during the responses of plants to pathogens. *Semin. Cell Dev. Biol.* **2016**, *56*, 174–189. [CrossRef] [PubMed]
77. Li, N.; Han, X.; Feng, D.; Yuan, D.Y.; Huang, L.J. Signaling crosstalk between Salicylic Acid and Ethylene/Jasmonate in plant defense: Do we understand what they are whispering? *Int. J. Mol. Sci.* **2019**, *20*, 671. [CrossRef]
78. Sembdner, G.; Parthier, B. The biochemistry and the physiological and molecular actions of jasmonates. *Annu. Rev. Plant Phys.* **1993**, *44*, 569–589. [CrossRef]
79. Kawamura, Y.; Takenaka, S.; Hase, S.; Kubota, M.; Ichinose, Y.; Kanayama, Y.; Nakaho, K.; Klessig, D.F.; Takahashi, H. Enhanced defense responses in *Arabidopsis* induced by the cell wall protein fractions from *Pythium oligandrum* require *SGT1, RAR1, NPR1* and *JAR1*. *Plant Cell Physiol.* **2009**, *50*, 924–934. [CrossRef]
80. Liu, Q.; Xu, C.; Wen, C.K. Genetic and transformation studies reveal negative regulation of ERS1 ethylene receptor signaling in *Arabidopsis*. *BMC Plant Biol.* **2010**, *10*, 60. [CrossRef]

microorganisms

Article

Bacillus velezensis BY6 Promotes Growth of Poplar and Improves Resistance Contributing to the Biocontrol of *Armillaria solidipes*

Ping Zhang, Guangqiang Xie, Lihai Wang * and Yanqiu Xing *

Heilongjiang Provincial Key Laboratory of Forest Sustainable Management and Environmental Microbial Engineering, Northeast Forestry University, Harbin 150040, China
* Correspondence: wanglihai@nefu.edu.cn (L.W.); yanqiuxing@nefu.edu.cn (Y.X.)

Abstract: To improve the application of endophyte *Bacillus velezensis* BY6 from the xylem of poplar, the effect of BY6 on the growth of diseased *Populus davidiana* × *Populus. alba var. pyramidalis* Louche (Pdpap poplar) seedlings and the biological control effect on the pathogen *Armillaria solidipes* were tested using a plant split-root experiment. After applying BY6 to the roots of diseased Pdpap poplar seedlings, the results show that plant growth indicators (dry mass, fresh mass, and plant height) were significantly increased ($p < 0.05$), and genes related to auxin hormone signal transcription were activated. BY6 indicated a surprising control effect after the inoculation of diseased Pdpap poplar seedlings. Compared to the infected control group, the treated disease index of the diseased Pdpap poplar seedlings in the treatment group were reduced by 49.53% on the 20th day. The relative staining areas of diaminobenzidine (DAB) and Trypan blue decreased by 3.37 and 7.31 times, respectively. The physiological indicators (soluble sugar and protein) and oxidase indicators were significantly increased ($p < 0.05$). The expression levels of defense genes related to salicylic acid (SA) and jasmonic acid (JA) signaling pathways were significantly increased ($p < 0.05$). Amazingly, the results indicate that BY6 simultaneously activates induced systemic resistance (ISR) and systemic acquired resistance (SAR) in diseased Pdpap poplar seedlings and promotes growth. The results indicate that BY6 is a promising candidate for developing forest tree biofertilizers and biopesticides.

Keywords: biological control; *Armillaria solidipes*; growth promotion; *Bacillus velezensis*; ISR; SAR; *Populus davidiana* × *Populus. alba var. pyramidalis* Louche

Citation: Zhang, P.; Xie, G.; Wang, L.; Xing, Y. *Bacillus velezensis* BY6 Promotes Growth of Poplar and Improves Resistance Contributing to the Biocontrol of *Armillaria solidipes*. *Microorganisms* **2022**, *10*, 2472. https://doi.org/10.3390/microorganisms10122472

Academic Editors: Beibei Ge, Tomislav Cernava and Kyungseok Park

Received: 27 October 2022
Accepted: 9 December 2022
Published: 14 December 2022

1. Introduction

The *Populus davidiana* × *Populus. alba var. pyramidalis* Louche (Pdpap poplar) on of the most commonly planted tree species in China. It is characterized by rapid growth, drought resistance, and low-temperature resistance [1,2]. Pdpap poplar is a female clone, and catkins can naturally fall off. It is also a model species for investigating woody plants [3]. With the continuous expansion of Pdpap poplar planting areas and the degradation of the environment, the incidence of root diseases caused by fungi is increasing, adversely affecting the growth of Pdpap poplar seedlings. Among them, Pdpap poplar *Armillaria* root rot (ARR) caused by *Armillaria solidipes* is one of the primary soil-borne diseases [4]. The chemical fungicides used to control the ARR of trees are of a single type, and the impact of use remains limited. It cannot meet the needs of the current nursery gardens. Some fungicides (methyl bromide) with better effects are prohibited from use because of environmental pollution [5,6]. Therefore, there is an urgent need to search for new green control agents for the ARR disease of Pdpap poplar.

Endophytes are often used to control root diseases in plants. Endophytes are a group of fungi and bacteria typically designated to colonize within the cells of healthy plants. Endophytes do not cause substantial damage to the host plant and establish a harmonious symbiotic relationship [7]. They can also inhibit various pathogenic fungi [8], promote plant growth [9], enhance plant tolerance to abiotic stress [10], and regulate and improve the soil environment [11]. Among these, *Bacillus velezensis* is an essential plant endophyte. Several studies have revealed that *B. velezensis* is a potent bacterium that produces lipopeptides (LPs), volatile compounds, and hydrolytic enzymes [12]. Huang et al. [13] isolated *B. velezensis* HYEB5-6 from boxwood leaves. They determined that this strain could significantly reduce the disease index of *Euonymus japonicus* anthracnoseon. Meng et al. [14] reported that a single strain of *B. velezensis* BAC03 could produce IAA, NH_3, and ACC deaminase and promote the growth of nine crops. To date, the literature lacks a relevant report on BY6 strain inducing disease resistance and growth promotion of Pdpap poplar.

To survive, pathogens constantly break plant defenses. Several studies have shown that pathogenic bacteria can indirectly inhibit the disease-resistance signaling pathway mediated by salicylic acid (SA) or jasmonic acid (JA) to invade plants by regulating the ethylene (ET) content [15]. *A. solidipes* infects the Pdpap poplar after breaking through their defenses. Whether *B. velezensis* can induce the diseased Pdpap poplar to continue to defend the *A. solidipes* requires further in-depth understanding. In a previous study, we isolated and identified the bacterial strain *B. velezensis* BY6 from the Pdpap poplar xylem [16]. This study investigates the effects of BY6 on the growth, physiology, and disease resistance of diseased Pdpap poplar seedlings. We also determine the structure-activity relationship between the growth promotion and disease resistance of BY6 on diseased Pdpap poplar and provide a theoretical basis for future application of BY6 in forestry.

2. Materials and Methods

2.1. Source of Strains and Plant Material

Strain source: *A. solidipes* (number A2001) strain was isolated from diseased Pdpap poplar roots. A2001 strain internal transcribed spacer information was stored in National Center for Biotechnology Information (NCBI) GenBank, with the accession number OP787670.1. The *B. velezensis* BY6 was isolated from Pdpap poplar xylem. Its whole genome information has been stored in NCBI GenBank, the accession number CP051011-CP051012. The above strains were preserved at the Heilongjiang Provincial Key Laboratory of the Forest Sustainable Management and Environmental Microbial Engineering Department, Northeast Forestry University.

Plant Material: The experiments were conducted under controlled greenhouse conditions. Sterile Pdpap poplar seedlings used in this study were planted in a confined and sterile environment throughout the trial. After the stems were changed into seedlings, they were transferred to the rooting medium ($\frac{1}{2}$ Murashige and Skoog's macronutrient ($\frac{1}{2}$ MS) define medium) at a temperature of 25 °C, with light for 16 h per day. After the Pdpap poplar grew 3~5 leaves, it was transplanted into the combined define culture medium of vermiculite + $\frac{1}{2}$ MS. Before transplanting, 100 g of vermiculite was prepared and placed into flower pots (26 × 35 × 18 cm). A total of 50 mL of $\frac{1}{2}$ MS define medium was added to each pot to ensure that the nutrient solution was evenly distributed in the soil and sterilized at 121 °C for 2 h. Each pot was watered with 50 mL of sterile water. The greenhouse temperature was maintained at 26 ± 1 °C and the relative humidity was 70 ± 2%, with a light period of 16 h and a dark period of 8 h.

2.2. Determination of Treated Disease Index and Relative Control Effect on Pdpap Poplar ARR after BY6 Induction

Healthy Pdpap poplar potted seedlings with a growth cycle of four weeks individuals were selected. The planted cultivar was highly susceptible to *A. solidipes*. The *A. solidipes* and BY6 were inoculated in a define PDA medium and cultured at 26 °C for 7 d. Hyphae from the edge of the colony were picked, transferred to a PD liquid medium, and rotary shaken (THZ-98A, Shanghai, China) at 160 rpm for 7 d at 28 °C. The concentration based on the ultraviolet spectrophotometer (Eppendorf D30, Berlin, Germany) OD_{600} reached approximately 1.9×10^8–2.0×10^8 CFU/mL.

After sterile Pdpap poplar seedlings were transplanted into vermiculite + $\frac{1}{2}$ MS aseptic soil, the roots were washed and divided into two equal parts (part 1 and part 2), and then transplanted separately into two adjacent pots for 20 d. Seven days in advance, the treatment and infected control groups were inoculated with *A. solidipes* fermentation broth by root irrigation, and each pot was inoculated with 20 mL. After the Pdpap poplar leaves had obvious disease symptoms, BY6 fermentation broth (20 mL) was inoculated into each pot in the treatment group by irrigating the roots. An equal volume of sterile water was used as the control. Subsequently, the control and treatment groups were collected at 0, 6, 12, 24, and 48 h (from the 5th to the 8th leaf from the stem base), wrapped in tin foil, quickly frozen in liquid nitrogen, and stored at −80 °C.

Referring to the research method described by Hernandez et al. [17], based on the magnitude of 0~4, the disease degree of the aboveground leaves in the control and treatment groups at 0, 5, 10, 15 and 20 d was reported (leaf yellowing or wilting number/total leaf number, were 0 = 0%, 1 = 1~33%, 2 = 34~66%, 3 = 67~100%, 4 = dead plants). Referring to the research method described by Tang et al. [18], the treated disease index and control effect were calculated using the following formula: treated disease index = [$\sum$ (number of disease-grade plants × representative value)/(total number of plants × representative value of highest disease grade)] × 100%, relative control effect = (control disease index-treated disease index)/(control disease index) × 100%.

2.3. Determination of Pdpap Poplar Enzyme Activity after BY6 Induction

To verify the BY6-induced resistance of Pdpap poplar, defense enzyme assays were performed on Pdpap poplar leaves of infection [18]. The leaves of the control and treatment groups at 0, 6, 12, 24, and 48 h (from the 5th to the 8th leaf from the stem base) were randomly mixed. They were then transferred to −20 °C in a cold mortar. After adding liquid nitrogen thoroughly grinding, the ground leaf tissue was transferred to a 1.5 mL centrifuge tube using a spatula. A solution of 0.02 M Phosphate buffer was added in a ratio of 1:9 weight of leave tissue to volume phosphate buffer. The mixture was vortexed for 1 min, followed by centrifugation at 3000 rpm for 4 min at 4 °C. A total of 50 μL of the supernatant was transferred into a 5 mL centrifuge tube. According into the kit's (NJJCBIO, Nanjing, China) instruction manual, the reaction reagents were introduced to the control and measurement tubes and mixed. The absorbance OD value was measured (Eppendorf D30, Berlin, Germany), and the enzyme activity was calculated according to the following formula.

2.4. Detection of Reactive Oxygen Species and Cell Death in Pdpap Poplar Leaves after BY6 Induction

The experiment was terminated 20 d post-inoculation with the BY6 strain. Referring to the method described by Lin et al. [19], the leaves were immersed in diaminobenzidine (DAB) (Sigma-Aldrich D8001, Berlin, Germany) staining solution (1% (*w*/*v*)) for 50 min. They were then removed from the solution and placed in a mixture of 75% ethanol, 5% glycerol, and 20% distilled water. They tubes were incubated in a water bath (CU-600, Shanghai, China) at 100 °C for 30 min. The stained leaves turned brown, suggesting the production of H_2O_2. Referring to the method described by Hung et al. [20], the leaves were immersed in trypan blue (1% (*w*/*v*)) (Sigma-Aldrich V900876, Berlin, Germany) for 50 min, removed, and then placed in a centrifuge tube containing a mixture of 75% ethanol, 5% glycerol, and 20% distilled water. They were then placed in a water bath at 100 °C for 30 min. The staining result turned blue, suggesting that the Pdpap poplar leaf cells were necrotic.

2.5. Determination of Pdpap Poplar Seedling Growth after BY6 Induction

Referring to the method described by Goicoechea et al. [21], the whole Pdpap poplar seedlings were taken out. The root soil was washed with clean water, placed in water, and then straightened. Vernier calipers (SATA91511, Shanghai, China) were used to measure root length and plant height. Plant height is defined as the distance from the ground position to the tip of the tallest leaf. The whole Pdpap poplar was placed in a sealed bag filled with distilled water to ensure that the entire seedling could fully absorb water. It was soaked in water for 24 h, and the fresh mass was measured after removing the excess water with filter paper. The whole Pdpap poplar seedling was cut into pieces for dry matter quality determination and placed in a glass petri plate. It was then placed in a blast drying oven (BPG-9206A, Shanghai, China) at 115 ± 10 °C for 30 min. The Pdpap poplar seedling was naturally cooled, placed in an oven at 65 °C for 4 h, and weighed using an electronic balance (Shimadzu ATX224R, Kyoto, Japan).

2.6. Determination of Soluble Sugar and Soluble Protein in Pdpap Poplar Seedlings after BY6 Induction

One gram of leaf tissue was washed with 0.9% normal saline to the grind. The Phenol and Coomassie brilliant blue G250 methods [21] were used to determine the soluble sugar and protein, respectively. The Pdpap poplar leaves were cut and placed in a 50 mL centrifuge tube. Five mL of deionized water was added to a water bath at 100 °C for 30 min. The tissue fluid was filtered through filter paper and diluted five times, yielding a total of 0.5 mL of the dilution. A total of 200 mg of Pdpap poplar leaves was used similarly. Five mL of deionized water was added to the ground sample, and the homogenate was centrifuged at 8000 rpm for 5 min (Eppendorf D30, Berlin, Germany). The supernatant was diluted five times with deionized water, and then 0.1 mL of the dilution was transferred to a centrifuge tube to determine soluble protein content in the Pdpap poplar. Simultaneously, a standard curve was prepared using bovine serum albumin (BSA) as the standard. Five mL of Coomassie Brilliant blue G250 was added to the standard and sample test tubes. It was inverted, mixed several times, and maintained for 5 min. A 1 mL was extracted and put in a micro cuvette. Absorbance was measured at a wavelength of 595 nm. The soluble protein content was then calculated.

2.7. Determination of Disease Resistance and Auxin Gene Expression in Pdpap Poplar after BY6 Induction

Disease resistance and growth-related gene expression in Pdpap poplar were determined after BY6 induction. The leaves of the control and treatment groups at 0, 6, 12, 24, and 48 h (counting from the 5th to the 8th leaf from the stem base) were randomly mixed. They were then transferred to −20 °C in a cold mortar. After adding liquid nitrogen thoroughly grinding, the ground leaf tissue was transferred to a 1.5 mL centrifuge tube using a spatula. Total RNA was extracted using RNA Extraction Kit (BIOTEKE PR3502, Beijing, China). The extracted RNA was assayed for RNA concentration using a UV spectrophotometer (BIOTEKE ND5000, Beijing, China) (OD260/280). The RNA final concentration of 1 ug, and cDNA was synthesized using a PrimeScript^{TM} RT reagent Kit with gDNA Eraser (Takara, Tokyo, Japan).

Primers were designed using the Primer 5.0 online tools (https://www.premierbiosoft.com/primerdesign/index.html accessed on 26 October 2022), and were the primers synthesized by Shanghai Sangon Bioengineering Co., Ltd. (Shanghai, China). *Act* (P)-L and *Tub* (P)-L were housekeeping genes. The sequences of housekeeping genes and PCR primers are illustrated in (Table 1). PCR reagents (SYBR Green Realtime PCR Master Mix 10 μL (Takara RR820A, Tokyo, Japan), cDNA 2 μL, Primer F (10 μM) 1 μL, Primer R (10 μM) 1 μL, and ddH_2O 6 μL) were added. The RT-qPCR (Agilent Mx3000P, Calif, USA) reaction program was as follows: (denaturation: at 94 °C, for 10 s, annealing at 60 °C, for 30 s, extension at 72 °C for 30 s, 30 cycles). The relative expression levels of the genes obtained by the reaction were calculated using the $2^{-\Delta\Delta CT}$ method [22]. Four biological replicates were used for each gene.

Table 1. Genes and primers used for the RT-qPCR analysis.

Target Gene	Primer Sequence (5′–3′)	Predicted Product Size (bp)
PR1	ACACCACCGTGCAACCTATG CGAGCAGAGTTCGCCAAACCA	240
NPR1	GGCCGACGATATTCCCTAGTT TGCCCTCATAGTTTCTGAGCT	261
AUX1	CAGTCGGTGCTCTTCGGTCA GTCCACCTGCCTACGAAGT	230
LAX2	TCATTCGGCCCTTCACCAAGT GCCATGATTGCATCACACCTT	212
MP	GCACATGAACGGCAGGGTTT ACCTGACCATGACGACACTT	213
TIR1	GTTGGTACGCAAAGGTAGAGA GGCTGACCATGCAATACTAGC	235
IAA8	CGGAGACTGGATGTTGTTGGT ACCTGAAACCTGATCGTGCTC	226
JAR1	AGTGGTGACCAAGCGAGGAG GGATTTGTGCACCTTGCTGTT	210
JAZ6571	GGAAGCTCATTGGCACAGATG CCGGAGTGGGTTGTTTGTCTG	221
MYC2	ACGAAGCGCAATCTGCTGAGT CCAGGCTCTCAAAGCCGACAT	223
COI	GAGGTATTGTGGGTGCATGGT ACGCGAACCTACCCTCGCT	245
ORCA3	AGAGGATGAGGCAAAGACCCT TCCGGCTCTCGCGCTTAGT	231
Actin(P)	AGCTGATCGAATAGCAAG CTAGAAGCACTCCTGTG	196
β-tubulin (P)	TACCGAGGCTGAGATAACAT GGACCCACAACTCATCACAT	210

Note: Pathogenesis-related 1 (*PR1*); Non expressor of pathogenesis-related 1 (*NPR1*); Auxin resistant 1 (*AUX1*); Auxin influx carrier protein (*LAX2*); Monopteros (*MP*); Transport inhibitor response protein 1 (*TIR1*); Auxin/indoleacetic acids proteins (*IAA8*); Jasmonate resistant 1 (*JAR1*); Jasmonate ZIM-domain (*JAZ6571*); V-myc avian myelocytomatosis viral oncogene homolog (*MYC-2*); Cytochrome c oxidase I (*COI*); Octadecanoid-derivative responsive Catharanthus AP2-domain protein 3 (*ORCA3*).

2.8. Statistical Analysis

All experimental data were subjected to analysis of variance (ANOVA) using the SPSS software v19.0 (IBM Corp., Armonk, NY, USA). The statistical significance of the difference between means was determined using Duncan's multiple-range test and an independent-sample T test ($p < 0.05$).

3. Results

3.1. Treated Disease Index and Relative Control on Pdpap Poplar Root Rot after BY6 Induction

On the 20th day after inoculation with *A. solidipes*, Pdpap poplar seedlings in the greenhouse demonstrated disease symptoms similar to those in the field. The leaves in the infected control group revealed yellowing and withering and then began to die one after another (Figure 1A). On day 20, the treated disease index reached 49.53%, an increase of 36.51% compared with 0 d (Figure 1C). However, after BY6 induction, only the leaves below the ground demonstrated wilting symptoms, suggesting that the disease symptoms of the plant were relieved. New leaves were successively grown (Figure 1B). On day 20, the treated disease index decreased by 42.01% compared to the infected control group, and the relative control effect indicated an upward trend. The control effect on the 20th day reached 84.31%, an increase of 89.22% compared to that at 0 day (Figure 1D). This result indicated that the diseased Pdpap poplar seedlings' disease index was significantly reduced after BY6 induction and had a certain biological control effect.

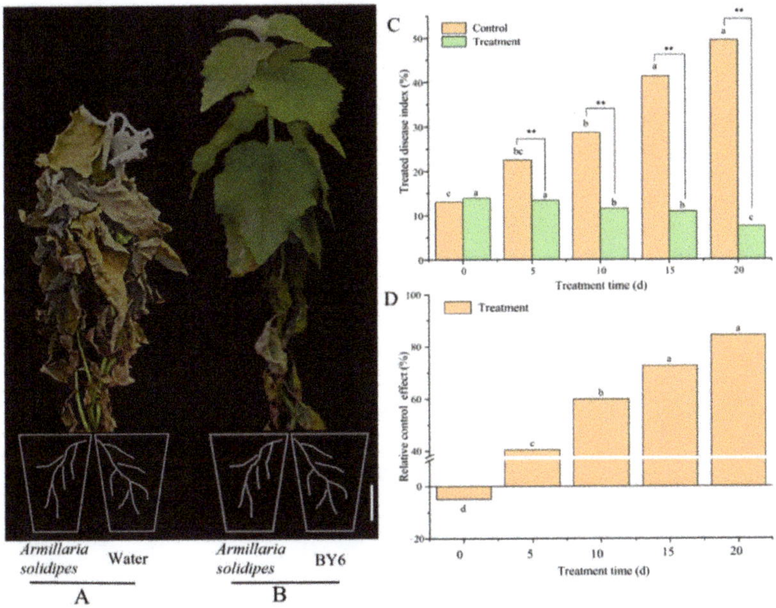

Figure 1. Pdpap poplar ARR symptoms and treated disease index. (**A**): Water control; (**B**): Inoculated BY6; (**A**,**B**): the inoculation of *A. solidipes* in advance 7 d, (**C**): the treated disease index; (**D**): Relative control effect; The roots of Pdpap poplar seedlings were divided into two equal parts (part 1 and part 2), and transplanted separately into two adjacent pots. Architect scale is 1 cm; The experiment was carried out in four biological replicates; ** indicate significant differences between the groups at the same time point (ANOVA, ** < 0.01), and the different lowercase letters indicate significant differences for the same treatment group at different times (ANOVA, $p < 0.05$).

3.2. Effects of BY6 Induction on the Activity of Defense Enzymes in Pdpap Poplar

According to the results (Figure 2A), Pdpap poplar catalase (CAT) activity in the treatment group exhibited a downward trend, from 0 to 48 h. It reached a minimum value at 48 h, which was 0.59 times lower than that at 0 h, while the difference between the infected control groups was insignificant. Phenylalanine ammonia-lyase (PAL) activity treatment group demonstrated an upward trend from 0 to 48 h (Figure 2B), reaching a maximum value at 48 h, which was 1.52 times higher than that at 0 h. In contrast, the differences between the values of the infected control groups at different points of time were insignificant. These results indicate that BY6-induced diseased Pdpap poplar decreased CAT activity, and increased PAL activity.

Peroxidase (POD) enzyme activity treatment results indicated that the treatment group had an upward trend from 0 h to 48 h (Figure 2C) and reached a maximum at 48 h. Compared with 0 h, the enzyme activity increased by 1.68 times. The infected control group was insignificant, and only a small increase was observed from 24 to 48 h. The superoxide dismutase (SOD) enzyme activity treatment group increased from 0 to 12 h (Figure 2D), reached the maximum value at 12 h, and increased by a factor of 1.12 compared to that at 0 h. It then stabilized from 24 to 48 h, while the difference between the groups was insignificant; this indicates that BY6 induced Pdpap poplar seedlings, increased POD and SOD activities in Pdpap poplar.

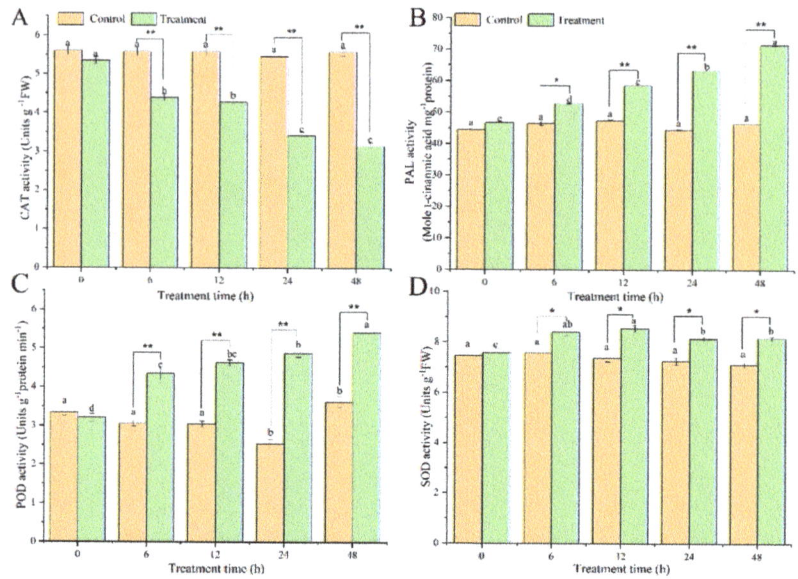

Figure 2. The effect of BY6 on Pdpap poplar oxidase indicators activity. (**A**): Catalase (CAT). (**B**): Phenylalanine ammonia-lyase (PAL). (**C**): Peroxidase (POD). (**D**): Superoxide dismutase (SOD). The experiment was carried out in four biological replicates. The graphical values were the average of four biological repeats. Error bars represent standard errors of the mean (n = 4) at each time point. Different * indicate significant differences between the groups at the same time point (ANOVA, * < 0.05; ** < 0.01), and the different lowercase letters indicate significant differences for the same treatment group at different times (ANOVA, $p < 0.05$).

3.3. Staining Results of Reactive Oxygen Species and Cell Death in Pdpap Poplar Leaves after BY6 Induction

On the 20th day after the infection experiment, the Pdpap poplar leaves were stained with DAB and trypan blue, respectively. The results shown in Figure 3A that the DAB staining treatment and infected control groups had different degrees of staining, and the relative staining area of the infected control group was 3.37 times higher than that of the treatment group. The results shown in Figure 3B were obtained using trypan blue staining, but the relative staining area of the infected control group was 7.31 times higher than that of the treatment group. Both indicate that BY6 induction significantly reduced the reactive oxygen species content and the degree of necrosis in the leaf cells of diseased Pdpap poplar.

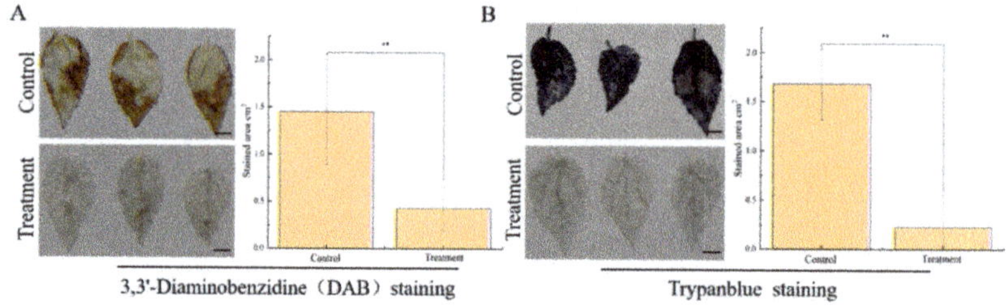

Figure 3. DAB and Trypan blue staining result. The results of Pdpap poplar leaves with DAB (**A**) and Trypan blue (**B**) 20 days after inoculation of treatment statistical analysis of DAB and trypan blue using Image J software; Biology repeated three times; ** indicate significant differences between the groups (ANOVA, ** < 0.01); Architect's scale is 1 cm.

3.4. Effects of BY6 Induction on Pdpap Poplar Growth

To analyze the impact of BY6 on the growth of diseased Pdpap poplar seedlings, the dry mass, fresh mass, plant height, and root length of the seedlings were all determined. On the 5th, 10th, 15th, and 20th d of inoculation with the BY6 strain, the dry mass index of the treated group exhibited an upward trend (Figure 4A) and finally reached the maximum value on the 20th day. The infected control group clarified a slight downward trend from 0 to 20 days. The fresh mass of the treated group exhibited an upward trend on the 10th, 15th, and 20th day (Figure 4B) and reached a maximum value on the 20th day, which was 2.07 times higher than that on the 0th day. It revealed a downward trend, reaching a minimum value on the 20th day, which was 1.77 times lower than 0 day. The plant height of the treated group demonstrated an upward trend from 0 to 20 days (Figure 4C), and the difference was significant. It reached a maximum value on the 20th day, which increased by 5.55 times compared to day 0, while the difference between the infected control groups was insignificant. The root length of the treated group demonstrated an upward trend from 0 to 20 days (Figure 4D), reaching a maximum value on the 20th day, 1.87 times higher than that on day 0. The infected control group stabilized after a slight increase from 0 to 10 days. In conclusion, after induction, the BY6 strain promoted the growth of diseased Pdpap poplar seedlings.

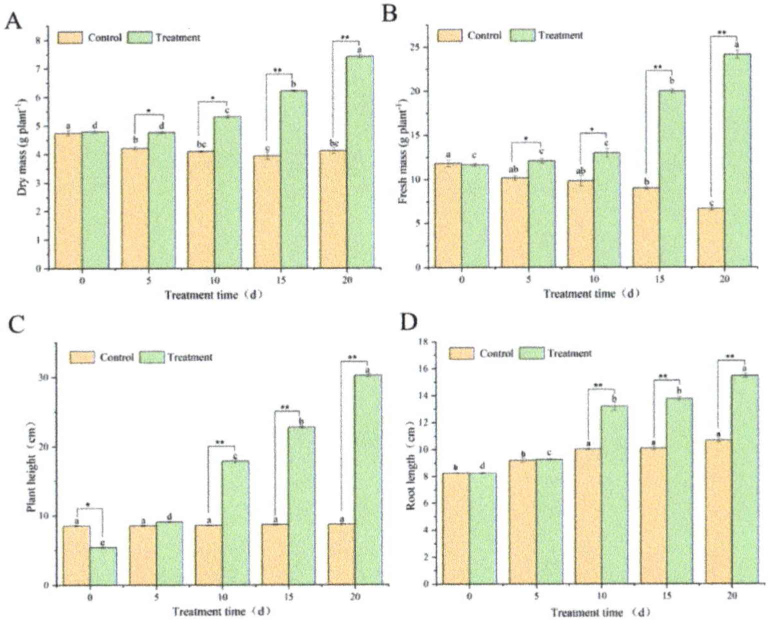

Figure 4. The effect of BY6 strain on the growth of Pdpap poplar seedlings. (**A**): Dry mass; (**B**): Fresh mass; (**C**): Plant height; (**D**): Root length. The graph value is the average of four biological replicates; the error bars represent the standard error of the average ($n = 4$) at each time point; The experiment was carried out in four biological replicates. Different * indicate significant differences between the groups at the same time point (ANOVA, * < 0.05; ** < 0.01), and the different lowercase letters indicate significant differences for the same treatment group at different times (ANOVA, $p < 0.05$).

3.5. Determination of Soluble Protein and Soluble Sugar in Pdpap Poplar after BY6 Induction

The results in Figure 5A show that the soluble protein content in the treated group had an upward trend from 0 to 15 days, reaching a maximum on day 15. Compared with day 0, it increased by 1.63 times, then decreased slightly from the 15th to the 20th day. The infected control group demonstrated a minor fluctuation, and the overall level was stable.

(Figure 5B) displays that the soluble sugar content of the treated group demonstrated an overall upward trend from 0 to 15 days and reached the maximum value on day 15, which increased by 5.72 times compared at day 0. A slight decrease was observed from 15 to 20 days, whereas the infected control group stabilized overall.

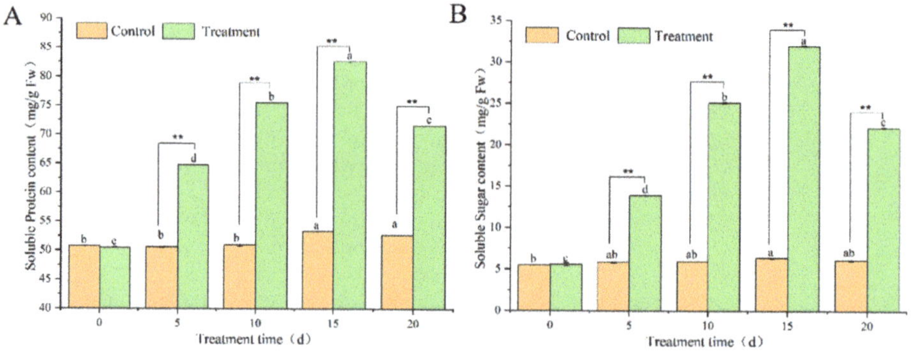

Figure 5. Effects of BY6 on soluble protein (**A**) and sugar (**B**) content in Pdpap poplar leaves. The experiment was carried out in four biological replicates; ** indicate significant differences between the groups at the same time point (ANOVA, ** < 0.01), and the different lowercase letters indicate significant differences for the same treatment group at different times (ANOVA, $p < 0.05$).

3.6. Determination of Expression Levels of Disease Resistance and Growth-Related Genes in Pdpap Poplar after BY6 Induction

The expression of genes related to the signal transduction pathway of disease resistance in Pdpap poplar was determined (Figure 6). The first was the SA pathway illustrateed. The *NPR1* genes expression levels in the BY6-treated group demonstrated an up-regulation trend, reaching a maximum transcriptional peak at 24 h. The expression level was 5.64 times that of the infected control group (Supplementary Table S1); The *PR1* genes-treated and infected control groups revealed an upward-regulated trend, reaching the maximum transcriptional peak at the 6th h.

The second is the JA pathway. The *JAR1* enzyme synthesizes JA-Ile genes treatment group revealed an up-regulation trend, reaching a maximum value at 24 h, while the infected control group demonstrated a down-regulation trend. The *COI* genes expression levels in the BY6-treated group exhibited an upward-regulated trend, reaching the maximum transcriptional peak at 48 h. In contrast, the infected control group was first up-regulated at 0~12 h, and then down-regulated after 12~48 h. The *JAZ6571* genes expression levels in the BY6-treated group displayed an up-regulation trend, reaching the maximum transcriptional peak at 48 h, whereas the infected control group revealed a down-regulation trend. The *MYC-2* genes in the treatment group was up-regulated and reached the maximum transcriptional peak at 48 h, while the infected control group indicated a down-regulation trend. The *ORCA3* genes in the control and genes expression levels in the BY6-treated group revealed a downward trend, and the difference between them was insignificant.

BY6 induces the transcription of Pdpap poplar auxin signaling-related genes. The *AUX1* genes in the control and genes expression levels in the BY6-treated group indicated an upward-regulated trend. The treatment group reached the maximum transcription peak at 24 h, and the infected control group was up-regulated at 0~6 h first. There was no significant change between 6~48 h. The *TIR1* genes in the infected control group indicated an up-regulation trend, reaching the maximum transcriptional peak at 48 h, and the treatment group demonstrated a down-regulation trend. The *LAX2* genes in the control and genes expression levels in the BY6-treated group exhibited the same performance; they indicated an upward-regulated trend and reached the maximum transcription peak at 72 h. In contrast, the *MP* genes in the control and gene expression levels in the BY6-treated group revealed a downward trend. The changing trend of the *IAA8* genes treatment and infected control groups was the same. The maximum transcription peak was reached at the 12th hour, and the maximum transcription peak of the genes expression levels in the BY6-treated group was 1.95 times that of the infected control group. This result indicated that BY6 promoted the expression of genes related to disease resistance signal transduction and the growth pathway in Pdpap poplar.

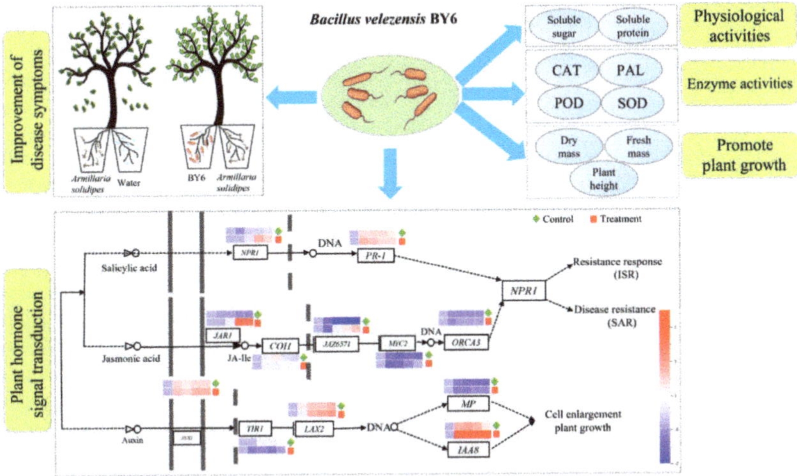

Figure 6. Schematic representation of the growth-promoting and disease-resistant responses of BY6 to diseased Pdpap poplar after induction. SA pathway genes (*NPR1*; *PR-1*). JA pathway genes (*JAR1*; *COI-1*; *JAZ6571*; *MYC-2*; *ORCA-3*). Auxin pathway genes (*AUX1*; *TIR1*; *LAX2*; *MP*; *IAA8*). Color bar: the red part represents genes with positive expression levels, and the blue represents negative expression levels. The roots of Pdpap poplar seedlings were divided into two equal parts (part 1 and part 2), and transplanted solid and dashed arrows indicate direct and indirect activation, respectively. Prohibition sign means prohibition.

4. Discussion

To further understand the biological control effect of the BY6 strain, its effect of BY6 on the ARR treated disease index and relative control of diseased Pdpap poplar was determined under indoor conditions. The results demonstrate that, compared with the infected control group, the treated disease index of Pdpap poplar root rot disease decreased by 42.02% after inoculation with BY6, and the relative control effect exhibited an upward trend; this confirms that BY6 has a therapeutic effect on the diseased Pdpap poplar and first report of *B. velezensis* BY6 for controlling ARR in Pdpap poplar. *Bacillus* strains are known for their metabolic capability and environmental versatility as well as for their ability to manage bacterial and fungal pathogens infecting plants. In some similar studies, Sarah Hong et al. [23] using *B. velezensis* CE 100 not only increased the fruit yield of strawberries by controlling the fungal diseases (*Macrophomina phaseolina and Fusarium oxysporum f.* sp. *fragariae*) but also through enhancing plant growth. Peng Hong et al. [24] also using *Bacillus amyloliquefaciens* HF-01, SBC control of postharvest decay on citrus while maintaining fruit quality after harvest. The strain BY6 of *B. velezensis* proved its antagonistic potential in controlled, reducing ARR on Pdpap poplar.

It remains unclear whether the symptomatic relief of ARR in Pdpap poplar is related to BY6 induction. To some extent, the interaction between plants and pathogens is a process by which the pathogen hydrolase system of the pathogen associates with the oxidase system of the host plant [25]. It has been reported that oxidase activity levels (POD, PAL, SOD, and CAT) increase or decrease after inoculation with inducers, further enhancing plant resistance to pathogen invasion and expansion [22]. PAL is the rate-limiting enzyme in the phenylpropane metabolic pathway, which regulates the accumulation of the main antibacterial substances. Therefore, it is often used to evaluate plant disease resistance [26]. POD is a terminal oxidase in the metabolic pathway of phenylpropane that strengthens the cell wall by catalyzing the polymerization of phenolic acids and the synthesis of lignin and enhances the resistance of plants to pathogenic infection. Under certain conditions, POD can also have a synergistic effect with PAL and CAT to eliminate the damage caused by excessive reactive oxygen species that are toxic to cells in the host [27]. PAL and POD activities in Pdpap poplar leaves were significantly increased after inoculation with BY6, BY6 induced the activation of Pdpap poplar defense enzymes and reduced the toxicity of pathogenic infection to its cells. *Bacillus* can increase the content of total phenols, flavonoids, and lignin in plants by activating plant PAL and POD enzyme activities [28]. These compounds can directly inhibit spore germination and mycelial growth of pathogenic fungi and the precursors for lignin synthesis. Lignin deposition on cell walls can improve mechanical tissue strength and prevent pathogen invasion. A same plant split into two parts was employed in this study, and the BY6 strain was not in contact with *A. solidipes*, which ruled out the bacteriostatic effect of BY6. Instead, BY6 induced Pdpap poplar cell wall enhancement, which reduced the damage caused by *A. solidipes* to invasion.

Prior studies [29] also revealed that the induction of Pdpap poplar with *B. amyloliquefaciens* AW3 could increase the CAT activity of Pdpap poplar leaves, thus increasing the root rot of Pdpap poplar caused by *Fusarium oxysporum* infection. CAT activity significantly decreased after BY6-induced Pdpap poplar and obtained the opposite results. We speculate that, on the one hand, the changes in various enzyme activities caused by pathogenic bacteria are related to the degree of plant damage. In contrast, BY6 induction alleviated the stimulation of pathogenic bacteria to Pdpap poplar, thus reducing CAT enzyme activity. To confirm the presumed results, we tested reactive oxygen species (ROS) and cell death in Pdpap poplar leaves. As a consequence of trypan blue staining, more necrosis was observed in the leaves of the infected control group. DAB staining also revealed increased H_2O_2 accumulation in the leaves of the infected control group. Both indicated that BY6 induction alleviated the pathogenic stimulation of Pdpap poplar and decreased CAT activity. Notably, the prevention and control of diseases by biocontrol bacteria result from multiple mechanisms (such as niche and nutrient competition, secretion of antimicrobial substances, host resistance induction, and colonization ability) [30]. It is also associated with other microbial community structures in plants [27]. These factors affect biological control and the multiple mechanisms involved require further investigation.

In addition to preventing and controlling plant diseases, some biocontrol bacteria have growth-promoting effects [29]. The dry mass of treated groups increased significantly on the 5th, 10th, 15th' and 20th d after inoculation with the BY6 strain, and the fresh mass had the same performance. Dry and fresh mass are important indicators of plant growth status and organic matter accumulation and are indirectly involved in plant disease resistance [31,32]. BY6 can significantly increase the fresh and dry mass of the seedlings after inducing diseased Pdpap poplar seedlings. The results of this study are consistent with those reported by Jin et al. [33]. This study also found that BY6 induction increased the plant height of diseased Pdpap poplar seedlings. The root length of the BY6-treated group revealed an overall increase from 0 to 20 days, while the infected control group indicated no significant increase. We speculate that BY6 could reactivate the expression of auxin-related genes in diseased Pdpap poplar. To test this speculation, verified by qRT-PCR, the results show that the transcription of four auxin-related genes (*AUX1*, *TIR1*, *LAX2*, and *IAA8*) was activated, which confirmed the above speculation. The above results indicate that BY6 secondary induced the growth of diseased Pdpap poplar seedlings, which could have a crucial role in controlling the upper and lower levels.

Soluble sugars and proteins content are critical indicators for measuring a plant's growth status and quality. They are involved in many metabolic and energy supply processes, and their content is an important indicator for characterizing the physiological activity of plants [34]. Soluble protein and sugar content levels demonstrated an upward trend, suggesting that both increased after BY6 induction. Previous studies have confirmed that soluble sugars and proteins are the main photosynthetic products of higher plants and are also the primary form of carbohydrate metabolism and temporary storage [35]. They also regulate oxidase indicators, and changes in their content can reflect plant health. To verify this relationship, oxidase indicators revealed that, post BY6 induction, this activated SOD, POD, and CAT enzymes (expressed or inhibited) in diseased Pdpap poplar; this indicates that BY6 promotes the production of soluble protein and sugar in the diseased Pdpap poplar, indirectly regulating the activity of oxidase indicators in diseased Pdpap poplar and improving disease resistance. These results are consistent with the reported results for Cf-10-mediated and other leaf mold samples [36].

This study determined whether BY6 triggers defense responses in Pdpap poplar seedlings in the presence of fungal pathogens and offers insights into its molecular basis. The results depicted that the relative expression levels of *NPR1*, *PR1*, *JAR1*, *JAZ6571*, *MYC2*, and *ORCA3* increased after BY6 induction, suggesting that BY6 activated the induced systemic resistance (ISR) of Pdpap poplar seedlings. However, SA and JA signals are antagonistic and cooperative in activating the plant response system for disease resistance [37,38]. It is generally believed that the SA signaling pathway is primarily involved in the resistance response to biotrophic pathogens (obligate parasites), and JA is involved in resistance responses to necrotrophic pathogens [39]. This study found that after the infection of the pathogen *A. solidipes*, BY6 continued to induce the SA and JA pathways. *NPR1*, *PR1*, *JAR1*, *JAZ6571*, *MYC2*, and *ORCA3* genes continued to be expressed, and the expression fold was higher than that of the infected control group under *A. solidipes* infection alone; this suggests that BY6 induced ISR and systemic acquired resistance (SAR) in diseased Pdpap poplar. In some related studies, rhizosphere-promoting bacterium *Bacillus subtilis* PTA-271 is an essential component of the *Arabidopsis thaliana* ISR, and the elicited ISR is JA/ET pathways dependent, and not pathogen-dependent SAR [40]. This study found that probiotic BY6 induced both SAR and ISR of the diseased Pdpap poplar, not a single ISR. It is an interesting phenomenon, and the specific mechanism will be examined in follow-up studies in the future.

5. Conclusions

This study used the endophyte *B. velezensis* BY6 to induce the mechanism of growth promotion and disease resistance in diseased Pdpap poplar. The results show that the application of BY6 on Pdpap poplar seedlings resulted in a significant increase in plant growth indicators (dry mass, fresh mass, and plant height), and also observed that the transcription of four auxin-related genes (*AUX1*, *TIR1*, *LAX2*, and *IAA8*) was significantly increased. BY6 revealed a remarkable control effect after the inoculation of diseased Pdpap poplar plants. On the 20th day, the treated disease index of the treatment group was reduced by 49.53%, compared with those of the infected control group. The relative staining area of DAB and Trypan blue decreased by 3.37 and 7.31 times, respectively, and the physiological indicators (soluble sugar and soluble protein) and oxidase indicators (CAT, PAL, POD and SOD) were significantly improved. The expression levels of defense-related genes (*NPR1*, *PR1*, *JAR1*, *JAZ6571*, *MYC2*, and *ORCA3*) in SA and JA signaling pathways were significantly increased. To the best of our knowledge, this is the first report in which BY6 controls ARR during diseased Pdpap poplar progression, involving the molecular mechanisms underlying growth promotion and disease resistance.

Supplementary Materials: The following supporting information can be downloaded at: https://www.mdpi.com/article/10.3390/microorganisms10122472/s1, Table S1: RT-qPCR analysis date.

Author Contributions: Conceptualization, investigation, writing—original draft preparation, P.Z. and G.X.; Formal analysis, writing—review and editing, supervision, L.W.; Project administration, funding acquisition, Y.X. All authors have read and agreed to the published version of the manuscript.

Funding: This work was financed by the Special Project for Double First-Class—Cultivation of Innovative Talents (000/41113102). Ping Zhang is grateful to the China Scholarship Council for the financial support (No. 202006600025).

Institutional Review Board Statement: Not applicable.

Informed Consent Statement: Not applicable.

Data Availability Statement: Not applicable.

Acknowledgments: All authors thank to Jian Diao, Chang Zhou and Ruiting Guo for providing experiment sample collect work on this article.

Conflicts of Interest: The authors declare that they have no competing interests. The manuscript does not contain experiments involving animal or human studies.

References

1. Ji, S.D.; Liu, Z.H.; Liu, B.; Wang, Y.C. Comparative analysis of biocontrol agent *Trichoderma asperellum* ACCC30536 transcriptome during its interaction with *Populus davidiana × P. alba var. pyramidalis*. *Microbiol. Res.* **2019**, *227*, 126294. [CrossRef] [PubMed]
2. Diao, J.; Li, M.R.; Zhang, P.; Zong, C.; Ma, W.; Ma, L. Overexpression of the Pdpap poplar ERF109 gene enhances resistance of *Populus davidiana × P. alba var. pyramidalis* to *Fusarium oxysporum* infection. *J. For. Res.* **2022**, *33*, 1925–1937. [CrossRef]
3. Sun, Y.S.; Chen, S.; Huang, H.J.; Jiang, J.; Bai, S.; Liu, G.F. Improved salt tolerance of *Populus davidiana × P. bolleana* overexpressed LEA from *Tamarix androssowii*. *J. For. Res.* **2014**, *25*, 813–818. [CrossRef]
4. Cleary, M.R.; van der Kamp, B.J.; Morrison, D.J. Effects of wounding and fungal infection with *Armillaria ostoyae* in three conifer species. II. Host response to the pathogen. *Forest Pathol.* **2012**, *42*, 109–123. [CrossRef]
5. Teklu, B.M.; Haileslassie, A.; Mekuria, W. Pesticides as water pollutants and level of risks to environment and people: An example from central rift valley of Ethiopia. *Environ. Dev. Sustain.* **2022**, *24*, 5275–5294. [CrossRef]
6. Munnecke, D.E.; Wilbur, W.D.; Kolbezen, M.J. Dosage response of *Armillaria-mellea* to methyl bromide. *Phytopathology* **1970**, *60*, 992. [CrossRef]
7. Ismail, M.A.; Amin, M.A.; Eid, A.M.; Hassan, S.E.; Mahgoub, H.; Lashin, I.; Abdelwahab, A.T.; Azab, E.; Gobouri, A.A.; Elkelish, A.; et al. Comparative study between exogenously applied plant growth hormones versus metabolites of microbial endophytes as plant growth-promoting for *Phaseolus vulgaris* L. *Cells* **2021**, *10*, 1059. [CrossRef]
8. Ren, Z.H.; Xie, L.; Okyere, S.K.; Wen, J.; Ran, Y.A.; Nong, X.; Hu, Y.C. Antibacterial activity of two metabolites isolated from endophytic bacteria *Bacillus velezensis* Ea73 in *Ageratina adenophora*. *Front. Microbiol.* **2022**, *13*, 9. [CrossRef]
9. Nanjani, S.; Soni, R.; Paul, D.; Keharia, H. Genome analysis uncovers the prolific antagonistic and plant growth-promoting potential of endophyte *Bacillus velezensis* K1. *Gene* **2022**, *836*, 146671. [CrossRef]
10. Ji, C.; Wang, X.H.; Song, X.; Zhou, Q.S.; Li, C.H.; Chen, Z.Z.; Gao, Q.X.; Li, H.Y.; Li, J.T.; Zhang, P.C.; et al. Effect of *Bacillus velezensis* JC-K3 on endophytic bacterial and fungal diversity in wheat under salt stress. *Front. Microbiol.* **2021**, *12*, 802054. [CrossRef]
11. Doan, H.V.; Hoseinifar, S.H.; Khanongnuch, C.; Kanpiengjai, A.; Unban, K.; Kim, V.V.; Srichaiyo, S. Host-associated probiotics boosted mucosal and serum immunity, disease resistance and growth performance of Nile tilapia (*Oreochromis niloticus*). *Aquaculture* **2018**, *491*, 94–100. [CrossRef]
12. Han, V.C.; Yu, N.H.; Yoon, H.; Ahn, N.H.; Son, Y.K.; Lee, B.H.; Kim, J.C. Identification, characterization, and efficacy evaluation of *Bacillus velezensis* for shot-hole disease biocontrol in flowering Cherry. *Plant Pathol. J.* **2022**, *38*, 115–130. [CrossRef]
13. Huang, L.; Li, Q.C.; Hou, Y.; Li, G.Q.; Yang, J.Y.; Li, D.W.; Ye, J.R. *Bacillus velezensis* strain HYEB5-6 as a potential biocontrol agent against anthracnose on *Euonymus japonicus*. *Biocontrol Sci. Technol.* **2017**, *27*, 636–653. [CrossRef]
14. Meng, Q.X.; Jiang, H.; Hao, J. Effects of *Bacillus velezensis* strain BAC03 in promoting plant growth. *Biol. Control* **2016**, *98*, 18–26. [CrossRef]
15. Constantin, M.E.; de Lamo, F.J.; Vlieger, B.V.; Rep, M.; Takken, F. Endophyte-mediated resistance in tomato to *Fusarium oxysporum* is independent of ET, JA, and SA. *Front. Plant Sci.* **2019**, *10*, 979. [CrossRef]
16. Zhang, P.; Diao, J.; Xie, G.Q.; Ma, L.; Wang, L.H. A complete genome sequence of the wood stem endophyte *Bacillus velezensis* BY6 strain possessing plant growth-promoting and antifungal activities. *Biomed. Res. Int.* **2021**, *2021*, 3904120. [CrossRef]
17. Hernandez, M.; Davila, A.R.; de Algaba, A.P.; Lopez, M.; Casas, A.T. Occurrence and etiology of death of young olive trees in southern Spain. *Eur. J. Plant Pathol.* **1998**, *104*, 347–357. [CrossRef]
18. Tang, Y.; Li, R.; Jiang, Z.C.; Cheng, Z.Y.; Li, W.; Shao, Y.Z. Combined effect of Debaryomyces hansenii and *Bacillus atrophaeus* on the physicochemical attributes, defense-related enzyme activity, and transcriptomic profile of stored litchi fruit. *Biol. Control* **2022**, *172*, 104975. [CrossRef]
19. Lin, C.W.; Chang, H.B.; Huang, H.J. Zinc induces mitogen-activated protein kinase activation mediated by reactive oxygen species in rice roots. *Plant Physiol. Biochem.* **2005**, *43*, 963–968. [CrossRef]
20. Hung, W.C.; Huang, D.D.; Chien, P.S.; Yeh, C.M.; Chen, P.Y.; Chi, W.C.; Huang, H.J. Protein tyrosine dephosphorylation during copper-induced cell death in rice roots. *Chemosphere* **2007**, *69*, 55–62. [CrossRef]

21. Goicoechea, N.; Aguirreolea, J.; Cenoz, S.; Garcia-Mina, J.M. *Verticillium dahliae* modifies the concentrations of proline, soluble sugars, starch, soluble protein and abscisic acid in pepper plants. *Eur. J. Plant. Pathol.* **2000**, *106*, 19–25. [CrossRef]
22. Livak, K.J.; Schmittgen, T.D. Analysis of relative gene expression data using real-time quantitative PCR and the $2-\Delta\Delta CT$ method. *Methods* **2001**, *25*, 402–408. [CrossRef]
23. Hong, S.R.; Kim, T.Y.; Won, S.J.; Moon, J.H.; Ajuna, H.B.; Kim, K.Y.; Ahn, Y.S. Control of fungal diseases and fruit yield improvement of strawberry using *Bacillus velezensis* CE 100. *Microorganisms* **2022**, *10*, 365. [CrossRef]
24. Hong, P.; Hao, W.N.; Luo, J.J.; Chen, S.H.; Hu, M.Y.; Zhong, G.H. Combination of hot water, *Bacillus amyloliquefaciens* HF-01 and sodium bicarbonate treatments to control postharvest decay of mandarin fruit. *Postharvest Biol. Technol.* **2014**, *88*, 96–102. [CrossRef]
25. Saijo, Y.; Loo, E.P.; Yasuda, S. Pattern recognition receptors and signaling in plant-microbe interactions. *Plant J.* **2018**, *93*, 592–613. [CrossRef]
26. Xu, J.; Zhang, Z.; Li, X.P.; Wei, J.; Wu, B. Effect of nitrous oxide against *Botrytis cinerea* and phenylpropanoid pathway metabolism in table grapes. *Sci. Hortic.* **2019**, *254*, 99–105. [CrossRef]
27. Zhang, X.T.; Kong, W.L.; Wu, X.Q.; Ye, J.R. *Bacillus velezensis* JK-XZ8 prevents and controls crown gall disease on *Prunus subhirtella* by colonizing and inducing resistance. *J. For. Res.* **2022**, *33*, 1019–1031. [CrossRef]
28. Kashyap, A.S.; Manzar, N.; Nebapure, S.M.; Rajawat, M.; Deo, M.M.; Singh, J.P.; Kesharwani, A.K.; Singh, R.P.; Dubey, S.C.; Singh, D. Unraveling microbial volatile elicitors using a transparent methodology for induction of systemic resistance and regulation of antioxidant genes at expression levels in chili against bacterial wilt disease. *Antioxidants* **2022**, *11*, 404. [CrossRef]
29. Zhang, P.; Hao, H.T.; Wang, L.H.; Liu, Z.H.; Ma, L. *Endophytes Bacillus* amyloliquefaciens AW3 (CGMCC1.16683) improves the growth of *Populus davidiana* × *Populus bolleana* (Pdpap poplar) and induces its resistance to wilt disease by *Fusarium oxysporum* Fox68 (CFCC86068). *Eur. J. Plant Pathol.* **2022**, *162*, 1–17. [CrossRef]
30. Wang, X.; Liang, L.Q.; Shao, H.; Ye, X.X.; Yang, X.B.; Chen, X.Y.; Shi, Y.; Zhang, L.H.; Xu, L.H.; Wang, J.X. Isolation of the Novel strain *Bacillus amyloliquefaciens* F9 and identification of lipopeptide extract components responsible for activity against *Xanthomonas citri* subsp. *citri*. *Plants* **2022**, *11*, 457. [CrossRef]
31. Ji, C.L.; Zhang, M.L.; Kong, Z.R.; Chen, X.; Wang, X.; Ding, W.; Lai, H.X.; Guo, Q. Genomic analysis reveals potential mechanisms underlying promotion of tomato plant growth and antagonism of soilborne pathogens by *Bacillus amyloliquefaciens* Ba13. *Microbiol. Spectr.* **2021**, *9*, e01615-21. [CrossRef]
32. Zhang, Y.; Qin, B.X.; Narisawa, K.; Nong, Q.; Qin, L.P.; Xie, L. The dark septate endophyte, *Phialocephala fortinii* J2PC4, Mitigates southern rice black-streaked dwarf disease and impacts the mortality of white back planthopper. *Biol. Control.* **2022**, *170*, 104911. [CrossRef]
33. Jin, W.; Peng, L.; Zhang, X.G.; Sun, H.J.; Yuan, Z.L. Effects of endophytic and ectomycorrhizal basidiomycetes on *Quercus virginiana* seedling growth and nutrient absorption. *J. Sustain. For.* **2019**, *38*, 457–470. [CrossRef]
34. Jang, J.H.; Kim, S.H.; Khaine, I.; Kwak, M.J.; Lee, H.K.; Lee, T.Y.; Lee, W.Y.; Woo, S.Y. Physiological changes and growth promotion induced in poplar seedlings by the plant growth-promoting rhizobacteria *Bacillus subtilis* JS. *Photosynthetica* **2018**, *56*, 1188–1203. [CrossRef]
35. Chen, Z.X.; Klessig, D.F. Identification of a soluble salicylic acid-binding protein that may function in signal transduction in the plant disease-resistance response. *Proc. Natl. Acad. Sci. USA* **1991**, *88*, 8179–8183. [CrossRef]
36. Liu, G.; Liu, J.F.; Zhang, C.L.; You, X.Q.; Zhao, T.T.; Jiang, J.B.; Chen, X.L.; Zhang, H.; Yang, H.H.; Zhang, D.Y.; et al. Physiological and RNA-seq analyses provide insights into the response mechanism of the Cf-10-mediated resistance to *Cladosporium fulvum* infection in tomato. *Plant Mol. Biol.* **2018**, *96*, 403–416. [CrossRef]
37. Yang, Y.N.; Qi, M.; Mei, C.S. Endogenous salicylic acid protects rice plants from oxidative damage caused by aging as well as biotic and abiotic stress. *Plant J.* **2004**, *40*, 909–919. [CrossRef]
38. De Vleesschauwer, D.; Gheysen, G.; Hofte, M. Hormone defense networking in rice: Tales from a different world. *Trends Plant Sci.* **2013**, *18*, 555–565. [CrossRef]
39. Kalsi, H.S.; Karkhanis, A.A.; Natarajan, B.; Bhide, A.J.; Banerjee, A.K. AUXIN RESPONSE FACTOR 16 (StARF16) regulates defense gene StNPR1 upon infection with necrotrophic pathogen in potato. *Plant Mol. Biol.* **2022**, *109*, 13–28. [CrossRef]
40. Nguyen, N.H.; Trotel-Aziz, P.; Villaume, S.; Rabenoelina, F.; Schwarzenberg, A.; Nguema-Ona, E.; Clement, C.; Baillieul, F.; Aziz, A. *Bacillus subtilis* and *Pseudomonas fluorescens* trigger common and distinct systemic immune responses in *Arabidopsis thaliana* depending on the pathogen lifestyle. *Vaccines* **2020**, *8*, 503. [CrossRef]

microorganisms

MDPI

Article

First Report of Olive Branch Dieback in Croatia Caused by *Cytospora pruinosa* Défago

Elena Petrović [1], Karolina Vrandečić [2], Dario Ivić [3,*], Jasenka Ćosić [2] and Sara Godena [1,*]

1 Institute of Agriculture and Tourism, Karla Huguesa 8, 52440 Poreč, Croatia; elena@iptpo.hr
2 Faculty of Agrobiotechnical Sciences Osijek, Josip Juraj Strossmayer University of Osijek, Vladimira Preloga 1, 31000 Osijek, Croatia; kvrandecic@fazos.hr (K.V.); jcosic@fazos.hr (J.Ć.)
3 Centre for Plant Protection, Croatian Agency for Agriculture and Food, Gorice 68b, 10000 Zagreb, Croatia
* Correspondence: dario.ivic@hapih.hr (D.I.); sara@iptpo.hr (S.G.)

Abstract: Olive (*Olea europaea* L.) is a very important crop grown in the Mediterranean part of Croatia. Olive branch and fruit dieback symptoms were observed in two olive orchards in Istria, Croatia. The samples from symptomatic trees were collected and brought to the laboratory for analysis. Based on their morphological characterization, isolated fungi were identified as *Cytospora* sp. Two representative isolates (one per orchard) were taken for molecular analysis, and based on DNA sequence data of the ITS and TUB gene regions, and phylogenetic analysis of the sequences, the isolates were identified as *Cytospora pruinosa* Défago. To determine pathogenicity, pathogenicity tests were conducted on detached olive branches and two-year-old olive trees in the greenhouse. This is the first report of *C. pruinosa* causing olive branch and fruit dieback in Croatia.

Keywords: canker; *Cytospora* sp.; fungal disease; *Olea europaea* L.

Citation: Petrović, E.; Vrandečić, K.; Ivić, D.; Ćosić, J.; Godena, S. First Report of Olive Branch Dieback in Croatia Caused by *Cytospora pruinosa* Défago. *Microorganisms* **2023**, *11*, 1679. https://doi.org/10.3390/microorganisms11071679

Academic Editors: Tomislav Cernava, Beibei Ge and Kyungseok Park

Received: 7 June 2023
Revised: 23 June 2023
Accepted: 26 June 2023
Published: 28 June 2023

1. Introduction

The olive (*Olea europaea* L.) is a medium-sized evergreen tree from the family *Oleaceae*, which integrates a unique set of morphological and developmental characteristics suited to the conditions of its Mediterranean origin [1]. The Mediterranean climate is characterized by an amount of rainfall ranging from 150 to 800 mm per year, and by the uneven distribution of rains, concentrated above all in winter and spring months [2,3]. The olive tree is very adapted to extreme environmental conditions, such as drought and high temperatures, and it is resistant to decay [2,4].

During the last decade, plantings and production of European olive (*Olea europaea* L.) have increased globally by about 10 and 20% [5]. According to the latest statistical data, worldwide production of olives is approximately 23 million tons, and it is cultivated on approximately 10 million ha [6]. In Croatia, olives are cultivated on almost 20 thousand ha with a production of 23 thousand tons of olives [6]. Olive is, along with vines, the most common crop grown in the Mediterranean part of Croatia [7]. In Istria, olive trees are known for about 2500 years [8]. Hundreds of named cultivars of both types of olives, table and oil, are grown. The most important domesticated and introduced olive cultivars in Istria (Croatia) are 'Bjelica', 'Buža', and 'Leccino'.

Olives are susceptible to different bacterial, viral, and fungal pathogens, which can cause severe diseases of the drupe, leaves, wood, and roots [9]. Trunk pathogens can infect olive trees through wounds, and cause dieback of twigs and branches, which can lead to a reduced fruit-bearing capacity and lifespan of olive trees [9,10]. Consequently, fungal trunk diseases can cause substantial economic losses [5]. Branches affected with cankers can show symptoms such as fruit rot or twig dieback. A number of different pathogens are reported as the causal agents associated with olive cankers and twig dieback [10]. One of the most harmful olive pathogens associated with the fungal canker of olives is species from the *Botryosphaeriaceae* family. The *Botryophaeriaceae* were found to be the most prevalent fungal family causing olive twig and branch dieback in California, Italy, and Spain [9–11]. Olive diseases are poorly studied in Croatia, with most research focusing on practical aspects of leaf spot, caused by *Venturia oleaginea* (Castagne) Rossman & Crous, and olive knot, caused by *Pseudomonas savastanoi* (Janse 1982) Gardan et al., management [12]. Thus far, the species *Botryosphaeria dothidea* (Moug. ex Fr.) Ces. & De Not., *Diaporthe neotheicola* (Sacc.) Udayanga & Castl., *Diplodia seriata* De Notaris, *Neofusicuccoum parvum* (Pennycook & Samuels) Crous, Slippers & A.J.L. Phillips, *Phaeoacremonium iranianum* L. Mostert, Gräfenhan, W. Gams & Crous, *Phoma incompta* Sacc. & Martelli (*Comoclathris incompta* (Sacc. & Martelli) Ariyaw. & K.D. Hyde), *Pleurostomophora richardsiae* (Nannf.) Reblova & Jaklitsch, and *Verticillium dahliae* Klebahn have been described as fungal pathogens of olive in Croatia [12–17].

Cytospora spp. (anamorphs of *Valsa* spp.) are common inhabitants of woody plants and they include important stem and branch canker pathogens [18]. Since many different pathogens could be associated with the same syndrome, the correct diagnosis of the diseases is difficult. Diseases associated with *Cytospora* spp. have been referred to as Cytospora-, Leucostoma-, Perennial-, or Valsa canker [5,18–20]. There have been several reports of *Cytospora* spp. on olive trees, but there is a lack of knowledge about the diversity and biology of *Cytospora* spp. affecting olive [5]. *Cytospora* genus belongs to the order *Diaporthales*, and was introduced in 1818 with four species, namely: *C. betulina*, *C. epimyces*, *C. resinae*, and *C. ribis* [21]. The genus *Cytospora* has asexual morphs in *Valsa*, *Leucostoma*, *Valsella*, and *Valseutypella* [19]. Recently, all sexual genera were synonymized with *Valsa* as a subgenus or species [19]. This order is well known to contain endophytes, phytopathogens, and saprobes, with worldwide distribution [22–24]. Currently, there are 3373 described species of *Diaporthales*, in 238 genera and 11 families [25]. Many different names and name changes within *Cytospora* species have caused confusion for plant pathologists. The Index Fungorum lists 672 species of *Cytospora* and 886 species of *Valsa*. Most species are believed to be synonyms [18]. Currently, there are 173 records of *Cytospora* species in GenBank.

In the past, species identification in *Cytospora* was largely based on host affiliation, morphological characters of pycnidia/perithecia, and spore dimensions, with morphological description [26–28]. This morphological species approach is confounded by many morphological characters' overlap among species and by the morphological plasticity of pycnidial locules which are affected by the host bark and cambium characteristics (e.g., stromatal arrangement in the host tissues, locular arrangement within pycnidia, locule division into chambers, independent or shared locular walls) [18,19,29–31]. The first molecular phylogeny of *Cytospora* was inferred from ITS sequences [29], and identifications were performed mostly based on the ITS gene region [19,32,33]. Recently, more *Cytospora* species were recognized when using analysis including multiple protein-coding loci, such as α-actin (ACT), beta-tubulin (TUB2), the RNA polymerase II second largest subunit (RPB2), or translation elongation factor 1-alpha (TEF1-a) [20,34].

Cytospora species are primarily wound pathogens and cause cankers and dieback on many genera of hardwoods and coniferous trees, but rarely on herbaceous plants [18]. On many tree species, these fungi are considered facultative wound parasites that attack weakened trees [18]. Petrini [35] describes *Cytospora* spp. as latent pathogens existing as symptomless endophytic infections. The pathogens infect the inner bark, which is also referred to as the bark periderm [18]. *Cytospora pruinosa* Défago may infect olive trees through pruning wounds and wounds caused by cold. Abiotic stress may also play a role in decline [36]. The climatological conditions and the availability of susceptible hosts may influence the distribution of *Cytospora* species [5]. Plant pathologists are concerned that pathogens such as species of *Cytospora* might move from an introduced host to a native host or vice versa [18]. Because *Cytospora* species cause disease in plants under stress, the movement of a pathogen to a new host may reflect the vulnerability of a host's defensive systems under stress or the severity of the stress [18]. Fruiting bodies of *Cytospora* spp. consist of stromata (conidiomata) that usually contain either labyrinthine chambers or clusters of pycnidia, having filamentous conidiophores and allantoid hyaline conidia [18].

Conidia oozing from pycnidia embedded in dead or dying host cortical tissues during humid or wet conditions are considered infectious propagules potentially initiating new infections. Conidia exude from the fruiting bodies in gelatinous matrices, usually as yellow, orange, red, or pallid tendrils [18]. They are dispersed to new plant tissues by rain-splash, where they germinate and infect the host through cracks and wounds to the bark created by breakage of shade-weakened twigs and branches, insect injuries, leaf scars, pruning wounds, winter-injured buds, twigs, and bark [31,37,38]. As plant pathogens, *Cytospora* species are primarily associated with canker diseases. Symptoms vary with host species and stage of development [26]. *Cytospora* species mainly impact branches, but they can cause more destructive infections in the trunk and larger scaffolds, consequently limiting the productivity of orchards [38,39]. The diseased inner bark and the bark above the infected cambium may appear sunken and yellow, brown, reddish-brown, grey, or black, becoming watery and odorous as the tissues deteriorate. The wood below the cambium is stained brown. Later, these fungi quickly girdle and kill branches and twigs, forming several black sporocarps [26].

The objective of this study was to identify the causal agent of canker disease on olive trees observed in two orchards in Istria, Croatia, based on morphological examination, phylogenetic analysis, and to confirm a pathogen is the cause of a particular disease using Koch's postulates.

2. Materials and Methods

2.1. Sampling and Fungal Isolation

In 2021, olive trees showing signs of dieback were observed in two olive orchards in Istria, Croatia. The first orchard was on the northern side of Istria, in Kaštelir (45°17′30″ N, 13°40′51″ E). The area of the orchard was 6.5 ha, with 44 different olive cultivars. Symptoms were observed on olive trees from the local cultivar 'Porečka rosulja'. The second orchard was on the western side, in Vodnjan (44°56′21″ N, 13°50′18″ E). The area of this orchard was 1 ha and it contained two different olive cultivars. Symptoms were observed on olive trees of the local cultivar 'Buža'. In both orchards, olive trees were between five and 20 years old. In total, five samples (one sample per tree) of branches from symptomatic trees were collected from each orchard and delivered to the Laboratory for Plant Protection at the Institute of Agriculture and Tourism in Poreč (Istria, Croatia) for analysis.

Whole olive fruits and small pieces of branches and leaves (5 × 5 mm) were rinsed under tap water, surface sterilized in 70% ethanol for one minute (two minutes for fruits), rinsed two times in sterile distilled water, and placed on a sterile paper sheet in a laminar flow cabinet until dry. Plant parts were plated on potato-dextrose agar (PDA) and incubated for seven days at 25 °C under laboratory conditions. After incubation, the growing tips of hyphae were transferred aseptically on the fresh PDA medium, for pure culture.

2.2. Morphological Identification

Pure fungal cultures of ten isolates on PDA were taken for examination. Species have been identified based on the spores (color, shape, presence or absence of septa, and dimensions) and colony characteristics (color, form, elevation, margin, surface, and opacity). A study of the fungal structure was performed with a Boeco BM-2000 microscope, a Boeco BCAM10 camera, and a B-View software (Boeckel + Co (GmbH + Co) KG, Hamburg, Germany). Morphometric values were compared with previously published data for the genus [26].

2.3. Molecular Identification and Phylogenetic Analyses

Two representative isolates, SL2 PRIV and V16 BIII (one per orchard), were chosen for molecular identification. To fully characterize the isolates, DNA sequences of internal transcribed spacer (ITS) and beta-tubulin (TUB) were determined. Fresh mycelia of fungal isolates grown on PDA for seven days at 25 °C were scraped with a sterile toothpick from the colony margins and used for genomic DNA extraction. Total genomic DNA from the isolate was extracted using a Maxwell® RSC Instrument (Promega, Madison, WI, USA) and Maxwell® RSC Plant DNA Kit (Promega, Madison, WI, USA). The nuclear ribosomal DNA repeats were amplified using ITS1 (5′ TCCGTAGGTGAACCTGCGG 3′) and ITS4 (5′ TCCTCCGCTTATTGATATGC 3′) pair of primers [40]. Oligonucleotide primers Bt2a (5′ GGTAACCAAATCGGTGCTGCTTTC 3′) and Bt2b (5′ ACCCTCAGTGTAGTGACCCTTGGC 3′) were used to amplify a portion of the TUB gene [41]. The polymerase chain reaction (PCR) mixture was composed of 12.5 μL of EmeraldAmp® GT PCR Master Mix, 0.5 μL of each primer, 6.5 μL of nuclease-free water, and 5 μL of genomic DNA. PCR was conducted in a SureCycler 8800 Thermal Cycler (Agilent Technologies, Santa Clara, CA, USA) under the following conditions for both gene regions: initial denaturation step for two minutes at 95 °C followed by 35 cycles for 30 s of denaturation at 95 °C, 30 s for annealing at 48 °C, one minute for extension at 72 °C, and a final extension step of eight minutes at 72 °C [26]. For both isolates, the amplification of the TUB region was not accomplished, so a second PCR was performed using 1 μL of the first PCR amplification as a template. PCR products were visualized on 1% agarose gel light using an iBright CL1000 Imaging System (Invitrogen, Thermo Fisher Scientific, Waltham, MA, USA). Purification and sequencing of PCR products was conducted by Macrogen Europe services (Amsterdam, The Netherlands). Nucleotide sequences were read and edited in Sequencher® (Gene Codes Corporation, Ann Arbor, MI, USA). Sequences were compared with those of *Cytospora* species from previous studies available in the National Center for Biotechnology Information database GenBank®. Consensus sequences were produced and deposited into GenBank®. Phylogenetic analysis was made using ITS sequence data from isolates used in this study and relevant isolates from GenBank®. Sequences were aligned using ClustalX2 (UCD Dublin, Dublin, Ireland) software, and a phylogenetic tree was made using MEGA11 (Pennsylvania State University, Center, PA, USA) software.

2.4. Pathogenicity Tests

To determine the pathogenicity of fungal species, the same two isolates selected for molecular analysis were chosen for the inoculation of olive branches. Two pathogenicity tests were performed: inoculation on detached branches and inoculation on olive trees.

2.4.1. Pathogenicity on Detached Branches

In total, ten segments of branches per cultivar, 10 cm long, were collected from healthy olive trees of cultivars 'Buža', 'Leccino', and 'Porečka rosulja' grown at the Institute of Agriculture and Tourism in Poreč. Branch segments were rinsed under tap water, surface sterilized in 10% sodium hypochlorite for 10 min, and rinsed with sterile distilled water for 10 min. Segments were placed in a laminar flow cabinet on a sterile paper tissue. After air drying, the branch segments were marked and sealed at both ends with Parafilm to reduce desiccation. Wounds 4 mm in diameter were made in the bark with a cork borer to remove the outer bark but to leave the inner bark intact. A 4 mm diameter mycelium plug from an 8-day-old PDA culture of isolates was placed in each wound. Inoculated wounds were sealed with Vaseline and protected with Parafilm. Two replicates were performed. PDA plugs without mycelium were used as a control. Inoculated branches were placed in a plastic bag on a sterile paper tissue soaked with sterile distilled water, and incubated under laboratory conditions at approximately 21 °C for 10 weeks.

2.4.2. Pathogenicity on Olive Trees

Branches of 2-year-old olive trees of cultivars 'Buža', 'Leccino', and 'Porečka rosulja' were inoculated in a greenhouse at the Institute of Agriculture and Tourism in Poreč. Four branches per replicate, per cultivar were inoculated for each isolate. Branches were disinfected with 70% ethanol, and 4 mm diameter wounds were made in the bark with a cork borer and inoculated the same way as previously described for detached branches. Inoculated plants had been kept in a greenhouse, at approximately 25 °C, for 6 months, from March to September 2022, and monitored for the presence of symptoms. After incubation, samples were collected. In an attempt to fulfill Koch's postulate, small pieces of necrotic tissue from the edge of the developed lesion were placed on a PDA medium to reisolate the inoculated fungus.

3. Results

3.1. Sampling and Fungal Isolation

The symptoms of the disease in the field can be described as the dieback of twigs and branches (Figure 1), brown internal necrosis on branches and under the bark, leaf necrosis, and fruit dieback. The symptoms were observed on trees from cultivars 'Porečka rosulja' and 'Buža'. Symptoms affect only part of the trees. Morphologically similar fungal colonies were retrieved from all 10 samples.

Figure 1. (**a**) Disease symptoms on olive tree in orchard. (**b**) Branch segment with bark discoloration taken for analysis.

3.2. Morphological Identification

Based on the colony and spore characteristics, 10 fungal isolates have been identified as *Cytospora pruinosa* Défago. Colonies developing on PDA had reached a nine cm diameter after 10 days at 25 °C on PDA. Colonies were irregularly shaped, with aerial, opaque, fluffy mycelium becoming flat with age. Initially, they were white becoming creamy white-beige with age with black pycnidia distributed on the surface (Figure 2). The reverse of the colonies was creamy white with an orange undertone and visible black pycnidia. Hyphae were septate, hyaline, and yellowish. Conidia were elongated, hyaline, aseptate, and smooth, and 5.6–6.8 × 1.1–1.3 µm diameter ($\overline{x}$ = 5.9 × 1.1 µm, n = 30).

3.3. Molecular Identification and Phylogenetic Analyses

Consensus sequences of representative isolates were produced and deposited in GenBank® under accession numbers: OQ642321 and OQ644501 for ITS, and OQ652101 and OQ694815 for the TUB gene region (available in Supplementary Materials). Blast analysis of the sequences from the SL2 PRIV isolate showed 100% similarity for ITS and TUB gene regions with *C. pruinosa*. Blast analysis of the sequence from the V16 BIII isolate showed 100% similarity for ITS and 99.79% similarity for the TUB gene region with *C. pruinosa*. The phylogenetic tree was constructed using the Neighbor-Joining method (Figure 3). DNA sequence analysis and phylogenetic analysis confirm the identity of SL2 PRI and V16 BIII isolates as *C. pruinosa*.

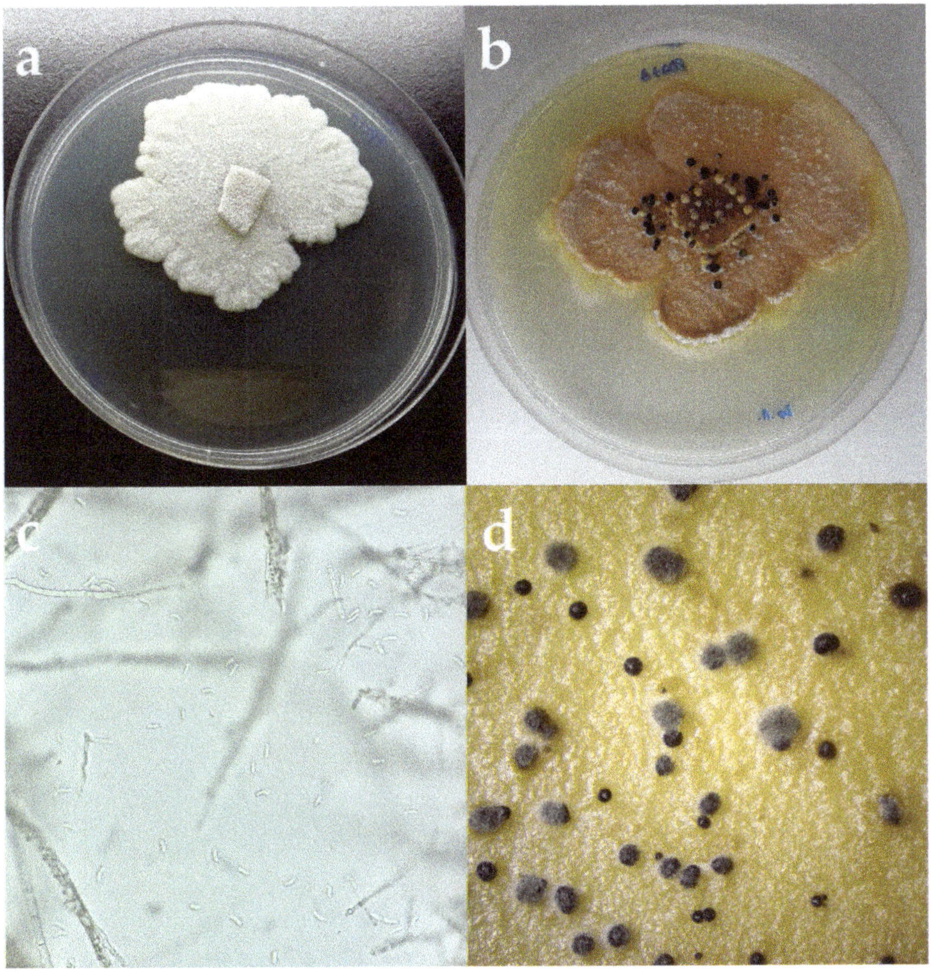

Figure 2. (**a**) *Cytospora pruinosa* colony on PDA after 5 days in the dark at 25 °C. (**b**) *C. pruinosa* colony on PDA after 2 weeks. (**c**) Hyphae and spores of *C. pruinosa* isolate under the microscope. Scale bar = 10 µm. (**d**) Conidiomata formed on PDA.

3.4. Pathogenicity Tests of Isolate

The symptoms of the disease on the olive branches tested in the laboratory and on the olive trees in the greenhouse showed the same symptoms as olive trees observed in the field survey. Dieback of twigs, brown internal necrosis on branches and under the bark, bark discoloration, and fruit collapse were detected (Figure 4). The pathogen had been consistently reisolated from all affected pieces of wood. Only saprobes were isolated from the control branches and trees.

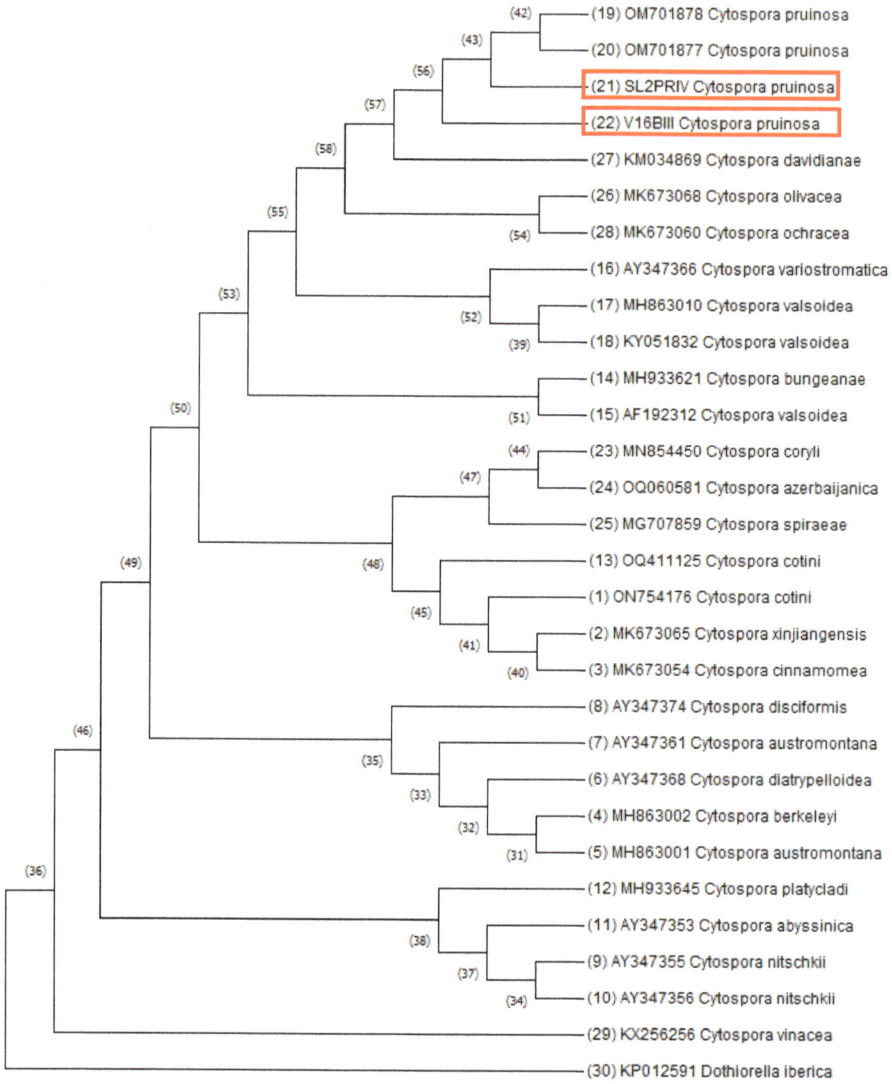

Figure 3. The evolutionary history was inferred using the Neighbor-Joining method [42]. The optimal tree is shown. The percentage of replicate trees in which the associated taxa clustered together in the bootstrap test (1000 replicates) is shown next to the branches [43]. The evolutionary distances were computed using the Maximum Composite Likelihood method [44] and are in the units of the number of base substitutions per site. This analysis involved 30 nucleotide sequences. *Dothiorella iberica* isolate 211 KP012591 was used as an outgroup. Sequences from this research are marked with red rectangles. All ambiguous positions were removed for each sequence pair (pairwise deletion option). There were a total of 769 positions in the final dataset. Evolutionary analyses were conducted in MEGA11 [45].

Figure 4. Disease symptoms on olive trees used in pathogenicity tests in the greenhouse after 6 months at 25 °C: (**a**) fruit collapse, (**b**) branch necrosis, (**c**) branch dieback, (**d**) bark discoloration. (**e**,**f**) Disease symptoms on olive branches used in pathogenicity tests in the laboratory: (**g**) Difference between the control branch inoculated with pure PDA plug (left) and the branch inoculated with *C. pruinosa* (right).

4. Discussion

In this research, *C. pruinosa* was identified based on morphological characteristics, molecular data of ITS and TUB gene region, and a phylogenetic tree made based on internal transcribed spacer sequence alignment. Pathogenicity tests were conducted on detached olive branches and whole plants from three different olive cultivars. Cultivars 'Buža' and 'Porečka rosulja' were chosen for the test because all isolates were derived from those olive cultivars. Cultivar 'Leccino' was chosen as one of the most distributed and resistant olive cultivars in Istria. Lesions appeared on all tested branches and plants, except for the controls. In addition, Koch's postulate was carried out on all infected and control plants. Disease symptoms of infection of olive trees inoculated with *C. pruinosa* comprised reddish-brown discoloration of bark, stained brown discoloration below the cambium, leaf necrosis, branch, fruit, and twig dieback.

Members of *Cytospora* genus are cosmopolitan and occur on a broad host range [26]. The first record of *C. pruinosa* on olives was in 2006 in South Africa [18]. In addition to Africa [18,46] it was found as a pathogen of olives in Spain [11]. Other *Cytospora* species known as olive pathogens are *C. oleicola* D.P. Lawr., L.A. Holland & Trouillas [5,31], *C. oleina* Berl. [47], *C. olivarum* Úrbez-Torr., D.P. Lawr., Peduto, Gubler & Trouillas [5], *C. plurivora* D.P. Lawr., L.A. Holland & Trouillas, and *C. sorbicola* Norphanph., Bulgakov, T.C. Wen & K.D. Hyde [31]. Except dieback of olives, *C. pruinosa* is associated with dieback of ash tree *Fraxinus excelsior* L. *C. pruinosa* can infect ash trees weakened after primary infection by fungus *Hymenoscyphis fraxineus* (T. Kowalski) Baral, Queloz & Hosoya, or emerald ash borer *Agrilus planipennis* Fairmaire [48,49].

Moral et al. [11] reinforce the idea that inoculation in vivo is essential for the characterization of fungal pathogens. In their trial, *C. pruinosa* did not cause symptoms when tested on 5-year-old potted 'Gordal Sevilliana' olive trees in a greenhouse at 25 to 30 °C. Contradictory to their study, *C. pruinosa* formed the largest average lesion length of all the isolates used in a study conducted in 2021 in South Africa [46]. However, of the six tested isolates of *C. pruinosa*, two isolates developed long lesions with values of 41.14 mm and 36.94 mm, one isolate with 10.53 mm, while the rest had values of less than 10 mm. In total, of 58 fungal isolates comprising 38 species, *C. pruinosa* produced not only the longest lesions but also one isolate was on the bottom 13 isolates with an average lesion length of only 3.20 mm [46]. In the research carried out by Úrbez-Torres et al. [5], six months after inoculation of olive branches, *C. oleicola* and *C. olivarum* caused lesions that averaged 26.7 mm in length. All tested isolates of *C. oleina* in spring inoculations resulted in the formation of a necrotic area around the inoculation point 43 to 56 mm long, while inoculation made in autumn resulted in the death of the whole twig [47]. In this trial, six months after inoculation of trees, *C. pruinosa* caused lesions that averaged 10.4 mm on cv. Porečka rosulja, 22.4 mm on cv. Leccino, and 31.36 mm on cv. Buža. Compared to other fungal pathogens, such as species from the *Botryosphaeriaceae* family, *Cytospora* species were shown as less aggressive pathogens.

Inability to form symptoms could be explained by *C. pruinosa* previously being part of a species complex with genetic variances expected to be high. These variances could create implications regarding the variation in virulence-related genes [18,46].

Control of *Cytospora* diseases is difficult and focusing management efforts against the most aggressive encountered *Cytospora* species will be essential [31]. Preventive practices, such as proper pruning and pruning of infected parts, clean tools, removal of infected plant material from orchards, treatment of pruning wounds, and selection of resistant cultivars may be an effective preventative strategy against infection with *Cytospora* species. Mechanical pruning results in multiple cuts on trees, which greatly surpass the number of pruning wounds produced in traditionally farmed low-density orchards, so further research needs to be carried out to investigate the impact of mechanized practices on infection by *Cytospora* spp. [5].

There is no data about protection measures for *C. pruinosa* exclusively; protection measures against other *Cytospora* species have been listed. Regarding fungicides, thiophanate-methyl (alone, amended in 50% latex paint, combined with VitiSeal, and combined with latex paint at 50 and 70%), captan, 50% latex paint, lime sulfur, and VitiSeal combined with lime sulfur has the potential for reducing the species *C. leucostoma*, *C. plurivora*, and *C. sorbicola* [50–52]. In addition, antagonistic fungal species *Trichoderma viride* SC1 provide excellent pruning wound protection against *C. sorbicola* [51]. Moreover, antagonistic assays in vitro showed that secondary metabolites of *Bacillus pumilus* Meyer and Gottheil strain JK-SX001 extracted using methylbenzene could also suppress the growth of *C. chrysosperma* (Pers.) Fr. [53].

Since there is a lack of information about the biology and pathogenicity of this species, further research needs to be carried out. Urbez-Torres [5] propose research based on spore-trapping studies combined with studies evaluating the effectiveness of pruning wound protectants for development of effective control strategies. *Cytospora* species has not been recorded on other plant species in Croatia so far. Given that spores of some of the *Cytospora* species, such as *C. leucostoma*, can be wind-blown to 76 m from the inoculum source [54], and given that *Cytospora* species are pathogens of numerous plant species, especially woods, the question arises as to the possibilities of their spread on other plants, mostly on vines, because vines and olives are often grown together in these areas.

5. Conclusions

To the best of our knowledge, this is the first report of *Cytospora pruinosa* Défago causing olive twig and branch dieback on olive trees in Croatia.

Supplementary Materials: The following supporting information can be downloaded at: https://www.mdpi.com/article/10.3390/microorganisms11071679/s1. Flatfile records of the isolates from NCBI.

Author Contributions: Conceptualization, E.P. and S.G.; methodology, E.P. and S.G.; investigation, E.P. and S.G.; writing—original draft preparation, E.P.; writing—review and editing, S.G., D.I., K.V. and J.Ć. All authors have read and agreed to the published version of the manuscript.

Funding: This research was funded by Croatian Science Foundation Installation Research Project "Natural bioactive compounds as a source of potential antimicrobial agents in the control of bacterial and other fungal pathogens of olives", Anti-Mikrobi-OL (AMO), UIP-2020-02-7413, and "Young Researchers' Career Development Project" DOK-2021-02-2882.

Institutional Review Board Statement: Not applicable.

Informed Consent Statement: Not applicable.

Data Availability Statement: All sequence data are available in NCBI GenBank following the accession numbers in the manuscript.

Conflicts of Interest: The authors declare no conflict of interest.

References

1. Rapoport, H.F.; Fabbri, A.; Sebastiani, L. *Olive Biology. The Olive Tree Genome*; Springer: Cham, Switzerland, 2016; pp. 13–25.
2. Fraga, H.; Moriondo, M.; Leolini, L.; Santos, J.A. Mediterranean olive orchards under climate change: A review of future impacts and adaptation strategies. *Agronomy* **2021**, *11*, 56. [CrossRef]
3. Ponti, L.; Gutierrez, A.P.; Ruti, P.M.; Dell'Aquila, A. Fine-scale ecological and economic assessment of climate change on olive in the Mediterranean Basin reveals winners and losers. *Proc. Natl. Acad. Sci. USA* **2014**, *111*, 5598–5603. [CrossRef]
4. Mafrica, R.; Piscopo, A.; De Bruno, A.; Poiana, M. Effects of climate on fruit growth and development on olive oil quality in cultivar Carolea. *Agriculture* **2021**, *11*, 147. [CrossRef]
5. Úrbez-Torres, J.R.; Lawrence, D.P.; Hand, F.P.; Trouillas, F.P. Olive twig and branch dieback in California caused by *Cytospora oleicola* and the newly described species *Cytospora olivarum* sp. nov. *Plant Dis.* **2020**, *104*, 1908–1917. [CrossRef]
6. Food and Agriculture Organization of the United Nations. Available online: https://www.fao.org/faostat/en/#data/QCL (accessed on 5 June 2023).
7. Gluhić, D. Jesenska gnojidba vinove loze i masline. *Glas. Zast. Bilja* **2020**, *37*, 97–110. [CrossRef]
8. Žužić, I. *Maslina i Maslinovo Ulje: Sa Posebnim Osvrtom na Istru*; Olea, udruga maslinara Istarske županije; TIPOMAT: Velika Gorica, Croatia, 2008; pp. 1–380.
9. Carlucci, A.; Raimondo, M.l.; Cibelli, F.; Phillips, A.J.l.; Lops, F. *Pleurostomophora richardsiae, Neofusicoccum parvum* and *Phaeoacremonium aleophilum* associated with a decline of olives in southern Italy. *Phytopathol. Mediterr.* **2013**, *52*, 517–527.
10. Úrbez-Torres, J.R.; Peduto, F.; Vossen, P.M.; Krueger, W.H.; Gubler, W.D. Olive twig and branch dieback: Etiology, incidence, and distribution in California. *Plant Dis.* **2013**, *97*, 231–244. [CrossRef] [PubMed]

11. Moral, J.; Agusti-Brisach, C.; Pérez-Rodríguez, M.; Xaviér, C.; Carmen-Raya, M.; Rhouma, A.; Trapero, A. Identification of fungal species associated with branch dieback of olive and resistance of table cultivars to *Neofusicoccum mediterraneum* and *Botryosphaeria dothidea*. *Plant Dis.* **2017**, *101*, 306–316. [CrossRef] [PubMed]
12. Ivić, D.; Ivanović, A.; Miličević, T.; Cvjetković, B. Shoot necrosis of olive caused by Phoma incompta, a new disease of olive in Croatia. *Phytopathol. Mediterr.* **2010**, *49*, 414–416.
13. Cvjetković, B. *Mikoze i Pseudomikoze Voćnjaka i Vinove Loze*; Zrinski: Čakovec, Croatia, 2010; pp. 5–534.
14. Godena, S.; Ivić, D.; Goreta Ban, S. *Uzročnici Djelomičnog ili Potpunog Sušenja Stabala Maslina*; Priručnik o rezultatima VIP projekta; Institut za Poljoprivredu i Turizam: Poreč, Croatia, 2019; pp. 1–49.
15. Kaliterna, J.; Miličević, T.; Ivić, D.; Benčić, D.; Mesić, A. First report of Diplodia seriata as causal agent of olive dieback in Croatia. *Plant Dis.* **2012**, *96*, 290. [CrossRef]
16. Petrović, E.; Vrandečić, K.; Ćosić, J.; Kanižai Šarić, G.; Godena, S. First Report of *Phaeoacremonium iranianum* causing olive twig and branch dieback. *Plants* **2022**, *11*, 3578. [CrossRef] [PubMed]
17. Kaliterna, J.; Miličević, T.; Benčić, D.; Mesić, A. First report of Verticillium wilt caused by *Verticillium dahliae* on olive trees in Croatia. *Plant Dis.* **2016**, *100*, 2526. [CrossRef]
18. Adams, G.C.; Roux, J.; Wingfield, M.J. *Cytospora* species (*Ascomycota, Diaporthales, Valsaceae*): Introduced and native pathogens of trees in South Africa. *Australasian Plant Pathol.* **2006**, *35*, 521–548. [CrossRef]
19. Adams, G.C.; Roux, J.; Wingfield, M.J.; Common, R.; Roux, J. Phylogenetic relationships and morphology of *Cytospora* species and related teleomorphs (*Ascomycota, Diaporthales, Valsaceae*) from Eucalyptus. *Stud. Mycol.* **2005**, *52*, 1–144.
20. Lawrence, D.P.; Travadon, R.; Pouzoulet, J.; Rolshausen, P.E.; Wilcox, W.F.; Baumgartner, K. Characterization of *Cytospora* isolates from wood cankers of declining grapevine in North America, with the description of two new *Cytospora* species. *Plant Pathol.* **2017**, *66*, 713–725. [CrossRef]
21. Ehrenberg, C.G. *Sylvae Mycologicae Berolinenses*; Formis Theophili Bruschcke: Berlin, Germany, 1818; pp. 1–52.
22. Castlebury, L.A.; Rossman, A.Y.; Jaklitsch, W.J.; Vasilyeva, L.N. A preliminary overview of the *Diaporthales* based on large subunit nuclear ribosomal DNA sequences. *Mycologia* **2002**, *94*, 1017–1031. [CrossRef]
23. Rossman, A.Y.; Farr, D.F.; Castlebury, L.A. A review of the phylogeny and biology of the *Diaporthales*. *Mycoscience* **2007**, *48*, 135–144. [CrossRef]
24. Fan, X.; Bezerra, J.D.P.; Tian, C.M.; Crous, P.W. Families and genera of Diaporthalean fungi associated with canker and dieback of tree hosts. *Persoonia* **2018**, *40*, 119–134. [CrossRef] [PubMed]
25. Encyclopedia of Life. 2023. Available online: https://eol.org/pages/5612 (accessed on 21 April 2023).
26. Fan, X.L.; Bezerra, J.D.P.; Tian, C.M.; Crous, P.W. *Cytospora* (*Diaporthales*) in China. *Persoonia* **2020**, *45*, 1–45. [CrossRef]
27. Grove, W. The British species of *Cytospora*. *Bull. Misc. Inf.* **1923**, *1923*, 1–30. [CrossRef]
28. Spielman, L.J. A monograph of *Valsa* on hardwoods in North America. *Can. J. Bot.* **1985**, *63*, 1355–1378. [CrossRef]
29. Adams, G.C.; Surve-Iyer, R.S.; Iezzoni, A.F. Ribosomal DNA sequence divergence and group I introns within the *Leucostoma* species *L. cinctum*, *L. persoonii*, and *L. parapersoonii* sp. nov., ascomycetes that cause Cytospora canker of fruit trees. *Mycologia* **2002**, *94*, 947–967.
30. Wang, X.; Wei, J.; Huang, L.; Kang, Z. Re-evaluation of pathogens causing *Valsa* canker on apple in China. *Mycologia* **2011**, *103*, 317–324. [CrossRef] [PubMed]
31. Lawrence, D.P.; Holland, L.A.; Nouri, M.T.; Tavadon, R.; Abramians, A.; Michailides, T.J.; Trouillas, F.P. Molecular phylogeny of Cytospora species associated with canker diseases of fruit and nut crops in California, with the descriptions of ten new species and one new combination. *IMA Fungus* **2018**, *9*, 333–369. [CrossRef] [PubMed]
32. Fan, X.; Hyde, K.D.; Liu, M.; Liang, Y.; Tian, C. *Cytospora* species associated with walnut canker disease in China, with description of a new species C. gigalocus. *Fungal Biol.* **2015**, *119*, 310–319. [CrossRef] [PubMed]
33. Zhu, H.Y.; Tian, C.M.; Fan, X. Multigene phylogeny and morphology reveal *Cytospora spiraeae* sp. nov. (*Diaporthales, Ascomycota*) in China. *Phytotaxa* **2018**, *338*, 49–62. [CrossRef]
34. Monkai, J.; Tibpromma, S.; Manowong, A.; Mapook, A.; Norphanphoun, C.; Hyde, K.D.; Promputtha, I. Discovery of three novel Cytospora species in Thailand and their antagonistic potential. *Diversity* **2021**, *13*, 488. [CrossRef]
35. Petrini, O. Fungal endophytes of tree leaves. In *Microbial Ecology of Leaves*; Andrews, J.H., Hirano, S.S., Eds.; Springer: New York, NY, USA, 1991; pp. 179–197.
36. Lazarevic, J.; Menkis, A. *Cytospora friesii* and *Sydowia polyspora* are associated with the sudden dieback of *Abies concolor* in Southern Europe. *Plant Prot. Sci.* **2022**, *58*, 258–263. [CrossRef]
37. Tekauz, A.; Patrick, Z. The role of twig infections on the incidence of perennial canker of peach. *Phytopathology* **1974**, *64*, 683–688. [CrossRef]
38. Biggs, A.R. Integrated approach to controlling Leucostoma canker of peach in Ontario. *Plant Dis.* **1989**, *73*, 869–874. [CrossRef]
39. Chang, L.S.; Iezzoni, A.F.; Adams, G.C.; Ewers, F.W. Hydraulic conductance in susceptible versus tolerant peach seedlings infected with *Leucostoma persoonii*. *J. Am. Soc. Hortic. Sci.* **1991**, *116*, 831–834. [CrossRef]
40. White, T.J.; Bruns, T.D.; Lee, S.B.; Taylor, J.W. 38—Amplification and direct sequencing of fungal ribosomal RNA Genes for phylogenetics. In *PCR—Protocols and Applications—A Laboratory Manual*; Innis, M.A., Gelfand, D.H., Sninsky, J.J., White, T.J., Eds.; Academic Press: Cambridge, MA, USA, 1990; pp. 315–322.

41. Glass, N.L.; Donaldson, G.C. Development of primer sets designed for use with the PCR to amplify conserved genes from filamentous ascomycetes. *Appl. Environ. Microbiol.* **1995**, *61*, 1323–1330. [CrossRef] [PubMed]
42. Saitou, N.; Nei, M. The neighbor-joining method: A new method for reconstructing phylogenetic trees. *Mol. Biol. Evol.* **1987**, *4*, 406–425.
43. Felsenstein, J. Confidence limits on phylogenies: An approach using the bootstrap. *Evolution* **1985**, *39*, 783–791. [CrossRef] [PubMed]
44. Tamura, K.; Nei, M.; Kumar, S. Prospects for inferring very large phylogenies by using the neighbor-joining method. *Proc. Natl. Acad. Sci. USA* **2004**, *101*, 11030–11035. [CrossRef]
45. Tamura, K.; Stecher, G.; Kumar, S. MEGA 11: Molecular Evolutionary Genetics Analysis Version 11. *Mol. Biol. Evol.* **2021**, *38*, 3022–3027. [CrossRef]
46. van Dyk, M.; Spies, C.F.J.; Mostert, L.; van der Rijst, M.; du Plessis, I.L.; Moyo, P.; van Jaarsveld, W.J.; Haleen, F. Pathogenicity testing of fungal isolates associated with olive trunk diseases in South Africa. *Plant Dis.* **2021**, *105*, 4060–4073. [CrossRef]
47. Rumbos, I.C. *Cytospora oleina* causing canker and dieback of olive in Greece. *Plant Pathol.* **1988**, *37*, 441–444. [CrossRef]
48. Kowalski, T.; Bilanski, P.; Kraj, W. Pathogenicity of fungi associated with ash dieback towards *Fraxinus excelsior*. *Plant Pathol.* **2017**, *66*, 1228–1238. [CrossRef]
49. Rajtar, N.N.; Held, B.W.; Blanchette, R.A. Fungi from galleries of the emerald ash borer produce cankers in ash trees. *Forests* **2021**, *12*, 1509. [CrossRef]
50. Miller, S.T.; Otto, K.L.; Sterle, D.; Minas, I.S.; Stewart, J.E. Preventive fungicidal control of *Cytospora leucostoma* in peach orchards in Colorado. *Plant Dis.* **2019**, *103*, 1138–1147. [CrossRef] [PubMed]
51. Holland, L.A.; Travadon, R.; Lawrence, D.P.; Nouri, M.T.; Trouillas, F. Evaluation of pruning wound protection products for the management of almond canker diseases in California. *Plant Dis.* **2021**, *105*, 3365–3375. [CrossRef] [PubMed]
52. Miller, S.T.; Sterle, D.; Minas, I.S.; Stewart, J.E. Exploring fungicides and sealants for management of *Cytospora plurivora* infections in western Colorado peach production systems. *Crop Prot.* **2021**, *146*, 105654. [CrossRef]
53. Ren, J.H.; Li, H.; Wang, Y.F.; Ye, J.R.; Yan, A.Q.; Wu, X.W. Biocontrol potential of an endophytic *Bacillus pumilus* JK-SX001 against poplar canker. *Biol. Control.* **2013**, *67*, 421–430. [CrossRef]
54. Bertrant, P.F.; English, H. Release and dispersal of conidia of Valsa leucostoma. *Phytopathology* **1976**, *66*, 987–991. [CrossRef]

MDPI AG
Grosspeteranlage 5
4052 Basel
Switzerland
Tel.: +41 61 683 77 34

Microorganisms Editorial Office
E-mail: microorganisms@mdpi.com
www.mdpi.com/journal/microorganisms

www.ingramcontent.com/pod-product-compliance
Lightning Source LLC
Chambersburg PA
CBHW061520210926
51850CB00008B/1980
9783725815562